DATE DUE

MODERN CONTROL SYSTEMS ANALYSIS AND DESIGN

MODERN CONTROL SYSTEMS ANALYSIS AND DESIGN

WALTER J. GRANTHAM
Mechanical and Materials Engineering
Washington State University

THOMAS L. VINCENT
Aerospace and Mechanical Engineering
University of Arizona

John Wiley & Sons, Inc.
New York • Chichester • Brisbane • Toronto • Singapore

Acquisitions Editor	Cliff Robichaud
Marketing Manager	Susan Elbe
Production Manager	Lucille Buonocore
Production Supervisors	Savoula Amanatidis
	and Marjorie Shustak
Copyediting Supervisor	Deborah Herbert
Designer	Pedro A. Noa
Illustration Coordinator	Jaime Perea
Manufacturing Manager	Andrea Price

This book was typeset in Times Roman by General Graphic Services and printed and bound by Hamilton Printing Company.

Recognizing the importance of preserving what has been written, it is a policy of John Wiley & Sons, Inc. to have books of enduring value published in the United States printed on acid-free paper, and we exert our best efforts to that end.

Library of Congress Cataloging in Publication Data:

Grantham, Walter J. (Walter Jervis), 1944-
    Modern control systems : analysis and design / Walter J. Grantham,
    Thomas L. Vincent.
        p.   cm.
    Includes bibliographical references and index.
    ISBN 0-471-81193-9 (acid-free paper)
    1. Feedback control systems.   2. Linear systems.   I. Vincent,
Thomas L.   II. Title.
TJ216.G73   1993
629.8'3--dc20   92-31072
                                                          CIP

Printed in the United States of America

10 9 8 7 6 5 4 3 2 1

To Tricia, Christopher, Thelma
and Peggie Jo, Tania, Tyrone

PREFACE

The purpose of this book is to provide an introduction to the analysis techniques used in the design of linear feedback control systems. The emphasis is on both classical and matrix methods. After introducing general state space systems and input-output systems, design and analysis techniques are then presented in a building-block sequence, with a thorough analysis of first-order systems, second-order systems, higher-order systems, and general state space systems. In addition to classical topics, such as Root Locus, Bode design methods, and Nyquist stability, this text also presents eigenvalue placement and observer design techniques for linear systems, as well as results on systems with uncertain inputs, systems with uncertain parameters, analogue computer implementations, and redesign for digital implementations.

This book is designed for one-semester introductory junior- or senior-level control systems courses and assumes a background on the student's part to include dynamics and differential equations. We recognize that many schools have introduced junior-level dynamic systems courses that focus on preliminary material dealing with system modeling (mechanical, thermal, electrical, etc.) and basic eigenvalue and Laplace transform analysis techniques for linear dynamical systems. These courses, necessarily, have time for at most a brief introduction to control systems. However, many schools do not have such courses, and the senior-level control systems course must provide the basic material. This text is designed to accommodate either situation. In particular, although this book's primary focus is on classical and modern control systems, it also contains basic background material associated with a Dynamic Systems course, so that such a course is not a prerequisite.

There are a variety of approaches to structuring an introductory one-semester control systems course using this text. Chapters 3 and 4 provide mathematical background on the free and forced response of linear systems, respectively, and should be a review for students who have already been exposed to linear systems. For such students these two chapters can be covered briefly, except for the Routh stability criteria in Section 3.5, which should be examined in detail. With this approach, the objective of the course should be to cover the remaining material throughout the text. For students with no previous exposure to linear systems modeling and analysis in a matrix setting, Chapters 3 and 4 should be covered in detail. In this circumstance, Sections 5.4, 6.2, 7.3, and 7.4 could be deemphasized.

We hope that this book will stimulate its readers to continue their study beyond this introduction to modern control systems analysis and design. Each chapter contains several examples and a variety of exercises to aid the student. In addition, a root locus numerical algorithm is given. It should also be noted that, because of our integrated coverage of both classical and modern control techniques, this book is well-suited for a course taught in conjunction with software which can be used for control systems analysis, such as Mathematica, MATLAB, and MathCAD. We have also emphasized the importance of uncertain inputs to a system and have provided an introduction to some methods for dealing with uncertain inputs. Mastering the material in this text will provide sufficient background for those students interested in more advanced texts dealing with nonlinear and optimal control design. As an additional aid, an Appendix presents answers to even-numbered exercises for each chapter.

Walter J. Grantham
Thomas L. Vincent

Pullman, Washington
Tucson, Arizona
March 1992

CONTENTS

Chapter 1

Automatic Control Systems

1.1 AUTOMATIC CONTROL APPLICATIONS

Automatic control applications are synonymous with modern technology. They are found in a complete spectrum of technological products ranging from robots to toasters. Automatic control deals with the problem of obtaining desired behavior from a **dynamical system** that runs by itself. Use of the phrase "dynamical system" implies an aggregate of time-varying quantities that identify the object(s) of interest. An autopilot is an example of an automatic control system for an airplane. Guided missiles are wholly dependent on automatic control. Less impressive, but more common, are constant-speed motors and temperature control systems.

Other examples of automatic control devices include speed and positioning controls for tape and disk drives, aircraft automatic landing systems, collision avoidance systems, magnetic bearings, artificial hearts and other prosthesis, pacemakers for the heart itself, all robots, and many biological systems (for example, fisheries). Even the U.S. economy is a control system. In the latter case, the Federal Reserve Board attempts to regulate the economy through its interest rate and money supply policies. The list of potential automatic control applications is limited only by one's imagination; almost any dynamical system may be subject to control.

For a given dynamical system there may be more than one possible automatic control application. For example, consider a stick, such as a billiard cue. The phrase "speak softly but carry a big stick" implies only one application. But, indeed, man is capable of manipulating a stick in several ways, over and above the implied threat. For example, he can balance it or use it to point to objects of interest. Both of these activities are very natural and perhaps intuitive reactions to having a stick placed in one's hand. The balancing act is an illustration of the basic control problem of stabilizing an unstable system, and the pointing act is typical of many automatic control problems, such as tracking a planet or comet with a telescope. In a more military setting, the "big stick" might be the guns on a battleship, which must be pointed accurately (to a fraction of a degree) in rolling seas.

Think of the stick as a system to be controlled and the human holding it as a controller. The combination makes up a complete control system. The eye plays an important role in sensing the orientation or "state" of the stick. This information "feeds back" via the brain to the hand for proper adjustment of the stick. Even though an object may move, as long as it can be tracked by the eye, one can still point a stick at it. The complexities of balancing a stick also usually present no problem, provided that the stick is of reasonable dimensions. Note that the human forms a "control loop." In other words, information about the state of the system is used to adjust the system via control action to achieve the desired state.

The objective of automatic control theory is to obtain this type of control regulation without the use of a human in the control loop. To do so, a similar type of operation is still required. Measurements related to the state of the system must be made and this information, in turn, must be used to supply control action to achieve a desired state. **Sensors** are devices used to make the measurements. **Actuators** are devices used to supply the control action. The conglomerate of sensors, actuators, and logic devices to implement the control actions makes up the **controller**.

If one can build a controller to balance a stick or point the stick at an object, then many other applications of this same basic problem are possible. For example, the precision pointing of a space telescope from an orbiting space shuttle or the balancing of a rocket on its engines as it carries the shuttle into orbit both represent the same basic control problem. Because of the reaction times required for each of these applications, a human controller could not be used. Indeed, one must understand and be able to apply automatic control theory in order to implement these and other "high-technology" applications.

1.2 CONTROL SYSTEM MODELS

General Definition of a Control System

We will be dealing with dynamical systems, consisting of a primary system to be controlled and a control mechanism (to be determined) for accomplishing the control. For such systems there are three fundamental concepts involved: state, output, and input. The **state** of a system is simply those dynamical components of a system that completely identify it at any moment in time. The **outputs** from a system are those functions of the state, such as some of the states themselves or some combination of states, that can be measured. The **inputs** to the dynamical system are quantities that can affect the evolution of either the state or output of the system.

Quantities associated with the inputs to the system may be either **control inputs** or **uncertain inputs**. Control inputs include both **command inputs** external to the system and **automatic control inputs** internal to the system. Command inputs are initiated by a human operator or some other external device. Automatic control inputs are determined by an automatic control algorithm within the system (usually computed by an analog or digital device) and are implemented by actuators considered to be part of the system. The uncertain inputs to the system are also of two types: those associated with uncertainties in the system dynamics (manufacturing defects, unpredictable forces) and those associated with uncertainties in the output measurements (noise, instrument error).

The general dynamical system **model** that we will consider in this text consists of a system of algebraic equations and first-order differential equations, written in vector form as

$$\dot{\mathbf{x}} = \mathbf{f}(\mathbf{x},\mathbf{u},\mathbf{v}) \tag{1.2-1}$$

$$\mathbf{y} = \mathbf{g}(\mathbf{x},\mathbf{u},\mathbf{v},\mathbf{w}), \tag{1.2-2}$$

where $(\dot{}) \triangleq d()/dt$ is used to denote differentiation with respect to time t.

Equations (1.2-1) and (1.2-2) are called the **state equations** and **output equations**, respectively. In general, $\mathbf{f}(\mathbf{x},\mathbf{u},\mathbf{v})$ and $\mathbf{g}(\mathbf{x},\mathbf{u},\mathbf{v},\mathbf{w})$ are vector functions of their arguments $\mathbf{x}$, $\mathbf{u}$, $\mathbf{v}$, and $\mathbf{w}$. These, in turn, are vector variables that vary with time. The variables $\mathbf{u}$, $\mathbf{v}$, and $\mathbf{w}$ as well as the initial state $\mathbf{x}(0)$ must be specified before $\mathbf{x}(t)$ and $\mathbf{y}(t)$ can be determined from (1.2-1) and (1.2-2). Both $\mathbf{v}$ and $\mathbf{w}$ will always be considered to be functions of time. However, $\mathbf{u}$ may be either a function of time, $\mathbf{u}(t)$, a direct function of the state, $\mathbf{u}(\mathbf{x})$, or an indirect function $\mathbf{u}(\mathbf{y})$ of the state as measured by the output $\mathbf{y}$. Also, Equations (1.2-1) and (1.2-2) may contain parameters (constants) that we do not display explicitly. We will assume that the functions $\mathbf{f}(\cdot)$ and $\mathbf{g}(\cdot)$ are differentiable in all of their arguments.

The variable $\mathbf{x}$ denotes the **state** of the system. The state is a vector represented in a chosen coordinate system by a column matrix with N_x components,

$$\mathbf{x} = [x_1 \cdots x_{N_x}]^T, \tag{1.2-3}$$

where $()^T$ denotes the transpose. The state $\mathbf{x}(t)$ evolves as a function of time according to (1.2-1), with the evolution depending on the input functions $\mathbf{u}(\cdot)$ and $\mathbf{v}(\cdot)$. The dimension N_x of the state vector defines the **order** of the dynamical system, the number of first-order differential equations required to describe the dynamic behavior of the system. Note that time itself does not appear explicitly in (1.2-1) or (1.2-2). This assures us that, for given inputs $\mathbf{u}(t)$ and $\mathbf{v}(t)$, there is only one solution $\mathbf{x}(t)$ to (1.2-1) passing through an initial state $\mathbf{x}(0)$. To handle systems in which time does appear explicitly, t can be incorporated in the state vector by including the differential equation, for example as $\dot{x}_{N_x} = 1$, where $t = x_{N_x}$.

The variable $\mathbf{u}$ represents the **control** input vector and has N_u components

$$\mathbf{u} = [u_1 \cdots u_{N_u}]^T. \qquad (1.2\text{-}4)$$

The basic control problem is to determine a functional relationship for $\mathbf{u}$ [$\mathbf{u}(t)$, $\mathbf{u}(\mathbf{x})$, or $\mathbf{u}(\mathbf{y})$] designated by $\mathbf{u}(\cdot)$, so that the system behaves in a suitable manner. The value of the control vector is generally required to satisfy a set of specified **control constraints** $\mathbf{u} \in \mathcal{U}$, where $\mathcal{U}$ is some given set of admissible control values. For example, in the case of a bounded scalar control u, the constraint set might be of the form $\mathcal{U} = \{u \mid |u| \leq 1\}$. Except for the constraints $\mathbf{u} \in \mathcal{U}$, the choice of a specific control function $\mathbf{u}(\cdot)$ is completely at the discretion of the control system designer.

The vector variables $\mathbf{v}$ and $\mathbf{w}$ are called the **system uncertainty** and **measurement uncertainty** vectors, respectively. The vector $\mathbf{v}$ has N_v components and $\mathbf{w}$ has N_w components

$$\mathbf{v} = [v_1 \cdots v_{N_v}]^T \qquad (1.2\text{-}5)$$

$$\mathbf{w} = [w_1 \cdots w_{N_w}]^T. \qquad (1.2\text{-}6)$$

We will require that the uncertain inputs $\mathbf{v}$ and $\mathbf{w}$ be bounded. That is, $\mathbf{v} \in \mathcal{V}$ and $\mathbf{w} \in \mathcal{W}$, where $\mathcal{V}$ and $\mathcal{W}$ are bounded sets in N_v and N_w space. These bounds may or may not be known. Both $\mathbf{v}$ and $\mathbf{w}$ are completely independent of the control input.

The variable $\mathbf{y}$ represents the **output** from the system, given by (1.2-2), and has N_y components

$$\mathbf{y} = [y_1 \cdots y_{N_y}]^T. \qquad (1.2\text{-}7)$$

The algebraic output equation (1.2-2) embodies the relationship between the actual state $\mathbf{x}$ of the system and the observations or measurements $\mathbf{y}$ that are made. In the case of complete, perfect information, $\mathbf{y} = \mathbf{x}$. But often it is not possible to measure all of the states directly or perfectly.

The differential equation given by (1.2-1) is a mathematical representation of the physical system without reference to the specific form of the inputs $\mathbf{u}(\cdot)$, $\mathbf{v}(t)$, and $\mathbf{w}(t)$ and with only general reference to how the system will be observed via the measurements. We will not attempt to distinguish here between a physical system and its mathematical model as embodied in (1.2-1) and (1.2-2). For analysis purposes, the mathematical model is the system! Of course, we do recognize a difference, and the extent of the difference is a measure of how well we have modeled the system. From this viewpoint, investigating the dynamical behavior of a system is equivalent to studying the solutions to (1.2-1) and (1.2-2). Since many different physical systems can be described by the same differential equations, strong motivation exists for this approach.

EXAMPLE 1.2-1   **A Model For a Toy Train**

To illustrate the process of developing a model of the form discussed above, consider the problem of designing an on-board controller for a toy train. The controller is to regulate all aspects of the dynamics of the train by adjusting the thrust produced by the engine. That is, it will be used to produce a smooth startup, constant-speed motion, and a smooth stop. To design such a controller, a model for the system must be first obtained.

The toy train as illustrated in Figure 1.2-1 consists of an engine (mass m_1) and one car (mass m_2) that is connected to the engine by a rather soft spring (constant k). The rolling resistance of the engine and car is given by $\mu m_1 g$ and $\mu m_2 g$, respectively, where μ is the coefficient of rolling resistance and g the acceleration of gravity. The engine can produce a thrust u in either direction that is bounded by the relation

$$|u| \leq \bar{u} \qquad (1.2\text{-}8)$$

where $\bar{u}$ is the maximum positive (and negative) thrust available. Note that the control constraint set $\mathcal{U}$ is defined by (1.2-8).

Let r_1 and r_2 denote the position of the engine and car, respectively. Then by applying Newton's second law to a free-body diagram of the engine and car, we obtain

$$m_1\ddot{r}_1 = u - k(r_1 - r_2) - \mu m_1 g \, \text{sgn}(\dot{r}_1) \qquad (1.2\text{-}9)$$

$$m_2\ddot{r}_2 = k(r_1 - r_2) - \mu m_2 g \, \text{sgn}(\dot{r}_2), \qquad (1.2\text{-}10)$$

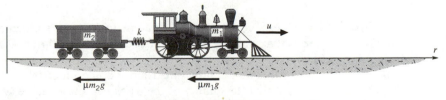

Figure 1.2-1 Modeling the toy train.

where the sgn (sign) function has been used to determine the proper sign for the rolling resistance. The sgn function is defined by

$$\text{sgn}(x) = \begin{cases} +1 & \text{if } x > 0 \\ 0 & \text{if } x = 0. \\ -1 & \text{if } x < 0 \end{cases} \qquad \text{(1.2-11)}$$

These equations can be put in the form of (1.2-1) by letting

$$x_1 = r_1 - r_2$$
$$x_2 = \dot{r}_1 \qquad \text{(1.2-12)}$$
$$x_3 = \dot{r}_2.$$

It follows from the above definitions and (1.2-9) and (1.2-10) that

$$\dot{x}_1 = x_2 - x_3 \qquad \text{(1.2-13)}$$

$$\dot{x}_2 = -\left(\frac{k}{m_1}\right) x_1 - \mu g \, \text{sgn}(x_2) + \frac{u}{m_1} \qquad \text{(1.2-14)}$$

$$\dot{x}_3 = \left(\frac{k}{m_2}\right) x_1 - \mu g \, \text{sgn}(x_3). \qquad \text{(1.2-15)}$$

Equations (1.2-13)–(1.2-15) are of the form of (1.2-1) and represent a mathematical model of this toy train subject to control. To design a controller for the train, some output(s) must be measured. A possible sensor for this case could be an accelerometer attached to the engine and another attached to the car. The accelerometers would measure the acceleration of the engine and car so that the output, in the form of (1.2-2), is given by

$$y_1 = -\left(\frac{k}{m_1}\right) x_1 - \mu g \, \text{sgn}(x_2) + \frac{u}{m_1} \qquad \text{(1.2-16)}$$

$$y_2 = \left(\frac{k}{m_2}\right) x_1 - \mu g \, \text{sgn}(x_3). \qquad \text{(1.2-17)}$$

Of course, other sensors could be used (such as tachometers attached to some of the wheels) that would yield output equations very different from those given above.

It is easy to see in this model how uncertainty could enter the system. For example, suppose that the quality of the track is quite variable so that it produces a variable coefficient of friction μ. We could then express

(1.2-14) and (1.2-15) in terms of some average μ and deviations from the average given by $\Delta\mu$. In this case, these equations could be written as

$$\dot{x}_2 = -\left(\frac{k}{m_1}\right) x_1 - (\mu g)\, \text{sgn}(x_2) + \frac{u}{m_1} + v_1 \qquad \text{(1.2-18)}$$

$$\dot{x}_3 = \left(\frac{k}{m_2}\right) x_1 - (\mu g)\, \text{sgn}(x_3) + v_2, \qquad \text{(1.2-19)}$$

where

$$v_1 = -\Delta\mu g\, \text{sgn}(x_2) \qquad \text{(1.2-20)}$$

$$v_2 = -\Delta\mu g\, \text{sgn}(x_3). \qquad \text{(1.2-21)}$$

Linear Control Systems

The state equations (1.2-1) for most real systems are nonlinear. They may contain products of the state variables, trigonometric or square root functions, or other types of nonlinear terms (such as the sgn function in the above example). As a result, we are not able to obtain closed-form mathematical solutions to many differential equation models. For specified initial conditions and specified input functions, we will need to employ numerical algorithms for generating the corresponding motion $\mathbf{x}(t)$. Fortunately, software is available that can be used to solve general equations in the form of (1.2-1). However, in spite of this fact, we will focus on linear systems, for which we are able to obtain closed-form solutions. Having closed-form solutions available will give us an obvious advantage in the analysis and design of controllers for such systems.

In this book we will be studying constant-coefficient **linear systems**. These systems are an important subclass of (1.2-1) and (1.2-2) that are completely characterized in terms of constant matrices as follows:

$$\dot{\mathbf{x}} = \mathbf{Ax} + \mathbf{Bu} + \mathbf{Rv} \qquad \text{(1.2-22)}$$

$$\mathbf{y} = \mathbf{Cx} + \mathbf{Du} + \mathbf{Ev} + \mathbf{Sw}, \qquad \text{(1.2-23)}$$

where $\mathbf{A}$ is a square $N_x \times N_x$ matrix, $\mathbf{B}$ is $N_x \times N_u$, $\mathbf{R}$ is $N_x \times N_v$, $\mathbf{C}$ is $N_y \times N_x$, $\mathbf{D}$ is $N_y \times N_u$, $\mathbf{E}$ is $N_y \times N_v$, and $\mathbf{S}$ is $N_y \times N_w$. Although the constraint set $\mathcal{U}$ for the control $\mathbf{u}$ always exists (and is usually known), it cannot be introduced in linear analysis. This is because "saturation" (caused by the control reaching the boundary of such a set) is a nonlinear phenomenon. Ignoring the constraint set $\mathcal{U}$ is reasonable as long as control actions do not violate the control constraints. However, the actual presence of $\mathcal{U}$ must be kept in mind for proper implementation and interpretation of the linear analysis applied to real systems. Similar arguments hold for the constraints on the uncertainty vectors $\mathbf{v}$ and $\mathbf{w}$. However, for the most part we will focus on the "deterministic" case, in which $\mathbf{v}(t) = \mathbf{w}(t) \equiv \mathbf{0}$.

EXAMPLE 1.2-2   **Scalar First-Order Systems**

A wide variety of dynamical systems can be represented by a single first-order linear differential equation. For example, consider the four systems illustrated in Figure 1.2-2. The first-order differential equation

$$\dot{x} = bu \tag{1.2-24}$$

describes the state of each system, where the state x, control input u, and constant b for each system are the scalars indicated in Table 1.2-1.

We observe that the dynamic behavior of four very different systems is expressed by the same first-order differential equation, depending on the value of the parameter b. Indeed, many more systems can be expressed by this same differential equation. In general, solving a given differential equation results in the solution to a large number of physical dynamical systems. The investigation of the dynamic behavior of (perhaps idealized) components or a collection of interacting components is equivalent to investigating the solution to the differential equation.

Note that for the systems represented by (1.2-24), the input $u(\cdot)$ is unspecified. A solution to the differential equation can be obtained only after specifying the input $u(\cdot)$ and initial conditions for the state x. Of course, (1.2-24) is just a special case of the more general first-order scalar equation

$$\dot{x} = ax + bu, \tag{1.2-25}$$

Table 1.2-1 Four Systems of the Form $\dot{x} = bu$ in Figure 1.2-2

Dynamical System	State x	Input u	Constant b
(a) Electrical inductance	I	E	$1/L$
(b) Electrical capacitance	E	I	$1/C$
(c) Mechanical damper	y	F	$1/\mu$
(d) Hydraulic tank	h	q	$1/A$

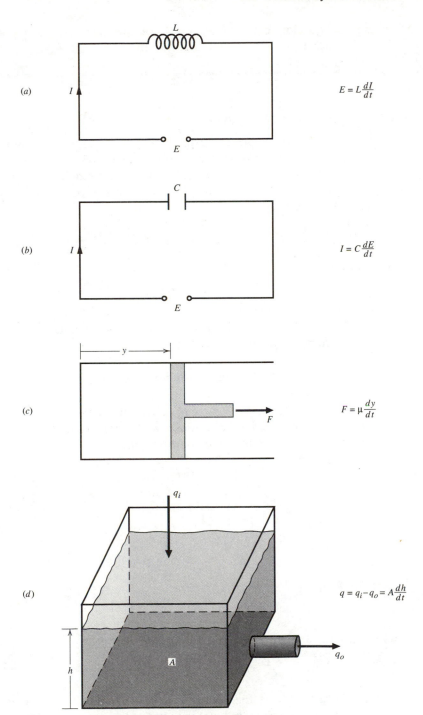

Figure 1.2-2 Examples of scalar first-order dynamic systems.

which includes an even larger class of systems. For example, see Exercises 1.5-1 through 1.5-3.

Linearization of Nonlinear Systems

Systems of interest are often nonlinear. However, control system design of such systems, within a limited range of operation, is still possible by using linear methods. We can easily show that any nonlinear system will behave linearly if it is sufficiently near a reference operating condition. Consider the system

$$\dot{\mathbf{X}} = \mathbf{f}(\mathbf{X},\mathbf{U},\mathbf{V})$$

with outputs

$$\mathbf{Y} = \mathbf{g}(\mathbf{X},\mathbf{U},\mathbf{V},\mathbf{W}),$$

and let $\overline{\mathbf{X}}(t)$ denote a solution generated by specified inputs $\overline{\mathbf{U}}(t)$, $\overline{\mathbf{V}}(t)$, and $\overline{\mathbf{W}}(t)$. The corresponding output is given by

$$\overline{\mathbf{Y}}(t) = \mathbf{g}[\overline{\mathbf{X}}(t),\overline{\mathbf{U}}(t),\overline{\mathbf{V}}(t),\overline{\mathbf{W}}(t)].$$

We will let the **perturbations** $\mathbf{x}(t)$, $\mathbf{u}(t)$, $\mathbf{v}(t)$, $\mathbf{w}(t)$, and $\mathbf{y}(t)$ denote small changes from the **reference conditions** $\overline{\mathbf{X}}(t)$, $\overline{\mathbf{U}}(t)$, $\overline{\mathbf{V}}(t)$, $\overline{\mathbf{W}}(t)$, and $\overline{\mathbf{Y}}(t)$, respectively. The corresponding solution generated by

$$\mathbf{U}(t) = \overline{\mathbf{U}}(t) + \mathbf{u}(t) \tag{1.2-26}$$

$$\mathbf{V}(t) = \overline{\mathbf{V}}(t) + \mathbf{v}(t) \tag{1.2-27}$$

and

$$\mathbf{W}(t) = \overline{\mathbf{W}}(t) + \mathbf{w}(t) \tag{1.2-28}$$

is of the form

$$\mathbf{X}(t) = \overline{\mathbf{X}}(t) + \mathbf{x}(t), \tag{1.2-29}$$

with outputs

$$\mathbf{Y}(t) = \overline{\mathbf{Y}}(t) + \mathbf{y}(t). \tag{1.2-30}$$

If the magnitudes of the perturbations are small for all times, then Taylor's theorem implies that $\mathbf{x}(t)$ is approximated by the solution to the

state perturbation equations

$$\dot{\mathbf{x}} = \left[\frac{\partial \mathbf{f}}{\partial \mathbf{X}}\right]\mathbf{x} + \left[\frac{\partial \mathbf{f}}{\partial \mathbf{U}}\right]\mathbf{u} + \left[\frac{\partial \mathbf{f}}{\partial \mathbf{V}}\right]\mathbf{v} \qquad \text{(1.2-31)}$$

and $\mathbf{y}(t)$ is approximately the solution to the **output perturbation equations**

$$\mathbf{y} = \left[\frac{\partial \mathbf{g}}{\partial \mathbf{X}}\right]\mathbf{x} + \left[\frac{\partial \mathbf{g}}{\partial \mathbf{U}}\right]\mathbf{u} + \left[\frac{\partial \mathbf{g}}{\partial \mathbf{V}}\right]\mathbf{v} + \left[\frac{\partial \mathbf{g}}{\partial \mathbf{W}}\right]\mathbf{w}, \qquad \text{(1.2-32)}$$

where the partial derivative matrices in (1.2-31) and (1.2-32) are evaluated along the reference conditions. If the reference conditions are all constants, as they frequently are, so that these matrices are constant, then (1.2-31) and (1.2-32) are of the form of (1.2-22) and (1.2-23). These equations accurately describe the motion of the nonlinear system in the neighborhood of the reference conditions. Provided that the system remains in this neighborhood, all of the linear control analysis presented in this book remains valid.

In (1.2-31) and (1.2-32) we have used a shorthand notation for matrices of partial derivatives. To illustrate this notation more fully, let $\mathbf{x}$ and $\delta\mathbf{x}$ be N_x vectors (that is, $N_x \times 1$ column matrices), $\psi(\mathbf{x})$ be a scalar-valued function, and $\mathbf{z}(\mathbf{x})$ be an N_z vector function. By convention we treat the **gradient operator** $\partial()/\partial\mathbf{x}$ as yielding a *row* matrix when applied to a scalar

$$\frac{\partial \psi}{\partial \mathbf{x}} \triangleq \left[\frac{\partial \psi}{\partial x_1} \cdots \frac{\partial \psi}{\partial x_{N_x}}\right], \qquad \text{(1.2-33)}$$

so that we can compactly write "dot product" expressions, such as

$$\left[\frac{\partial \psi}{\partial \mathbf{x}}\right]\delta\mathbf{x} = \frac{\partial \psi}{\partial x_1}\delta\mathbf{x}_1 + \cdots + \frac{\partial \psi}{\partial x_{N_x}}\delta\mathbf{x}_{N_x}.$$

The partial derivative of a (column) vector with respect to a vector is obtained by applying the operator to each element of the column vector

$$\frac{\partial \mathbf{z}}{\partial \mathbf{x}} \triangleq \begin{bmatrix} \dfrac{\partial z_1}{\partial \mathbf{x}} \\ \vdots \\ \dfrac{\partial z_{N_z}}{\partial \mathbf{x}} \end{bmatrix} = \begin{bmatrix} \dfrac{\partial z_1}{\partial x_1} & \dfrac{\partial z_1}{\partial x_2} & \cdots & \dfrac{\partial z_1}{\partial x_{N_x}} \\ \vdots & & \ddots & \vdots \\ \dfrac{\partial z_{N_z}}{\partial x_1} & \dfrac{\partial z_{N_z}}{\partial x_2} & \cdots & \dfrac{\partial z_{N_z}}{\partial x_{N_x}} \end{bmatrix}. \qquad \text{(1.2-34)}$$

To find the partial derivative of a row matrix, first take the transpose of the row to create a column matrix. For example, the partial derivative of (1.2-33) is given by

$$\frac{\partial^2 \psi}{\partial \mathbf{x}^2} \triangleq \frac{\partial}{\partial \mathbf{x}} \left\{ \left[\frac{\partial \psi}{\partial \mathbf{x}} \right]^T \right\} = \begin{bmatrix} \dfrac{\partial^2 \psi}{\partial x_1^2} & \dfrac{\partial^2 \psi}{\partial x_2 \partial x_1} & \cdots & \dfrac{\partial^2 \psi}{\partial x_{N_x} \partial x_1} \\ \vdots & \vdots & \ddots & \vdots \\ \dfrac{\partial^2 \psi}{\partial x_1 \partial x_{N_x}} & \dfrac{\partial^2 \psi}{\partial x_2 \partial x_{N_x}} & \cdots & \dfrac{\partial^2 \psi}{\partial x_{N_x}^2} \end{bmatrix}. \qquad \textbf{(1.2-35)}$$

EXAMPLE 1.2-3   Automobile Cruise Control

A familiar device to many motorists today is the automatic cruise control. After the driver sets the desired speed, the car is able to maintain that speed in spite of changes in wind conditions and slope of the highway (provided, of course, that these changes are not too great).

Figure 1.2-3 illustrates the forces acting on an automobile as it cruises at a velocity X along a highway of some unknown slope V. The forces opposing motion are a component of the car's weight ($mg \sin V$), rolling friction ($\mu mg \cos V$), and wind resistance (kX^2). The engine provides power to the wheels, which results in a net thrust U. The equation of motion is thus given by

$$m\dot{X} = -mg \sin V - \mu mg \cos V - kX^2 + U. \qquad \textbf{(1.2-36)}$$

Note that a negative slope will provide a positive accelerating force due to weight and that the engine alone can provide both positive and negative

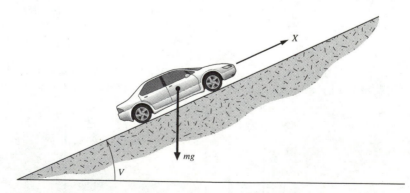

Figure 1.2-3 An automobile traveling at a velocity X along a highway of slope V.

values for U. That is, if we assume that the car is in high gear, the engine compression can provide negative thrust without the use of brakes. As the throttle is moved from its minimum to its maximum position, the thrust is varied from a minimum negative value to a maximum positive value.

Let us assume that we wish to design a cruise controller that will maintain the car's speed at some fixed value $\overline{X} = 50$ mph (the car is assumed to remain in high gear). Since we are interested in motion only in the neighborhood of $\overline{X}$, we will linearize (1.2-36) about this nominal value. Assume that when the cruise control is engaged, the slope of the highway is $\overline{V}$. We obtain the corresponding value for $\overline{U}$ from (1.2-36) by setting $\dot{X} = 0$

$$\overline{U} = mg \sin \overline{V} + \mu mg \cos \overline{V} + k\overline{X}^2. \tag{1.2-37}$$

Linearizing (1.2-36) by means of (1.2-31) yields

$$\dot{x} = -(g \cos \overline{V})v + (\mu g \sin \overline{V})v - \left(\frac{2k\overline{X}}{m}\right) x + \frac{u}{m} \tag{1.2-38}$$

or equivalently

$$\dot{x} = Ax + Bu + Rv, \tag{1.2-39}$$

where $A = -2k\overline{X}/m$, $B = 1/m$, and $R = \mu g \sin \overline{V} - g \cos \overline{V}$. This is exactly of the form of (1.2-22). Note that for this problem, all variables are scalars. The only uncertain input considered here is the slope of the road. Since the change in speed from the reference condition can easily be measured, it would be used for the output. That is,

$$y = x. \tag{1.2-40}$$

1.3   SOME APPLICATION MODELS

In order to design a controller for a particular system, a model for that system must first be developed. In this section we will develop models for a number of dynamical systems that require automatic control for operation, accuracy, or ease of use.

Magnetic Suspension System

A simplified model for a magnetically suspended rotor is obtained by assuming one-dimensional motion along an unstable axial direction as shown in Figure 1.3-1. It is further assumed that an inverse square force field exists between the gaps. Thus, a rotor of length 2ℓ, whose center of

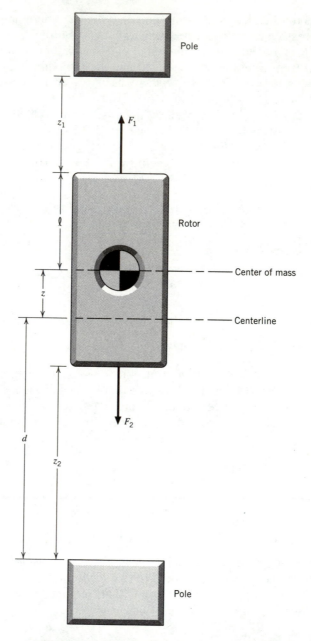

Figure 1.3-1 Magnetic suspension system.

mass is displaced by a distance z from the centerline between the poles, will experience a force of attraction given by

$$F_1 = \frac{k_1}{(k_2 + z_1)^2} \qquad (1.3\text{-}1)$$

to the upper pole and a force of attraction given by

$$F_2 = \frac{k_1}{(k_2 + z_2)^2} \qquad (1.3\text{-}2)$$

to the lower pole, where z_1 is the gap between the top of the rotor and the upper pole, z_2 is the gap between the bottom of the rotor and the lower pole, and k_1 and k_2 are constants associated with the magnetic system.

Let $2d$ be the distance between the poles; then

$$z_1 = d - \ell - z \qquad (1.3\text{-}3)$$

and

$$z_2 = d - \ell + z. \qquad (1.3\text{-}4)$$

If we let $k_3 = k_2 + d - \ell$, the net force toward the upper pole is a nonlinear function given by

$$F_1 - F_2 = \frac{4k_1 k_3 z}{(k_3^2 - z^2)^2}. \qquad (1.3\text{-}5)$$

For small z, from a Taylor series expansion about $z = 0$, the net force is approximately

$$F_1 - F_2 = kz, \qquad (1.3\text{-}6)$$

where $k = 4k_1/k_3^3$ represents the "stiffness" of the unstable rotor. In addition to this magnetic force, it is assumed that an additional force u can be created by applying current to control coils. Hence, the axial acceleration of the rotor is determined from

$$\ddot{z} - \beta_1 z = \beta_2 u, \qquad (1.3\text{-}7)$$

where

$$\beta_1 = \frac{k}{m} \qquad (1.3\text{-}8)$$

$$\beta_2 = \frac{1}{m} \qquad (1.3\text{-}9)$$

and m is the mass of the rotor.

For this system suppose that the displacement z can be measured. By letting $y = z$, $x_1 = z$, and $x_2 = \dot{z}$, we obtain the equivalent system of equations

$$\dot{x}_1 = x_2 \tag{1.3-10}$$

$$\dot{x}_2 = \beta_1 x_1 + \beta_2 u \tag{1.3-11}$$

$$y = x_1, \tag{1.3-12}$$

which are of the form of (1.2-22) and (1.2-23). Note that the two first-order equations (1.3-10) and (1.3-11) are equivalent to the single second-order equation (1.3-7), with the output y identified by means of (1.3-12).

Line-of-Sight Missile Guidance

Consider the constant-speed line-of-sight-seeking guided missile shown in Figure 1.3-2. This missile is to be guided so that it follows an established line of sight (LOS) from the starting point to the target. All motion is in the horizontal plane so that gravitational effects do not need to be modeled. Figure 1.3-2 is drawn looking down on the missile as it moves in the horizontal plane.

The rate at which the missile deviates from the LOS for small angles γ is given by

$$\dot{z} = V\gamma, \tag{1.3-13}$$

where V is the missile speed and γ the direction the velocity vector makes with the LOS. The time rate of change of flight path angle γ satisfies

$$mV\dot{\gamma} = \tfrac{1}{2}\rho V^2 SC_L = \tfrac{1}{2}\rho V^2 SC_{L_\alpha}\alpha, \tag{1.3-14}$$

provided that the angle of attack α remains small so that the lateral lift coefficient C_L can be approximated by the linear relation $C_L \approx C_{L_\alpha}\alpha$. The other parameters in (1.3-14) are air density ρ, lifting surface area S, missile mass m, and the change in lift coefficient due to α, C_{L_α}. Similarly, the approximate time rate of change of angular momentum is given by

$$I\ddot{\theta} = \tfrac{1}{2}\rho V^2 SdC_M = \tfrac{1}{2}\rho V^2 Sd(C_{M_\alpha}\alpha + C_{M_\delta}\delta), \tag{1.3-15}$$

where I is the moment of inertia of the missile about the center of mass, d is a characteristic distance (such as the average cord length of the fins, or the length of the missile), δ is the control tab deflection, C_{M_α} is the change in moment coefficient C_M due to α, and C_{M_δ} is the change in moment coefficient due to δ, with $C_M \approx C_{M_\alpha}\alpha + C_{M_\delta}\delta$. The other parameters are as previously defined.

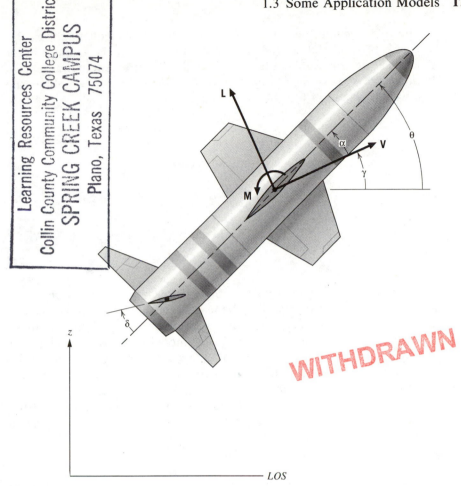

Figure 1.3-2 Line-of-sight missile.

If we use the fact that $\theta = \alpha + \gamma$, Equations (1.3-14) and (1.3-15) can be combined to yield

$$\ddot{\alpha} + \beta_2\dot{\alpha} + \beta_1\alpha = \beta_3\delta, \tag{1.3-16}$$

where

$$\beta_1 = -\frac{\rho S d V^2 C_{M_\alpha}}{2I} \tag{1.3-17}$$

$$\beta_2 = \frac{\rho S V C_{L_\alpha}}{2m} \tag{1.3-18}$$

$$\beta_3 = \frac{\rho S d V^2 C_{M_\delta}}{2I}. \tag{1.3-19}$$

The total system, described by the two first-order equations (1.3-13) and (1.3-14) and the second-order equation (1.3-16), is a fourth-order dynamical system. Suppose that z is the only measurement (output). Then letting $x_1 = \alpha$, $x_2 = \dot{\alpha}$, $x_3 = \gamma$, $x_4 = z$, $u = \delta$, and $y = z$, we obtain the equivalent system of four first-order equations

$$\dot{x}_1 = x_2 \tag{1.3-20}$$

$$\dot{x}_2 = -\beta_1 x_1 - \beta_2 x_2 + \beta_3 u \tag{1.3-21}$$

$$\dot{x}_3 = \beta_2 x_1 \tag{1.3-22}$$

$$\dot{x}_4 = V x_3 \tag{1.3-23}$$

$$y = x_4. \tag{1.3-24}$$

Note that the constant-speed assumption, the small angle assumptions, and other assumptions implied with the use of C_{L_α}, and so on, allowed us to develop a linear model for this system directly. If any of these assumptions is not valid, then the linear model is no longer valid. For example, if the speed is not constant, the order of the system will be increased by the addition of a differential equation for the speed change and the total system becomes nonlinear in V (see Exercise 1.5-12).

Notice that (1.3-20) and (1.3-21) do not contain x_3 or x_4, and that the evolution of x_3 and x_4 is completely governed by $x_1(t) = \alpha(t)$. Thus if α can be measured, we need only (1.3-20) and (1.3-21), with (1.3-22)–(1.3-24) replaced by

$$y = x_1, \tag{1.3-25}$$

to design a controller for the system. That is, (1.3-22) and (1.3-23) can be used after the fact to compute flight path angle and deviation from the LOS. This is an example of a case in which it is possible to reduce the order of the system model, provided that the new output α rather than z represents the dynamical variable of interest. Otherwise [as in the case of the originally stated objective of $z(t) \to 0$], we must deal with the fourth-order system.

Longitudinal Motion of an Aircraft

Consider the aircraft depicted in Figure 1.3-3. Motion will be determined with respect to a set of **body-fixed coordinates** x, y, and z. The angle Γ between the horizon and the velocity vector is called the flight path angle, and the angle A between the x axis and the velocity vector is called the

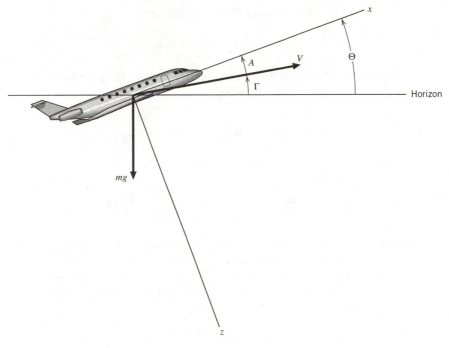

Figure 1.3-3 Longitudinal motion of an aircraft.

angle of attack. The angle Θ between the x axis and the horizon is called the pitch attitude angle.

In order to use Newton's laws to formulate the equations of motion, we need expressions for the velocity and acceleration vectors of the aircraft relative to a nonrotating coordinate system. These can be found using **Chasle's theorem,** which states that the general motion of a rigid body is the vector sum of a pure translation and pure rotation. A corollary of Chasle's theorem is that for any vector $\mathbf{a}(t)$, the time rate of change in a nonrotating coordinate system is related to the rate of change seen by a rotating observer by the formula

$$\frac{d\mathbf{a}}{dt} = \frac{d\mathbf{a}}{dt}\bigg|_{\text{rot}} + \mathbf{\Omega} \times \mathbf{a}, \qquad (1.3\text{-}26)$$

where $\times$ denotes the "cross product" vector and $\mathbf{\Omega}$ is the angular velocity of the rotating coordinate system.

For this problem the angular velocity of the rotating coordinate system is given by $\dot{\Theta}\mathbf{j}$, where $\mathbf{j}$ is a unit vector directed outward and perpendicular to the x and z axes. If we let U be the component of velocity directed along the x axis and W be the component of velocity directed along the z

axis, then the total velocity vector is given by

$$\mathbf{V} = U\mathbf{i} + W\mathbf{k},$$

where $\mathbf{i}$ and $\mathbf{k}$ are unit vectors in the direction of the positive x and z axes. Applying the corollary, we obtain

$$\frac{d\mathbf{V}}{dt} = \frac{d}{dt}(U\mathbf{i} + W\mathbf{k})\Big|_{\text{rot}} + \dot{\Theta}\mathbf{j} \times (U\mathbf{i} + W\mathbf{k})$$

or

$$\frac{d\mathbf{V}}{dt} = (\dot{U} + \dot{\Theta}W)\mathbf{i} + (\dot{W} - \dot{\Theta}U)\mathbf{k},$$

where "dot" now denotes time derivatives in the rotating coordinate system. In addition to the gravitational force shown in Figure 1.3-3, thrust and aerodynamic forces are also acting on the aircraft. Let X denote the x component and Z denote the z component of all these additional forces. From Newton's second law we obtain

$$m(\dot{U} + \dot{\Theta}W) = X - mg \sin \Theta$$
$$m(\dot{W} - \dot{\Theta}U) = Z + mg \cos \Theta.$$

In addition, rotational motion is determined from

$$I\ddot{\Theta} = M,$$

where I is the moment of inertia of the aircraft about the origin of the body-fixed coordinates and M the sum of all moments about the same point. If we define

$$\dot{\Theta} = Q, \tag{1.3-27}$$

then the above equations can be written as

$$m\dot{U} = X - mg \sin \Theta - mQW \tag{1.3-28}$$

$$m\dot{W} = Z + mg \cos \Theta + mQU \tag{1.3-29}$$

$$I\dot{Q} = M. \tag{1.3-30}$$

Equations (1.3-27)–(1.3-30) define the longitudinal motion of the aircraft in terms of the four state variables Θ, U, W, and Q. The state variable Q is called the pitch rate. In general, control forces can enter through X, Z, and M. We will simplify the system here by assuming that control will enter through the M term only. Specifically, we will assume that the forces and moments can be expressed functionally as

$$X = X(U,W)$$

$$Z = Z(U,W)$$

$$M = M(U,W,Q,N),$$

where N, the control input, corresponds to a deflection of the elevator on the horizontal stabilizer. In general, X, Z, and M are rather complex functions of the thrust, lift, drag, and moments acting on the aircraft. The lift, drag, and moment will all depend on the angle of attack A. Since

$$\sin A = \frac{W}{\sqrt{U^2 + W^2}}, \tag{1.3-31}$$

we see that such a dependence has been included. However, we have neglected the dependence on other quantities such as rate of change of A and Θ that for some aircraft may be important.

Suppose that our objective is to design a controller that will maintain the aircraft in the neighborhood of the nominal condition corresponding to steady level flight. That is, the nominal condition corresponds to

$$\dot{U} = \dot{W} = \dot{Q} = \dot{\Theta} = \Gamma = 0.$$

Substituting these requirements into (1.3-27)–(1.3-30) yields

$$Q = 0$$

$$X(U,W) - mg \sin \Theta = 0$$

$$Z(U,W) + mg \cos \Theta = 0$$

$$M(U,W,Q,N) = 0.$$

It follows from Figure 1.3-3 that when $\Gamma = 0$, $\Theta = A$, which from (1.3-31) is a function of U and W. Thus, the above equations, involving only four unknowns Q, U, W, and N, may be solved to obtain nominal values for $\overline{V}$, $\overline{W}$, and $\overline{N}$ along with $\overline{Q} = 0$. If the body-fixed coordinates are chosen so that the nominal value for $\overline{W} = 0$ [which from (1.3-31) implies $\overline{A} = \overline{Q} = 0$], then these coordinates are called **stability axes**. We make this choice not only because it is a common one but also because it will simplify the following analysis. The relationship between actual, nominal, and perturbation values [see (1.2-26)–(1.3-30)] is given by

$$U = \overline{U} + u$$
$$W = \overline{W} + w$$
$$Q = \overline{Q} + q$$
$$\Theta = \overline{\Theta} + \theta$$
$$N = \overline{N} + \eta,$$

where the lowercase letters are used to denote perturbations. Applying the state perturbation equations (1.2-31) to the system (1.3-27)–(1.3-30) evaluated at the reference conditions yields

$$\dot{\theta} = q \qquad\qquad\qquad (1.3\text{-}32)$$

$$m\dot{u} = X_u u + X_w w - mg\theta \qquad\qquad (1.3\text{-}33)$$

$$m\dot{w} = Z_u u + Z_w w + m\overline{U}q \qquad\qquad (1.3\text{-}34)$$

$$I\dot{q} = M_u u + M_w w + M_q q + M_\eta \eta, \qquad\qquad (1.3\text{-}35)$$

where X_u denotes the partial derivative of X with respect to U evaluated at the reference conditions. The same notation is used to define X_w, and so on. The system is now a linear system. However, it is still not quite in the form favored by aircraft stability and control engineers. It follows from (1.3-31) that for small A and W

$$A \approx \frac{W}{U}.$$

Since

$$A = \overline{A} + \alpha = \frac{\overline{W} + w}{\overline{U} + u} \approx \frac{\overline{W} + w}{\overline{U}} = \overline{A} + \frac{w}{\overline{U}},$$

if follows that

$$\alpha = \frac{w}{\overline{U}}.$$

Replacing w with $\overline{U}\alpha$ in (1.3-33)–(1.3-35) along with some rearrangement, we obtain

$$\dot{u} = \frac{X_u}{m} u + \frac{\overline{U} X_w}{m} \alpha - g\theta \qquad (1.3\text{-}36)$$

$$\dot{\alpha} = \frac{Z_u}{m\overline{U}} u + \frac{Z_w}{m} \alpha + q \qquad (1.3\text{-}37)$$

$$\dot{\theta} = q \qquad (1.3\text{-}38)$$

$$\dot{q} = \frac{M_u}{I} u + \frac{\overline{U} M_w}{I} \alpha + \frac{M_q}{I} q + \frac{M_\eta}{I} \eta, \qquad (1.3\text{-}39)$$

which is of the form

$$\dot{\mathbf{x}} = \mathbf{A}\mathbf{x} + \mathbf{B}\eta,$$

where

$$\mathbf{x} = [u, \alpha, \theta, q]^T,$$

$$\mathbf{A} = \begin{bmatrix} \dfrac{X_u}{m} & \dfrac{\overline{U} X_w}{m} & -g & 0 \\ \dfrac{Z_u}{\overline{U}m} & \dfrac{Z_w}{m} & 0 & 1 \\ 0 & 0 & 0 & 1 \\ \dfrac{M_u}{I} & \dfrac{\overline{U} M_w}{I} & 0 & \dfrac{M_q}{I} \end{bmatrix}, \qquad (1.3\text{-}40)$$

and

$$\mathbf{B} = \left[0, 0, 0, \frac{M_\eta}{I} \right]^T. \qquad (1.3\text{-}41)$$

The elements of the **A** matrix are known as **stability derivatives**. Values for a particular aircraft are determined from flight test results or are estimated from wind tunnel data.

Rotational Speed of a DC Motor

The basic elements of an armature-controlled dc motor are illustrated in Figure 1.3-4. A voltage E applied to the input leads produces a current I_A through the armature that is of resistance R_A and inductance L_A. This current produces a torque that is proportional to the product of the magnetic flux Φ of the field and the armature current I_A. That is,

$$\Gamma = k_1 \Phi I_A,$$

where k_1 is a motor constant.

The magnetic flux Φ may be produced either by a constant field current I_F through the coil as illustrated in Figure 1.3-4, with Φ being proportional to I_F, or it may be produced by using a permanent magnet. In either case, Φ is constant for an armature-controlled dc motor. Thus, the torque is proportional to the armature current as given by

$$\Gamma = k_\Gamma I_A, \tag{1.3-42}$$

where $k_\Gamma = k_1 \Phi$. The voltage E_B is the back emf induced by the rotation of the armature windings in the magnetic field. The back emf is proportional to the armature rotational speed $\dot{\theta}$ and field strength Φ. That is, $E_B = k_2 \Phi \dot{\theta}$, where k_2 is another motor constant. Since the field strength is constant, we can replace the product of constants by $k_B = k_2 \Phi$ to obtain

$$E_B = k_B \dot{\theta}. \tag{1.3-43}$$

Applying Kirchoff's voltage law around the armature loop yields

$$E = R_A I_A + L_A \dot{I}_A + k_B \dot{\theta}, \tag{1.3-44}$$

where R_A and L_A are the armature resistance and inductance, respectively. Using (1.3-42) to eliminate I_A in (1.3-44), we obtain

$$E = \frac{R_A}{k_\Gamma} \Gamma + \frac{L_A}{k_\Gamma} \dot{\Gamma} + k_B \dot{\theta}. \tag{1.3-45}$$

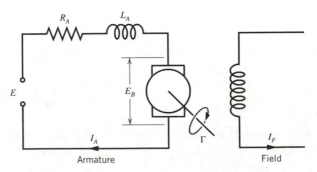

Figure 1.3-4 Armature-controlled dc motor.

Consider now the rotor and shaft of the motor. The angular acceleration of the rotor and shaft will be proportional to the torque Γ provided by the motor plus any external torques Γ_{ex} (such as the reaction torque from an external device attached by a pulley and belt) as given by

$$J\ddot{\theta} = \Gamma + \Gamma_{ex}, \tag{1.3-46}$$

where J is the moment of inertia of the rotor and shaft plus all attachments. Thus,

$$\Gamma = J\ddot{\theta} - \Gamma_{ex}.$$

Substituting into (1.3-45) and using $\theta^{(3)}$ to denote the third time derivative of θ yields

$$E = \frac{R_A}{K_\Gamma}(J\ddot{\theta} - \Gamma_{ex}) + \frac{L_A}{K_\Gamma}[J\theta^{(3)} - \dot{\Gamma}_{ex}] + K_B\dot{\theta}.$$

If we denote the rotation rate $\dot{\theta}$ by ω, then we can write the above as a second-order system

$$\frac{L_A J}{k_\Gamma}\ddot{\omega} + \frac{R_A J}{k_\Gamma}\dot{\omega} + k_B\omega = E + \frac{R_A}{k_\Gamma}\Gamma_{ex} + \frac{L_A}{k_\Gamma}\dot{\Gamma}_{ex}. \tag{1.3-47}$$

For motors of small inductance, $L_A \approx 0$, so this equation is often approximated by

$$\dot{\omega} + \frac{k_B k_\Gamma}{R_A J}\omega = \frac{k_\Gamma}{R_A J}E + \frac{1}{J}\Gamma_{ex}. \tag{1.3-48}$$

By setting $\omega = x$, $E = u$, $\Gamma_{ex} = v$, $-k_B k_\Gamma/(R_A J) = A$, $k_\Gamma/(R_A J) = B$, and $1/J = R$, we observe that (1.3-48) is in the scalar form of (1.2-22).

Finally, we will show that k_Γ and k_B are the same constant, expressed in different units. In the steady state, with $\Gamma_{ex} = 0$, we see from (1.3-48) that

$$E = k_B\omega = E_B.$$

That is, in the absence of external torques the voltage supplied to the motor equals the back emf. In this steady-state situation, the power input to the motor is given by

$$\text{Power in} = E_B I_A = k_B\omega I_A$$

and the power output of the motor is given by

$$\text{Power out} = \Gamma\omega = k_\Gamma\omega I_A.$$

In the absence of Γ_{ex}, the steady-state power input equals the power output, and we conclude that

$$k_B = k_\Gamma.$$

For example, if k_Γ is expressed in terms of ft-lb/amp and k_B is expressed in terms of V-sec/rad, then the conversion factor is given by

$$\frac{\text{V-sec}}{\text{rad}} \times \frac{2655}{3600} = \frac{\text{ft-lb}}{\text{amp}},$$

where 1 kW-hr $= 2.655 \times 10^6$ ft-lb.

Inverted Pendulum

Consider a version of the stick balancing problem mentioned earlier. Figure 1.3-5 illustrates a uniform cylindrical rod pinned at the origin and operating under the influence of gravity and a torque Γ applied at the origin.

The moment of inertia of the cylindrical rod about the center of rotation is given by $4m\ell^2/3$, where ℓ is the half-length and m is the total mass. Let

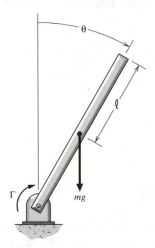

Figure 1.3-5 Inverted pendulum.

J_m be the moment of inertia of the motor shaft and rotor so that

$$J = J_m + \frac{4m\ell^2}{3}$$

is the total rotational moment of inertia of the system. Equating the rate of change of angular momentum about the origin to the applied torques yields

$$J\ddot{\theta} = \Gamma + mg\ell \sin\theta, \qquad (1.3\text{-}49)$$

where θ is the displacement angle from the vertical.

If we considered the torque Γ as the control input, we could stop here, and (1.3-49) would yield a second-order system. The controls analysis would be to determine the function $\Gamma(\cdot)$ for controlling the system. This would allow us not only to employ a low-order model but also to focus on the quantity Γ that directly controls the system. However, some actuator will be required to produce the control torque. In general, actuators are also dynamical subsystems, so the actual problem is to determine the actuator input to produce the desired torque. Suppose that for our inverted beam, the torque is generated by an armature-controlled dc motor, as illustrated in Figure 1.3-4. Combining (1.3-49) and (1.3-45) yields the following model for the rod-motor combination:

$$\frac{L_A}{k_\Gamma} J\theta^{(3)} + \frac{R_A}{k_\Gamma} J\ddot{\theta} + \left(k_B - \frac{L_A}{k_\Gamma} mg\ell \cos\theta\right)\dot{\theta} - \frac{R_A}{k_\Gamma} mg\ell \sin\theta = E.$$

$$(1.3\text{-}50)$$

The nonlinear differential equation (1.3-50) describes the general motion of the rod-motor system for the case where the actuator torque is generated by an armature-controlled dc motor. This nonlinear equation could be used, for example, to investigate the control problem of returning the rod to the upward vertical position $\theta = 0$, starting from some arbitrary position, such as $\theta = \pi$, or from some position possibly more than one revolution away from $\theta = 0$.

However, suppose the control problem is simply to restore the rod to the $\theta = 0$ position given that it is in some neighborhood of the desired position. If we make the small angle assumption $\sin\theta \approx \theta$ and $\cos\theta \approx 1$, then (1.3-50) after multiplying through by β_3 defined below reduces to the linear differential equation

$$\epsilon_1\theta^{(3)} + \ddot{\theta} + \epsilon_2\dot{\theta} + \beta_2\dot{\theta} + \beta_1\theta = \beta_3 E, \qquad (1.3\text{-}51)$$

where

$$\beta_1 = -\frac{mg\ell}{J} \tag{1.3-52}$$

$$\beta_2 = \frac{k_B k_\Gamma}{JR_A} \tag{1.3-53}$$

$$\beta_3 = \frac{k_\Gamma}{JR_A} \tag{1.3-54}$$

$$\epsilon_1 = \frac{L_A}{R_A} \tag{1.3-55}$$

$$\epsilon_2 = -\frac{L_A}{R_A}\frac{mg\ell}{J} = \epsilon_1\beta_1. \tag{1.3-56}$$

Suppose that the angular displacement θ is the only quantity that is measured ($y = \theta$). Letting $x_1 = \theta$, $x_2 = \dot{\theta}$, $x_3 = \ddot{\theta}$, and $u = E$, we see that (1.3-51) is equivalent to the system of equations

$$\dot{x}_1 = x_2 \tag{1.3-57}$$

$$\dot{x}_2 = x_3 \tag{1.3-58}$$

$$\epsilon_1\dot{x}_3 = -\beta_1 x_1 - (\beta_2 + \epsilon_2)x_2 - x_3 + \beta_3 u \tag{1.3-59}$$

$$y = x_1. \tag{1.3-60}$$

This third-order linear system can be put into the form of (1.2-22) and (1.2-23) by dividing (1.3-59) by $\epsilon_1 = L_A/R_A$, providing that $\epsilon_1 \neq 0$. But suppose that the armature inductance L_A is very small, which is frequently true. Then after we divide by ϵ_1, the right-hand side of the new equation would generally be very large compared with (1.3-57) and (1.3-58), which means that x_3 changes very rapidly. This is not a major problem, since we will be able to develop closed-form solutions for this linear system. However, if we were to use numerical integration techniques, we would have to either take very small time steps to accurately integrate (1.3-59) or to employ some special technique designed to handle **stiff systems**, that is, systems of differential equations having two or more time scales (for example, fast and slow) present.

On the other hand, the rapid (practically instantaneous) change in x_3 means that x_3 must essentially satisfy an algebraic equation ($\epsilon_1 = 0$), rather than a differential equation ($\epsilon_1 \neq 0$). Indeed, if L_A really is zero, then so are ϵ_1 and ϵ_2 and we can solve (1.3-59) for x_3, yielding a second-order system

$$\dot{x}_1 = x_2 \tag{1.3-61}$$

$$\dot{x}_2 = -\beta_1 x_1 - \beta_2 x_2 + \beta_3 u \tag{1.3-62}$$

$$y = x_1. \tag{1.3-63}$$

For small inductance, then, we are able to reduce the order of the system model and still obtain a reasonable representation of the real system.

Geosynchronous Satellite

Consider the problem of designing an on-board automatic thrust controller for a satellite, with the control task being to maintain the satellite in a geosynchronous circular orbit, directly above a certain point on the Earth's equator. We strive here to obtain the simplest possible dynamical system model that captures the important features of the problem.

Figure 1.3-6 shows an orthogonal rotating coordinate system with unit vectors $\mathbf{e}_r$, $\mathbf{e}_\theta$, $\mathbf{e}_z$, where $\mathbf{e}_z$ points out of the paper and coincides with the Earth's axis through the North Pole. The coordinate system is Earth-centered and rotates so that $\mathbf{e}_r$ always points at the satellite. Together the unit vectors $\mathbf{e}_r$ and $\mathbf{e}_\theta$ define the satellite's orbit plane, which contains the

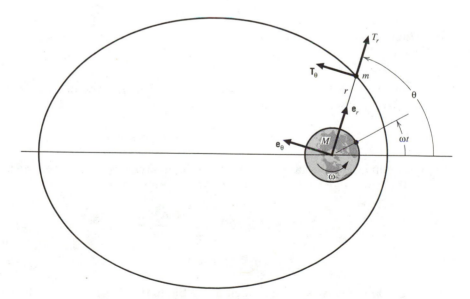

Figure 1.3-6 Satellite in Earth orbit.

Earth's equator. The Earth is rotating counterclockwise about the $\mathbf{e}_z$ axis with an angular speed ω and the satellite moves counterclockwise in its orbit, but it can change its orbit by thrusting. We let r be the distance from the center of the Earth to the satellite and let θ be the angle to the satellite from some fixed baseline direction, specifically, the direction from the Earth's center through the target position on the equator at time $t = 0$.

We will assume the mass of the satellite to be constant under the influence of the Earth's gravitational force only. The Earth is assumed to be a homogeneous sphere rotating at a constant angular velocity, with the center of the Earth assumed to be a fixed point in space. The control problem is to maintain the satellite in a circular orbit over a desired position on the surface of the Earth subject to all perturbations tending to move the satellite away from such a position (e.g., consider the assumptions employed in the model). That is, we seek $T_r(\cdot)$ and $T_\theta(\cdot)$ such that $r(t) \to$ constant and $\theta(t) \to \omega t$.

The position vector for the satellite is given by

$$\mathbf{r} = r\mathbf{e}_r \tag{1.3-64}$$

and the angular velocity of the rotating coordinate system is

$$\Omega = \dot{\theta}\mathbf{e}_z. \tag{1.3-65}$$

Using Chasle's corollary (1.3-26), we obtain the velocity vector $\mathbf{V}$ in a nonrotating coordinate system as

$$\mathbf{V} = \dot{r}\mathbf{e}_r + \dot{\theta}\mathbf{e}_z \times r\mathbf{e}_r,$$

which reduces to

$$\mathbf{V} = \dot{r}\mathbf{e}_r + r\dot{\theta}\mathbf{e}_\theta.$$

Applying Chasle's corollary again, to the velocity vector, yields the acceleration vector

$$\mathbf{A} = (\ddot{r} - r\dot{\theta}^2)\mathbf{e}_r + (r\ddot{\theta} + 2\dot{r}\dot{\theta})\mathbf{e}_\theta.$$

For an inverse square gravitational field, applying Newton's second law in the r and θ directions results in the equations of motion

$$m(\ddot{r} - r\dot{\theta}^2) = T_r - \frac{\mu M m}{r^2} \tag{1.3-66}$$

$$m(r\ddot{\theta} + 2\dot{r}\dot{\theta}) = T_\theta, \tag{1.3-67}$$

where m is the mass of the satellite, M the mass of the Earth, and μ the universal gravitational constant. The inputs to the system, which control the orbit of the satellite, are the thrust components $T_r(\cdot)$ and $T_\theta(\cdot)$ in the r and θ directions, respectively.

We make note of uncertain inputs that are implicit in the equations of motion. For example, the gravitational term in (1.3-66) is not exact. This term neglects the fact that the Earth is not a uniform sphere and neglects the motion of the Earth about the Sun. More important, it neglects the motion of the Moon about the Earth. For simplicity, we will model all of these effects by treating the mass of the Earth as $M + \Delta M$, where ΔM is a fluctuating uncertain value whose range is assumed to be some fraction of M, say, $\pm 10^{-6} M$. Then (1.3-66) becomes

$$m(\ddot{r} - r\dot{\theta}^2) = T_r - \frac{\mu(M + \Delta M)m}{r^2}. \tag{1.3-68}$$

Because these equations of motion consist of two second-order differential equations, any state-space representation will require four first-order differential equations, involving four state variables. Since we wish to maintain the satellite in a geosynchronous orbit, it is convenient to define state variables $\mathbf{x}$, control variables $\mathbf{u}$, and a scalar system uncertainty v by

$$x_1 = r, \qquad u_1 = \frac{T_r}{m}$$

$$x_2 = \dot{r}, \qquad u_2 = \frac{T_\theta}{m}$$

$$x_3 = \theta - \omega t, \qquad v = \Delta M$$

$$x_4 = \dot{\theta},$$

where ω is the angular rotation rate of the Earth. The state equations are

$$\dot{x}_1 = x_2 \qquad\qquad\qquad = f_1(\mathbf{x},\mathbf{u},v) \tag{1.3-69}$$

$$\dot{x}_2 = x_1 x_4^2 - \frac{\mu(M + v)}{x_1^2} + u_1 = f_2(\mathbf{x},\mathbf{u},v) \tag{1.3-70}$$

$$\dot{x}_3 = x_4 - \omega \qquad\qquad = f_3(\mathbf{x},\mathbf{u},v) \tag{1.3-71}$$

$$\dot{x}_4 = -\frac{2x_2 x_4}{x_1} + \frac{u_2}{x_1} \qquad = f_4(\mathbf{x},\mathbf{u},v). \tag{1.3-72}$$

Now suppose we use radar to make measurements. In general, we will only be able to measure a portion of the state or some functions that involve only a portion of the state. For example, an Earth-based radar might measure range, azimuth, and elevation, which are trigonometric functions of r and θ. A Doppler radar might also measure range rate, but probably not very accurately at orbital distances. For simplicity, assume that the radar, after some calculations, reports exact measurements of r and $\theta - \omega t$ (the angular deviation from the desired location). Then the observation equations would be

$$y_1 = x_1$$

$$y_2 = x_3.$$

To model "noisy" measurements, we could use

$$y_1 = x_1 + w_1 \tag{1.3-73}$$

$$y_2 = x_3 + w_2, \tag{1.3-74}$$

where $\mathbf{w} = [w_1 \quad w_2]^T$ is the measurement uncertainty vector.

In general, the input variables $\mathbf{u}$, $\mathbf{v}$, and $\mathbf{w}$ to a dynamical system have certain constraints associated with them. For our satellite example, the thrust magnitude might be limited by some maximum value, say, T_{max}, so that the control vector $\mathbf{u} = [T_r/m, T_\theta/m]^T$ would have a constraint set of the form

$$\mathbf{u} \in \mathcal{U} = \left\{ (u_1, u_2) \mid u_1^2 + u_2^2 \le \left(\frac{T_{max}}{m} \right)^2 \right\}. \tag{1.3-75}$$

The system uncertainty $v = \Delta M$ was assumed to have constraints of the form

$$v \in \mathcal{V} = \{v \mid |v| \le 10^{-6}\text{M}\}. \tag{1.3-76}$$

Finally, if each component of the measurement uncertainty vector $\mathbf{w}$ is bounded in absolute value, then the corresponding constraints would be of the form

$$\mathbf{w} \in \mathcal{W} = \{(w_1, w_2) \mid |w_i| \le W_i, \quad i = 1,2\}. \tag{1.3-77}$$

We will now illustrate the linearization process in terms of the control vector $\mathbf{U} = [T_r/m, T_\theta/m]^T$ and state vector $\mathbf{X} = [r, \dot{r}, \theta - \omega t, \dot{\theta}]^T$. Ignoring any uncertainty in the mass of the Earth, we can write the equations of motion (1.3-69)–(1.3-72) as

$$\dot{X}_1 = X_2 \qquad\qquad = f_1(\mathbf{X},\mathbf{U}) \qquad\qquad (1.3\text{-}78)$$

$$\dot{X}_2 = X_1 X_4^2 - \frac{\mu M}{X_1^2} + U_1 = f_2(\mathbf{X},\mathbf{U}) \qquad\qquad (1.3\text{-}79)$$

$$\dot{X}_3 = X_4 - \omega \qquad\qquad = f_3(\mathbf{X},\mathbf{U}) \qquad\qquad (1.3\text{-}80)$$

$$\dot{X}_4 = -\frac{2X_2 X_4}{X_1} + \frac{U_2}{X_1} \qquad = f_4(\mathbf{X},\mathbf{U}). \qquad\qquad (1.3\text{-}81)$$

For $\overline{\mathbf{U}} = \mathbf{0}$ and a geosynchronous circular orbit, we have the reference solution

$$\theta(t) = \omega t$$

$$r(t) = R = \text{constant},$$

where R is determined from (1.3-66) as

$$R = \left[\frac{\mu M}{\omega^2}\right]^{1/3}.$$

The corresponding constant reference state $\overline{\mathbf{X}} = [R, 0, 0, \omega]^T$ is a solution to the **equilibrium** equations $\dot{\mathbf{X}} = \mathbf{f}(\mathbf{X},\mathbf{U}) = \mathbf{0}$, with $\mathbf{U} = \mathbf{0}$. Note that, as in this case, an "equilibrium" (constant state) does not necessarily imply an absence of motion in the underlying physical system corresponding to a state-space dynamical system model.

Let x_1, x_2, x_3, and x_4 denote small changes in r, $\dot{r}$, θ, and $\dot{\theta}$, respectively, with $\mathbf{X}(t) = \overline{\mathbf{X}}(t) + \mathbf{x}(t)$. Then the linearized equations are

$$\begin{bmatrix} \dot{x}_1 \\ \dot{x}_2 \\ \dot{x}_3 \\ \dot{x}_4 \end{bmatrix} = \begin{bmatrix} 0 & 1 & 0 & 0 \\ 3\omega^2 & 0 & 0 & 2R\omega \\ 0 & 0 & 0 & 1 \\ 0 & \dfrac{-2\omega}{R} & 0 & 0 \end{bmatrix} \begin{bmatrix} x_1 \\ x_2 \\ x_3 \\ x_4 \end{bmatrix} + \begin{bmatrix} 0 & 0 \\ 1 & 0 \\ 0 & 0 \\ 0 & \dfrac{1}{R} \end{bmatrix} \begin{bmatrix} u_1 \\ u_2 \end{bmatrix}. \qquad (1.3\text{-}82)$$

1.4   PRINCIPLES OF AUTOMATIC CONTROL

Open-Loop and Closed-Loop Controls

Figure 1.4-1 illustrates a **block diagram** of a dynamical system. A block diagram is simply a pictorial representation of the dynamical process

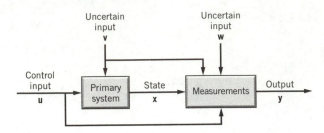

Figure 1.4-1 Dynamic system block diagram.

described by (1.2-1) and (1.2-2). The arrows represent the flow of information into and out of the blocks. Since this system is quite general, the "blocks" are also quite general. Later we will become more specific when we deal with linear systems. In the block diagram the "primary system" (also called the "plant") corresponds to (1.2-1) and is the object to be controlled.

The basic problem in the design of automatic control systems is to determine an algorithm for the control input $\mathbf{u}(\cdot)$. The objective is to achieve an acceptable behavior of the state $\mathbf{x}$ when the overall system is subjected to command inputs, such as a desired position or speed, despite any uncertain inputs that may be present in the dynamics and the measurements. That is, we wish to determine the functional relationship for the control input $\mathbf{u}$ so that integration of (1.2-1), subject to command inputs and possibly uncertain inputs, will yield a response $\mathbf{x}(t)$ with specified dynamical properties. This functional relationship for $\mathbf{u}(\cdot)$ is called the **controller**.

We will use the notation $\mathbf{r}(t)$ and the phrase **command input** to denote external inputs to the controller, such as operator-generated commands, that are available to the controller, but are not selectable by the controller. For example, $\mathbf{r}(t)$ could be the angular position of an automobile's steering wheel or the stick position on an aircraft. Indeed, $\mathbf{r}(t)$ can be a vector of many external command inputs.

The objective of automatic control design is to find a functional form for a controller (to be implemented in terms of physical components in the real system) that will produce acceptable system behavior. This functional form may vary depending on the application. For example, in some applications the controller may simply be equal to the command signal [that is, $\mathbf{u} = \mathbf{r}(t)$], or it may be a more complicated function of the command signal and time {that is, $\mathbf{u} = \mathbf{u}[\mathbf{r}(t),t]$}, or it may be a function of the command input and the output {that is, $\mathbf{u} = \mathbf{u}[\mathbf{r}(t),\mathbf{y}]$}. In the first two situations, where the control input does not depend explicitly on the output, the control is called **open-loop control**. The third situation, where the control input depends on the output rather than time, is called **closed-loop control**. The expression "closed-loop" refers to the fact that the output is used to *feed back* the measurement information into the system as part of the control algorithm.

Closed-loop control is further classified according to the exact nature of the feedback signal. If the closed-loop control input **u** depends explicitly on the output **y**, {that is, **u** = **u**[**r**(t),**y**]}, it is referred to as **output feedback**. Closed-loop control is called **state variable feedback** if all of the states are measured (or estimated by some other process) and are fed back into the system, so that the input **u** depends explicitly on the state **x** {that is, **u** = **u**[**r**(t),**x**]}.

The contrast between open-loop and closed-loop controls is a powerful basic concept in automatic controls. Imagine driving your car blindfolded (open-loop control). This is virtually impossible! But without the blindfold (closed-loop control), it's easy. In fact, you don't even need much of a model for the dynamical system to keep the car in its lane. You just monitor the car's position and turn the steering wheel until the car is where you want it. The task is even easier if the problem of simultaneously controlling the speed of the car is turned over to an automatic cruise control. Of course, the steering could also be done by an automatic controller, but it would require expensive instrumentation, both in the car and probably in the road, such as a buried electromagnetic guidewire. For these and other reasons, it is doubtful that we will ever delegate automobile steering completely to an automatic control system, as is done with an aircraft's autopilot.

Open-loop controllers exist primarily because they are cheaper to build and easier to implement than closed-loop controllers. This is because they do not require sensors to monitor the output, nor do they require much in the way of high-technology hardware to implement the control algorithm. Open-loop controllers require only a clock. As a consequence, they ignore what is happening to the response of the system and can never adapt.

In order for an open-loop control to be effective, the control designer must be able to predict perfectly the future behavior of the dynamical system. Hence, the model must be perfect, and the analysis task is essentially one of dynamics, not controls. However, no model is perfect and even if it were, all dynamical systems are subject to external disturbances, such as a gust of wind, which can cause the system to move in other than the predicted fashion. Nevertheless, there are many examples of highly effective open-loop control systems. Robotic devices that do "pick and place" operations work fine, as long as all of the parts are in the right place at the right time. But it is comic to watch such a robot "going through the motions" of spot welding thin air just because the part to be welded is not exactly where the robot "thinks" it ought to be. On the other hand, progress in robot vision systems, touch-sensitive systems, and other sensor systems enhances the prospects for closed-loop feedback control applications in robotics and several other areas.

Control System Structures

Our major concern will be with the problem of designing closed-loop feedback control systems. Figure 1.4-2 shows a block diagram of an output feedback closed-loop control system. If the dashed line were not present, the system would be an open-loop control system.

The inputs to the primary dynamical system block in Figure 1.4-2 fall into two categories: the **control inputs u** and **r**, and **uncertain inputs v** and **w**. The command input **r** comes from an operator (human or otherwise) external to the system. The automatic control input **u** is determined entirely by the controller internal to the system. The uncertain inputs **v** and **w** are generally unknown and are independent of the controller. However, their bounds may be known. The command function $\mathbf{r}(t)$ is assumed known for design purposes (often it is assumed constant), whereas the uncertain inputs $\mathbf{v}(t)$ and $\mathbf{w}(t)$ are always considered unknown. Uncertain inputs include random events, manufacturing defects, a jerky operator, or other generally unknown inputs that affect the system.

The inputs to the measurements block of Figure 1.4-2 are the state **x** and possibly the control **u** and the uncertain inputs **v** and **w**. Generally, **w** is either noise or measurement error or both. Various sensors are used to measure certain functions, components, or combinations of components of the state known as the output **y**, in accordance with (1.2-2). These measurements will not be exact because of instrument error and signal noise. Any such effects are summarized by the external input $\mathbf{w}(t)$ as indicated in Figure 1.4-2. Whether the system uncertainties **v** or control inputs **u** carry over to the measurements block will depend on the quantities being measured. This will be examined in more detail later.

If the dashed line in Figure 1.4-2 is present, the output feeds back through a block that defines an internal input to the controller block. This path "closes the loop," and satisfactory performance of the overall

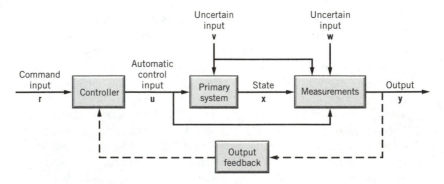

Figure 1.4-2 An output feedback control system.

system can usually be obtained through proper design of the feedback block and controller block.

We may also use a state variable feedback controller of the form $\mathbf{u} = \mathbf{u}(\mathbf{r},\mathbf{x})$, rather than an output feedback controller of the form $\mathbf{u} = \mathbf{u}(\mathbf{r},\mathbf{y})$. Figure 1.4-3 shows the structure of a state variable feedback control system. If $\mathbf{y} \neq \mathbf{x}$, the state must be estimated, using $\mathbf{x}(t) \approx \hat{\mathbf{x}}(t)$, as indicated by the estimator block. This will be true, for instance, when the output $\mathbf{y}$ contains only some of the components of the state vector $\mathbf{x}$. Generally, an estimator will require as inputs both the output $\mathbf{y}$ and control $\mathbf{u}$. Based on the control input $\mathbf{u}(t)$, the measured output $\mathbf{y}(t)$, and the model for the dynamical system, the estimator yields a state estimate $\hat{\mathbf{x}}(t)$, which is used in the state feedback system in lieu of the actual state $\mathbf{x}(t)$ to produce an input to the controller. If the full state is measured exactly, $\mathbf{y} = \mathbf{x}$, then an estimator is not required.

Fundamental Topics in Control Systems Analysis

The fundamental control problem is associated with transferring the system state $\mathbf{x}(t)$ to some given **target** set in the state space or maintaining the system state at or near the target. In the automobile cruise control Example 1.2-3 the speed X is a state variable subject to control. In this case, the target set is some preset speed $\overline{X}$. The automatic cruise control should be able to increase or decrease the speed to the preset speed and then maintain it at some acceptable tolerance about the desired speed. For an antiaircraft missile, the missile's target set might correspond to the position of the aircraft or possibly the tailpipe of one of the aircraft's engines. As another example, consider a swimmer in a river being swept rapidly toward a waterfall. The target in this instance may be a small island upstream of the waterfall.

If the target set is a constant set (usually a fixed point) in state space, the control problem is called a **regulator** control problem. If the target set

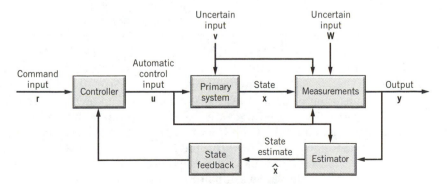

Figure 1.4-3 State feedback and estimation.

is time-varying, as specified by a time-varying command input, then the problem is called a **tracker** or **servomechanism** control problem. We will investigate both types of problems.

The topic of **controllability** deals with whether a control $u(\cdot)$ exists that will transfer the state to a specified target. The existence of such a control depends on the current state, the equations of motion, the control constraints $u \in \mathcal{U}$, any possible uncertain inputs, and the target itself. For example, if our swimmer cannot swim upstream, then there are clearly certain positions in the river from which he or she cannot reach the island.

A primary requirement in the design of an automatic control system is to have controllable systems. Depending on the target set specified, this may or may not require significant changes in the structure of the dynamical system itself. In sports, a high jumper who wants to clear a 15-ft height ought to switch to pole-vaulting. Later we will present some constructive tests for controllability that will not only determine whether the system is controllable but will also indicate the structural changes required to make the system controllable.

The specification of a target set may simply embody an objective of hitting the target, as with a rifle bullet. More often, the objective is also to stay in some neighborhood of the target after the target has been reached. In particular, the target set often corresponds to a set of desired operating conditions. These may change from time to time, because of operator inputs, and they may be time-varying. The command input $r(t)$ shown in Figures 1.4-2 and 1.4-3 expresses a desired operating condition to the controller. In this case, the objective is to make the output $y(t)$ approach, or "track," the command input $r(t)$.

Suppose that some desired operating condition has been specified that corresponds to a nominal reference motion for the dynamical system. This is frequently some constant-state condition. Furthermore, suppose that the system is not currently at the desired operating condition. The task of the controller is to move the system to the desired condition. If the dynamical system is controllable, the design objective corresponds to making the overall system **stable** about the operating condition. If the state is near the operating conditions, we not only want it to remain near (stability) but also to ultimately approach the reference operating conditions (asymptotic stability).

The topic of **observability** focuses on the problem of determining the state $x(t)$ from the measurements $y(t)$. Frequently, the measurements contain only some of the states. A system is said to be observable if it is possible to infer an initial state $x(0)$, given perfect measurements $y(t)$ over some finite time interval $[0,T]$. For example, radar position measurements on a moving object, coupled with a model of the dynamical system, allow us also to determine the velocity of the object. For an observable system, we will be able to design an estimator for computing an estimate $\hat{x}(t)$ of the state $x(t)$ based on the measurements $y(t)$ and the model of the dynamical system.

We will return to these topics of controllability, stability, and observability in later chapters. However, we will begin by discussing the problem of obtaining a solution to the differential equations that govern the motion of a dynamical system.

1.5 EXERCISES

1.5-1 Consider a mass attached to a shock absorber, with an external force *F* as shown in Figure 1.5-1. The state of the system is to be represented by the velocity of the mass. Obtain a first-order equation of the form of (1.2-25) that will model the system. The shock absorber yields a force proportional to velocity.

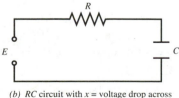

Figure 1.5-1 System for Exercise 1.5-1.

1.5-2 The voltage drop across a resistor, capacitor, or inductor is given by IR, $\int I\, dt/C$, or $L\, dI/dt$, respectively, where $I =$ the current flow through the element, $R =$ resistance, $C =$ capacitance, or $L =$ inductance of the element. The charge Q is related to the current flow by $I = dQ/dt$. For the state x and input u indicated, show that for appropriate constants a and b, the systems defined in Figure 1.5-2 satisfy a differential equation of the form of (1.2-25).

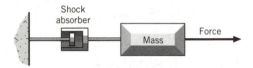

(a) RC circuit with $x =$ charge Q, $u =$ voltage E

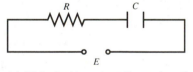

(b) RC circuit with $x =$ voltage drop across capacitor C, $u =$ voltage E

Figure 1.5-2 Circuits for Exercise 1.5-2.

1.5-3 The heat transferred to a liquid of temperature T in a stirred insulated container of temperature T_u is given by

$$Q = hA(T_u - T)$$

and the change in temperature due to heat transfer is given by

$$Q = mC\frac{dT}{dt},$$

where Q = rate of heat flow, h = coefficient of heat transfer, A = surface area, m = mass of liquid, and C = specific heat. Show that this system can be modeled by (1.2-25).

1.5-4 Let x_1 = charge Q, x_2 = current I, u = voltage E. Show that for an appropriate matrix **A** and vector **B** the system in Figure 1.5-3 can be expressed in the form of (1.2-22). Use the information contained in Exercise 1.5-2 and assume that $v = 0$.

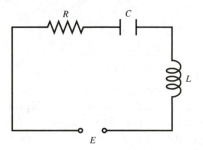

Figure 1.5-3 *RCL* circuit for Exercise 1.5-4.

1.5-5 Let **x** be an $N_x \times 1$ vector and consider the scalar-valued function (a quadratic form)

$$\psi(\mathbf{x}) = \mathbf{x}^T\mathbf{Px},$$

where **P** is a square constant matrix. Show that

$$\frac{\partial\psi(\mathbf{x})}{\partial\mathbf{x}} = \mathbf{x}^T[\mathbf{P} + \mathbf{P}^T] = 2\mathbf{x}^T\mathbf{P} \qquad \text{if } \mathbf{P} \text{ is symmetric.}$$

Hint: Apply the chain rule to differentiate the scalar bilinear form $\varphi(\mathbf{x},\mathbf{z}) = \mathbf{z}^T\mathbf{Px}$, with $\mathbf{z}(\mathbf{x}) = \mathbf{x}$. It might also help you to write the function as a sum, take the partials, and then write the results in matrix form.

1.5-6 From the calculus, under what conditions is the matrix in Equation (1.2-35) symmetric, that is, so that the sequence of differentiations can be interchanged?

1.5-7 For the system

$$\dot{X}_1 = X_1^2 - U$$
$$\dot{X}_2 = -X_1 + X_2^2$$

(a) Determine all (real) equilibrium points (constant-state points) when the input $U = 1$.

(b) In matrix form, write the linearized state-space equations of motion in terms of perturbations in X and U from the positive equilibrium points found in (a).

1.5-8 Assume that a nominal reference operating condition for the toy train in Example 1.2-1 is given by the constant-velocity condition, $\overline{X}_2 = \overline{X}_3 = $ constant > 0. (a) Determine $\overline{X}_1$ and $\overline{U}$ from the equations of motion, with all time derivatives set equal to zero. (b) Linearize the equations of motion as given by (1.2-13)–(1.2-15) about this nominal condition.

1.5-9 Reformulate Example 1.2-3 including wind V_w as an uncertain input. Linearize the resulting set of equations about a nominal operating speed of 65 mph on a level road with no wind. Assume $mg = 2000$ lb, $\mu = 0.01$, $k = 0.005$ lb-sec^2/ft^2.

1.5-10 As illustrated in Figure 1.5-4a, a ball of mass m, radius R, and moment of inertia J about its center of mass rolls (without sliding) on the top of a rectangular beam, under the influence of gravity and the motion of the beam. The beam has moment of inertia I about its center of mass, where it is pinned and is free to rotate due to an applied torque Γ.

(a) Write the position vector $\mathbf{r}$ for the ball's center of mass and use Chastle's corollary (1.3-26) to show that acceleration of the ball is given by

$$\ddot{\mathbf{r}} = [\ddot{r} - (R + h)\ddot{\theta} - r\dot{\theta}^2]\mathbf{e}_r + [r\ddot{\theta} + 2\dot{r}\dot{\theta} - (R + h)\dot{\theta}^2]\mathbf{e}_\theta.$$

(b) Using Newton's laws and the free-body diagrams in Figure 1.5-4, show that, after we eliminate the "internal" forces F_r and F_θ, the equations of motion can be written as

$$\left[m + \frac{J}{R^2} \right] \ddot{r} - \left[\frac{J}{R} + m(R + h) \right] \ddot{\theta} - mr\dot{\theta}^2 + mg \sin \theta = 0$$

$$\frac{hJ}{R^2} \ddot{r} + \left[I + mr^2 - \frac{hJ}{R} \right] \ddot{\theta} + 2mr\dot{r}\dot{\theta} - mr(R + h) \dot{\theta}^2$$
$$+ mgr \cos \theta = \Gamma.$$

(c) Assuming that $(r, \dot{r}, \theta, \dot{\theta})$ are small quantities, write the results in (b) as two linearized second-order differential equations. Convert these results to the linearized equations of motion in

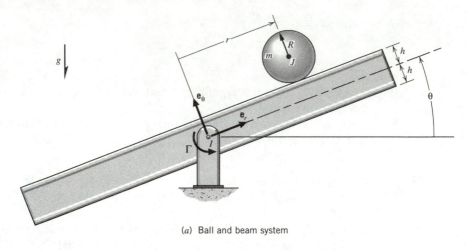

(*a*) Ball and beam system

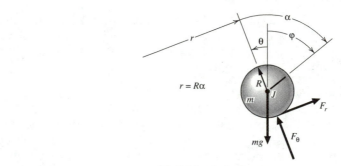

(*b*) Ball free body diagram

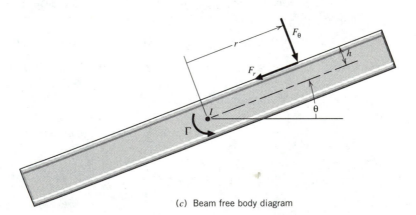

(*c*) Beam free body diagram

Figure 1.5-4 Ball and beam apparatus.

the form (1.2-22) and (1.2-23), with $\mathbf{x} = [r, \dot{r}, \theta, \dot{\theta}]^T$, $u = \Gamma$, $\mathbf{y} = [r, \theta]^T$, and with no uncertain inputs to the system ($\mathbf{v} = \mathbf{w} = \mathbf{0}$).

1.5-11 An elevator is moving upward in a building and a mass m is connected to the roof of the elevator through a nonlinear spring and shock absorber as shown in Figure 1.5-5. The nonlinear spring produces a force given by $-ky^2$, where y is the displacement of the mass from its position in the elevator in free fall. The shock absorber is linear and will produce a force given by $-\beta\dot{y}$.

(a) Obtain a state-space representation of this system of the form

$$\dot{X}_1 = f_1(X_1, X_2, U, V)$$

$$\dot{X}_2 = f_2(X_1, X_2, U, V),$$

where $X_1 = y$, $X_2 = \dot{y}$, U = control force, $V = g + a$, g = acceleration of gravity, and a = upward acceleration of the elevator with respect to the building.

(b) Linearize this system about the steady state given by $X_2 = 0$, X_1 = equilibrium position of the mass when $a = U = 0$, to obtain a system of the form

$$\dot{\mathbf{x}} = \mathbf{Ax} + \mathbf{Bu} + \mathbf{Rv}.$$

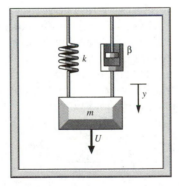

Figure 1.5-5 Elevator system for Exercise 1.5-11.

1.5-12 Obtain the state equations, corresponding to (1.3-20)–(1.3-24), for a variable speed line-of-sight missile. Assume a constant thrust along the velocity vector, with drag, opposite to the velocity vector, given by

$$D = \frac{\rho S C_D V^2}{2},$$

where C_D is a constant drag coefficient. Assume that all angles are small and that the aerodynamic lift and moment coefficients are linear as in Section 1.3. That is, $C_L(\alpha) = C_{L_\alpha}\alpha$, $C_M(\alpha,\delta) = C_{M_\alpha}\alpha + C_{M_\delta}\delta$.

1.5-13 A voice-coil actuator produces a force by supplying current to a coil in a field produced by a permanent magnet. The force is proportional to the coil current as given by

$$F = k_F I_C,$$

where F is the force on the coil, k_F is a constant associated with a particular design, and I_C is the current in the coil. Movement of the coil through the magnetic field produces a back emf according to

$$E_B = k_B \dot{y},$$

where E_B is the back emf voltage, k_B a constant associated with the design, and $\dot{y}$ the velocity of the coil through the field. If the coil has a resistance R_C and an inductance L_C, show that the displacement y is related to the voltage input across the coil E by

$$E = m\frac{L_C}{k_F}y^{(3)} + m\frac{R_C}{k_F}\ddot{y} + k_B\dot{y},$$

where m is the mass of the coil and $y^{(3)} \triangleq d^3y/dt^3$.

1.5-14 For the inverted pendulum in Section 1.3, rederive the equation of motion for the case where the torque is applied by a field-controlled dc motor, shown in Figure 1.5-6, in which the armature current is constant and the control is the voltage applied to the field of resistance R_F and inductance L_F, causing the magnetic flux to vary. Within the motor's operating range, the magnetic flux Φ is proportional to the field current.

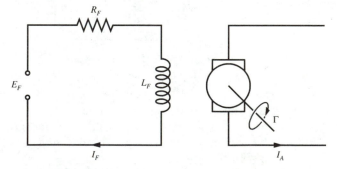

Figure 1.5-6 Field-controlled dc motor for Exercise 1.5-14.

1.5-15 For the geosynchronous satellite in Section 1.3, verify that Equation (1.3-82) follows from Equations (1.3-78)–(1.3-81) at equilibrium. The Westar satellite (launched April 13, 1974) was a geosynchronous communication satellite with a 7-yr station-keeping lifetime. It had an in-orbit mass $m = 297$ kg. Using $M = 5.976 \times 10^{24}$ Kg, $\mu = 6.673 \times 10^{-11}$ m³/(Kg·sec²), and $\omega = 0.7292 \times 10^{-4}$ rad/sec, determine the circular orbit radius and write the actual linearized equations of motion (1.3-82) for this satellite.

1.5-16 An armature-controlled motor is to be used as an actuator for the inverted pendulum. The manufacturer has provided us with the following data:

$$R_A = 7.7\,\Omega$$

$$L_A = 0.011 \text{ H } (\Omega \text{ sec}).$$

However, they did not include the motor constants K_Γ and K_B. Design two laboratory experiments that will independently determine these parameters.

Chapter 2

Linear Dynamical Systems

2.1 LINEAR SYSTEM REPRESENTATIONS

In control system applications, a model for a dynamical system can be presented in a variety of ways. Since "classical" control theory was developed prior to the advent of digital computers and the concept of the "state" of a dynamical system, its methods focus on dynamical systems that are viewed as **input-output** systems. This approach concentrates on a single-input, single-output, scalar nth-order differential equation describing how the output evolves due to the input. A variety of analysis techniques have been developed for such single-input single-output linear systems. Most of the techniques rely on the Laplace transform to convert the differential equation to an algebraic equation. The algebraic formulation lends itself to a graphical representation of the dynamical system. In particular, **block diagrams** (using Laplace transform "transfer functions") provide a visual description of the interactions of the various elements that make up a dynamical control system. These diagrams are also well-suited for use with analog computers; in many instances, they are essentially the wiring diagrams for an analog computer simulation of the dynamical system.

The more recently developed "modern" control theory focuses on matrix methods and on **state-space** models for describing a more general class of dynamical systems than those considered in classical control methods, including systems that may have many inputs and outputs. The analysis techniques of modern control theory are numerical rather than graphical, to handle large systems with perhaps hundreds of state variables, control variables, and output variables. These analysis techniques are designed to be independent of the dimension of the system, whenever possible, and they usually employ eigenvalue–eigenvector methods rather than Laplace transform methods.

To provide an introduction to both classical and modern control techniques, we first examine the various types of models that are used, including state-space models, input–output models, and block diagrams. We will focus, however, on the state-space representation, since the others are all special cases of this representation.

State-Space Systems

In this chapter we will consider linear, constant-coefficient dynamical systems that can be represented in **state-space** form by a system of first-order differential equations and a system of algebraic output equations of the form

$$\dot{\mathbf{x}} = \mathbf{A}\mathbf{x} + \mathbf{B}\mathbf{u} + \mathbf{R}\mathbf{v} \qquad (2.1\text{-}1)$$

$$\mathbf{y} = \mathbf{C}\mathbf{x} + \mathbf{D}\mathbf{u} + \mathbf{E}\mathbf{v} + \mathbf{S}\mathbf{w}, \qquad (2.1\text{-}2)$$

with real-valued constant matrices, where $\mathbf{A}$ is a square $N_x \times N_x$ matrix, $\mathbf{B}$ is an $N_x \times N_u$ matrix, $\mathbf{R}$ is an $N_x \times N_v$ matrix, $\mathbf{C}$ is an $N_y \times N_x$ matrix, $\mathbf{D}$ is an $N_y \times N_u$ matrix, $\mathbf{E}$ is an $N_y \times N_v$ matrix, and $\mathbf{S}$ is an $N_y \times N_w$ matrix. If the uncertain inputs are absent ($\mathbf{v} = \mathbf{w} = \mathbf{0}$), leaving the control $\mathbf{u}$ as the only input, then we have a **deterministic state space** system

$$\dot{\mathbf{x}} = \mathbf{A}\mathbf{x} + \mathbf{B}\mathbf{u} \qquad (2.1\text{-}3)$$

$$\mathbf{y} = \mathbf{C}\mathbf{x} + \mathbf{D}\mathbf{u}. \qquad (2.1\text{-}4)$$

For the special case of a scalar input and a scalar output ($N_u = N_y = 1$) with no uncertain inputs ($\mathbf{v} = \mathbf{w} = \mathbf{0}$), we have a deterministic, **single-input single-output (SISO)** state space system

$$\dot{\mathbf{x}} = \mathbf{A}\mathbf{x} + \mathbf{B}u \qquad (2.1\text{-}5)$$

$$y = \mathbf{C}\mathbf{x} + Du, \qquad (2.1\text{-}6)$$

where u and y are scalars, $\mathbf{B}$ is a one column $N_x \times 1$ matrix (i.e., a column vector), $\mathbf{C}$ is a one-row $1 \times N_x$ matrix (i.e., a row vector), and D is a scalar.

The choice of state variables in a dynamical system is generally not unique. The initial selection may be guided by physical considerations, but mathematical transformations are often employed for various analyses. Two canonical forms of the state equations occur frequently, depending on the choice of the coordinate system, that is, on the choice of state variables. One representation decouples the state equations and the other readily leads to a description of the state in terms of a single N_x-order differential equation.

The simplest representation of a state-space system occurs when the state equations are **decoupled,** that is, when $\mathbf{A}$ is in the **diagonalized** form

$$\mathbf{A} = \text{diag}[\lambda_1, \ldots, \lambda_{N_x}] \triangleq \begin{bmatrix} \lambda_1 & & & \\ & \lambda_2 & & \mathbf{0} \\ & & \cdot & \\ & & & \cdot \\ \mathbf{0} & & & \cdot \; \lambda_{N_x} \end{bmatrix}, \qquad (2.1\text{-}7)$$

so that the evolution of each state variable depends only on itself and the inputs, but not on the other state variables. We can often achieve this diagonalization by using a coordinate transformation composed of the eigenvectors of the original $\mathbf{A}$ matrix. Under this transformation the elements of the diagonalized matrix may be complex numbers. However, since the coordinates of the original system are real-valued, whatever intervening analysis is used, the inverse transformation back to the original coordinate system will again produce a real-valued system.

For a particular state-space system it may not be possible to completely diagonalize the $\mathbf{A}$ matrix so that each state variable is decoupled from the others. In this case, a generalization of the diagonal structure occurs when the $\mathbf{A}$ matrix is in a **block diagonal** form

$$\mathbf{A} = \text{diag}[\mathbf{A}_1, \mathbf{A}_2, \ldots, \mathbf{A}_n],$$

in which the $\mathbf{A}_i$ are square matrices on the diagonal of $\mathbf{A}$. This form decouples the system into a collection of noninteracting subsystems. It is always possible to achieve the block diagonal form. The advantage of the block diagonal form (with a completely diagonal matrix being a special case) becomes clear in applications involving large-scale systems, which can have several hundred state variables and control variables.

For a special class of SISO state-space systems, a representation exists that can lead to a description of the system in terms of a single higher-order differential equation. An SISO system is said to be in **companion form** if the $\mathbf{A}$ matrix is a **companion matrix**

$$\mathbf{A} = \begin{bmatrix} 0 & 1 & 0 & . & . & . & 0 \\ 0 & 0 & 1 & . & . & . & 0 \\ \vdots & & & \cdot & & & \vdots \\ & & & & \cdot & & \\ 0 & 0 & 0 & . & . & \cdot & 1 \\ -a_1 & -a_2 & -a_3 & . & . & . & -a_{N_x} \end{bmatrix} \qquad (2.1\text{-}8)$$

and the scalar control input u enters only in the $\dot{x}_{N_x}$ equation, with the column matrix $\mathbf{B}$ of the form

$$\mathbf{B} = \begin{bmatrix} 0 \\ 0 \\ \vdots \\ 1 \end{bmatrix}. \qquad (2.1\text{-}9)$$

Input–Output Systems

Equations (2.1-5) and (2.1-6) are the state variable representation of a general deterministic SISO system. Another common representation for a special class of SISO systems is referred to as an **input-output (IO)** representation. Using the notation $y^{(n)} \triangleq d^n y/dt^n$, we call an SISO system an IO system if the output y depends on the input u, and the system can be represented by a single linear N_x-order differential equation of the form

$$y^{(N_x)} + p_{N_x-1}y^{(N_x-1)} + \cdots + p_1\dot{y} + p_0 y$$
$$= q_0 u + q_1 \dot{u} + \cdots + q_{N_x-1}u^{(N_x-1)} + q_{N_x}u^{(N_x)}. \qquad (2.1\text{-}10)$$

Note that the left-hand side of (2.1-10) is expressed in terms of the output $y(t)$ and the right-hand side is expressed in terms of the input $u(t)$. In order for the output to depend on the input, not all of the right-hand coefficients $q_0, \ldots, q_{N_x}$ can be zero.

There is a close relationship between the state-space representation of an SISO system in companion form and the IO representation of the same system, provided that certain conditions are satisfied. For example, if the output is just the first state component, that is, the $\mathbf{C}$ vector is of the form

$$\mathbf{C} = [1 \quad 0 \ldots 0]$$

and $D = 0$, then the last state equation from (2.1-8) can be written as

$$\dot{x}_{N_x} + a_{N_x}x_{N_x} + \cdots + a_2 x_2 + a_1 x_1 = bu. \qquad (2.1\text{-}11)$$

Since $x_1 = y$, it follows from the other state equations of (2.1-8) that

$$x_2 = \dot{x}_1 = \dot{y}$$
$$x_3 = \dot{x}_2 = \ddot{y}$$
$$\vdots$$
$$x_{N_x} = \dot{x}_{N_x-1} = y^{(N_x-1)}.$$

Thus, (2.1-11) becomes

$$y^{(N_x)} + a_{N_x}y^{(N_x-1)} + \cdots + a_2\dot{y} + a_1 y = bu, \qquad (2.1\text{-}12)$$

which is in the IO format. If the output is not just the first state component, then it may not be possible to convert an SISO system in companion form to an equivalent IO representation. We will discuss later a general requirement that, when satisfied, will guarantee that a state-space SISO system (whether in companion form or not) can be converted to an equivalent IO representation.

In (2.1-10) note that N_x is the order of the state-space SISO system and that $y^{(N_x)}$ must appear as the highest-order y derivative. Also, the order of the highest u derivative must be less than or equal to N_x. Some textbooks treat this requirement as a "cause-and-effect" condition, ensuring that $y(t)$ is the output (the effect) and $u(t)$ is the input (the cause). This potentially misleading interpretation stems from the fact that IO formulations do not display the actual state of a dynamical system.

EXAMPLE 2.1-1   **Pseudo-Derivative System**

Consider the first-order state-space system

$$\dot{x} = -ax + au$$

with the output y equal to the velocity $\dot{x}$, so that

$$y = -ax + au.$$

The system is in companion form and can be converted to IO format by differentiating the output equation and then replacing $\dot{x}$ with y to obtain

$$\dot{y} + ay = a\dot{u}. \tag{2.1-13}$$

Let $\epsilon = 1/a$ and multiply both sides of (2.1-13) by ϵ to obtain

$$\epsilon\dot{y} + y = \dot{u}.$$

If $\epsilon << 1$, then the output y is approximately equal to the derivative of the input u. For this reason, when ϵ is small, y is called the **pseudo-derivative** of the input. This approximation will play an important role in the design of control systems in Chapter 6.

Example 2.1-1 also illustrates some mathematical differences between the two representations. With the state-space representation, the input is undifferentiated. It follows that a solution exists for both $x(t)$ and $y(t)$ as long as the input $u(t)$ is bounded and piecewise continuous with time. At points of discontinuity in the control, the state will be continuous but the output will be discontinuous if $D \neq 0$ in (2.1-6).

With the IO representation, derivatives of the control may appear in the

IO formulation (2.1-10). Since a discontinuous control input will result in an undefined derivative in the control at points of discontinuity, special difficulties are introduced with this formulation. However, the Laplace transformation method is particularly applicable for handling this situation. Clearly, with the same initial conditions and the same input both the state variable formulation and an equivalent IO formulation of the same problem will yield the same solution for the output.

Transforming System Representations

Certain analysis techniques are facilitated by choosing a particular representation for the control system model. We will first discuss general state-space transformations and then transformations of SISO systems.

In general, the choice of state variables in a state-space representation is not unique and any nonsingular **coordinate transformation**

$$z = M^{-1}x \tag{2.1-14}$$

applied to a state-space system will yield another state-space system. The matrix M^{-1} must be nonsingular (have an inverse M) so that a one-to-one correspondence exists between x and z, from which the original $x = Mz$ can be recovered. Applying such a transformation to (2.1-3) and (2.1-4) yields

$$\dot{z} = \hat{A}z + \hat{B}u \tag{2.1-15}$$

$$y = \hat{C}z + Du, \tag{2.1-16}$$

where

$$\hat{A} = M^{-1}AM \tag{2.1-17}$$

$$\hat{B} = M^{-1}B \tag{2.1-18}$$

$$\hat{C} = CM. \tag{2.1-19}$$

Recall that for a matrix $M = [m_{ij}]$ the inverse M^{-1} can be calculated from

$$M^{-1} = \frac{[adj\, M]^T}{|M|},$$

where $adj\, M = [\mu_{ij}]$ and μ_{ij} is the cofactor (signed minor) of m_{ij}.

Decoupled State-Space Form

We can convert a state-space system to decoupled form by choosing the columns of the M matrix to be the eigenvectors of A. For $i = 1, \ldots, N_x$, let λ_i denote the scalar **eigenvalues** of **A**, with corresponding **eigenvectors** $\xi_i \neq \mathbf{0}$. By definition, an eigenvector ξ is any nonzero vector such that multiplication by **A** yields the original vector itself multiplied by a scalar constant. That is, $\mathbf{A}\xi = \lambda\xi$. Alternately, we can write

$$[\lambda_i \mathbf{I} - \mathbf{A}]\xi_i = \mathbf{0}, \qquad i = 1, \ldots, N_x. \tag{2.1-20}$$

In order for these equations to have nonzero solutions for the eigenvectors ξ_i, the eigenvalues λ_i must be the roots of the N_x-order polynomial **characteristic equation**

$$|\lambda \mathbf{I} - \mathbf{A}| = 0, \tag{2.1-21}$$

with $|\cdot|$ denoting the determinant. Since (2.1-21) is an N_x-order polynomial equation, it will have exactly N_x roots (eigenvalues). Because the elements of **A** are real, so are coefficients in the characteristic equation. Thus, any complex roots will occur in conjugate pairs since (2.1-21) is real-valued. Similarly, the N_x eigenvectors may be complex-valued. With real elements in **A**, any complex eigenvectors will also occur in conjugate pairs.

For each eigenvalue λ_i, $i = 1, \ldots, N_x$, there are $N_x - \text{rank}\,[\lambda_i \mathbf{I} - \mathbf{A}]$ ≥ 1 linearly independent eigenvector solutions of (2.1-20). If an eigenvalue is repeated, with multiplicity m, (2.1-20) may not yield m linearly independent solutions. The complete set of N_x eigenvectors may not be linearly independent if any of the eigenvalues are repeated. On the other hand, one can show that for distinct eigenvalues ($\lambda_i \neq \lambda_j$ for $i \neq j$), the corresponding eigenvectors are linearly independent. Thus, a sufficient condition for a complete set of N_x linearly independent eigenvectors is that all the eigenvalues be distinct.

Suppose that the set of eigenvectors resulting from (2.1-20) are linearly independent. Then the **eigenvector matrix** (also called the **modal matrix**)

$$\mathbf{M} = [\xi_1, \ldots, \xi_{N_x}], \tag{2.1-22}$$

whose columns are the eigenvectors of **A**, has an inverse. The scaling of the eigenvectors is arbitrary, since any nonzero multiple of an eigenvector is also an eigenvector.

Under the modal matrix transformation $\mathbf{z} = \mathbf{M}^{-1}\mathbf{x}$, the resulting matrix $\hat{\mathbf{A}} = \mathbf{M}^{-1}\mathbf{A}\mathbf{M}$ will be diagonal, with the eigenvalues of **A** along the diagonal in the same order as the eigenvectors in **M**. To verify this, we note that the N_x eigenvector equations (2.1-20) can be written together as

$$[\lambda_1\xi_1, \ldots, \lambda_{N_x}\xi_{N_x}] = \mathbf{A}[\xi_1, \ldots, \xi_{N_x}],$$

which is equivalent to

$$M \begin{bmatrix} \lambda_1 & & \mathbf{0} \\ & \cdot & \\ & & \cdot \\ \mathbf{0} & & \lambda_{N_x} \end{bmatrix} = AM$$

and the result follows directly after premultiplying by M^{-1}.

In diagonal form the transformed state equations (2.1-15) are given in terms of components by the decoupled equations

$$\dot{z}_i = \lambda_i z_i + \hat{\mathbf{b}}_i^T \mathbf{u}, \qquad i = 1, \ldots, N_x, \tag{2.1-23}$$

and the output equations (2.1-16) can be written as

$$\mathbf{y} = z_1 \hat{\mathbf{c}}_1 + \cdots + z_{N_x} \hat{\mathbf{c}}_{N_x} + D\mathbf{u}, \tag{2.1-24}$$

where $\hat{\mathbf{b}}_i^T$ is the ith row of the new control input matrix $\hat{\mathbf{B}} = M^{-1}\mathbf{B}$ and $\hat{\mathbf{c}}_i$ the ith column of the new measurement output matrix $\hat{\mathbf{C}} = \mathbf{C}M$. Before discussing these results further, we give an example of the procedure.

EXAMPLE 2.1-2   **Magnetic Suspension System**

Consider diagonalizing the linear magnetic suspension system discussed in Section 1.3. The state-space equations (1.3-10)–(1.3-12) can be written as

$$\begin{bmatrix} \dot{x}_1 \\ \dot{x}_2 \end{bmatrix} = \begin{bmatrix} 0 & 1 \\ \beta_1 & 0 \end{bmatrix} \begin{bmatrix} x_1 \\ x_2 \end{bmatrix} + \begin{bmatrix} 0 \\ \beta_2 \end{bmatrix} u$$

$$y = \begin{bmatrix} 1 & 0 \end{bmatrix} \begin{bmatrix} x_1 \\ x_2 \end{bmatrix},$$

where the positive constants β_1 and β_2 are defined by (1.3-8) and (1.3-9). For convenience let $\omega^2 = \beta_1$. From (2.1-20) each eigenvector $\boldsymbol{\xi} = [\eta_1, \quad \eta_2]^T$ satisfies

$$\begin{bmatrix} \lambda & -1 \\ -\omega^2 & \lambda \end{bmatrix} \begin{bmatrix} \eta_1 \\ \eta_2 \end{bmatrix} = \begin{bmatrix} 0 \\ 0 \end{bmatrix},$$

where the corresponding eigenvalue λ satisfies the characteristic equation

$$0 = |\lambda \mathbf{I} - \mathbf{A}| = \lambda^2 - \omega^2.$$

Thus, the eigenvalues are $\lambda_1 = \omega$ and $\lambda_2 = -\omega$. The corresponding eigenvectors are unique in direction but not in magnitude. For convenience any nonzero component can be scaled to unity. Choosing $\eta_1 = 1$ we obtain $\boldsymbol{\xi}_1 = [1 \quad \omega]^T$ and $\boldsymbol{\xi}_2 = [1 \quad -\omega]^T$. Applying the coordinate transformation

$$
\begin{bmatrix} z_1 \\ z_2 \end{bmatrix} = \begin{bmatrix} 1 & 1 \\ \omega & -\omega \end{bmatrix}^{-1} \begin{bmatrix} x_1 \\ x_2 \end{bmatrix} = \tfrac{1}{2} \begin{bmatrix} 1 & \dfrac{1}{\omega} \\ 1 & \dfrac{-1}{\omega} \end{bmatrix} \begin{bmatrix} x_1 \\ x_2 \end{bmatrix} = \tfrac{1}{2} \begin{bmatrix} x_1 + \dfrac{x_2}{\omega} \\ x_1 - \dfrac{x_2}{\omega} \end{bmatrix}
$$

and using (2.1-5) yields the diagonalized system

$$
\begin{bmatrix} \dot{z}_1 \\ \dot{z}_2 \end{bmatrix} = \begin{bmatrix} \omega & 0 \\ 0 & -\omega \end{bmatrix} \begin{bmatrix} z_1 \\ z_2 \end{bmatrix} + \begin{bmatrix} \dfrac{\beta_2}{2\omega} \\ \dfrac{-\beta_2}{2\omega} \end{bmatrix} u,
$$

with the output equation

$$
y = [1 \quad 1] \begin{bmatrix} z_1 \\ z_2 \end{bmatrix}.
$$

SISO Companion Form

A state-space SISO system can be converted to a unique companion form if, and only if, the **controllability matrix**

$$
\mathbf{P} = [\mathbf{B}, \mathbf{AB}, \mathbf{A}^2\mathbf{B}, \ldots, \mathbf{A}^{N_x-1}\mathbf{B}], \tag{2.1-25}
$$

is of maximum rank (that is, $|\mathbf{P}| \neq 0$ for a single-input system). If this **controllability condition** is satisfied, then the system is said to be **controllable.** Use of this terminology will be clarified later in this chapter when we discuss complete controllability. For now, we examine the role that the controllability matrix plays in the process of converting a general SISO system to companion form.

We seek a nonsingular coordinate transformation $\mathbf{z} = \mathbf{M}^{-1}\mathbf{x}$, yielding a transformed system

$$
\dot{\mathbf{z}} = \hat{\mathbf{A}}\mathbf{z} + \hat{\mathbf{B}}u
$$
$$
y = \hat{\mathbf{C}}\mathbf{z} + Du,
$$

in which $\hat{\mathbf{A}}$ is a companion matrix and $\hat{\mathbf{B}} = [0 \quad 0 \ldots 1]^T$, where

$$\hat{\mathbf{A}} = \mathbf{M}^{-1}\mathbf{A}\mathbf{M}$$

$$\hat{\mathbf{B}} = \mathbf{M}^{-1}\mathbf{B}$$

$$\hat{\mathbf{C}} = \mathbf{C}\mathbf{M}.$$

Note that in this case, $\mathbf{M}$ is a transformation matrix to be determined and it is not the modal matrix used to diagonalize the system.

In companion form, the state variables all follow from z_1 in a cascade. Thus, we can construct the transformation matrix by finding an N_x-dimensional vector $\boldsymbol{\rho}$ such that, by choosing $z_1 = \boldsymbol{\rho}^T\mathbf{x}$, repeated differentiation starting from the condition $\dot{z}_1 = \boldsymbol{\rho}^T\dot{\mathbf{x}} = \boldsymbol{\rho}^T(\mathbf{A}\mathbf{x} + \mathbf{B}u)$ yields a system in companion form. In particular, we choose

$$
\begin{aligned}
z_1 &= \boldsymbol{\rho}^T\mathbf{x} \\
z_2 &= \dot{z}_1 = \boldsymbol{\rho}^T\mathbf{A}\mathbf{x} && \text{with} && \boldsymbol{\rho}^T\mathbf{B} = 0 \\
z_3 &= \dot{z}_2 = \boldsymbol{\rho}^T\mathbf{A}^2\mathbf{x} && \text{with} && \boldsymbol{\rho}^T\mathbf{A}\mathbf{B} = 0 \\
&\vdots \qquad\quad \vdots && && \vdots \\
z_{N_x} &= \dot{z}_{N_x-1} = \boldsymbol{\rho}^T\mathbf{A}^{N_x-1}\mathbf{x} && \text{with} && \boldsymbol{\rho}^T\mathbf{A}^{N_x-2}\mathbf{B} = 0 \\
\dot{z}_{N_x} &= \boldsymbol{\rho}^T\mathbf{A}^{N_x}\mathbf{x} + u && \text{with} && \boldsymbol{\rho}^T\mathbf{A}^{N_x-1}\mathbf{B} = 1.
\end{aligned}
\tag{2.1-26}
$$

From the left-hand sides of (2.1-26) and $\mathbf{z} = \mathbf{M}^{-1}\mathbf{x}$, we have

$$
\mathbf{M}^{-1} =
\begin{bmatrix}
\boldsymbol{\rho}^T \\
\boldsymbol{\rho}^T\mathbf{A} \\
\boldsymbol{\rho}^T\mathbf{A}^2 \\
\vdots \\
\boldsymbol{\rho}^T\mathbf{A}^{N_x-1}
\end{bmatrix}
\tag{2.1-27}
$$

From the right-hand sides of (2.1-26), $\boldsymbol{\rho}$ is the solution to

$$\boldsymbol{\rho}^T[\mathbf{B}, \mathbf{A}\mathbf{B}, \ldots, \mathbf{A}^{N_x-1}\mathbf{B}] = [0 \quad 0 \ldots 1].$$

For SISO systems, the controllability matrix $\mathbf{P}$ is square and the controllability condition assures the existence of $\mathbf{P}^{-1}$. Thus,

$$\boldsymbol{\rho}^T = [0 \quad 0 \ldots 1]\mathbf{P}^{-1}. \tag{2.1-28}$$

Hence, $\boldsymbol{\rho}^T$ is the last row of $\mathbf{P}^{-1}$, and we construct $\mathbf{M}^{-1}$ from $\boldsymbol{\rho}$ as shown above. The matrix $\mathbf{M}^{-1}$ has an inverse, since the right-hand sides of

(2.1-26) imply that the rows in (2.1-27) are linearly independent (Luenberger, 1979, p. 292).

Substituting the result for ρ into the $\dot{z}_{N_x}$ equation in (2.1-26) yields the state equations for the desired SISO system in companion form, with the corresponding output

$$y = \hat{C}z + Du. \tag{2.1-29}$$

EXAMPLE 2.1-3   **Two Carts**

To illustrate the process of converting a state-space system to companion form, consider two carts on frictionless wheels, as shown in Figure 2.1-1. The carts are connected by a damper and a force F is applied to one of the carts.

In terms of the speeds V_1 and V_2 of the two carts, Newton's second law yields

$$m_1 \dot{V}_1 = \beta(V_2 - V_1)$$
$$m_2 \dot{V}_2 = F - \beta(V_2 - V_1).$$

We want to control the system so that the carts move at the same (unspecified) speed, so we choose the measurement output as the speed difference

$$y = V_2 - V_1.$$

For numeric values, suppose $m_1 = 1$ kg, $m_2 = 0.5$ kg, and $\beta = 1$ Ns/m. Choosing the state variables as V_1 and V_2 and the control input as $u = F$ yields

$$\begin{bmatrix} \dot{V}_1 \\ \dot{V}_2 \end{bmatrix} = \begin{bmatrix} -1 & 1 \\ 2 & -2 \end{bmatrix} \begin{bmatrix} V_1 \\ V_2 \end{bmatrix} + \begin{bmatrix} 0 \\ 2 \end{bmatrix} u,$$

$$y = \begin{bmatrix} -1 & 1 \end{bmatrix} \begin{bmatrix} V_1 \\ V_2 \end{bmatrix}.$$

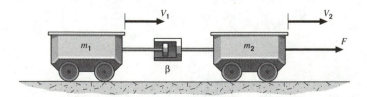

Figure 2.1-1 Cart system for Example 2.1-3.

To convert to companion form, we first compute the controllability matrix

$$\mathbf{P} = [\mathbf{B}, \mathbf{AB}] = \begin{bmatrix} 0 & 2 \\ 2 & -4 \end{bmatrix}.$$

Since $|\mathbf{P}| \neq 0$, an inverse exists for $\mathbf{P}$ that is given by

$$\mathbf{P}^{-1} = \begin{bmatrix} 1 & 0.5 \\ 0.5 & 0 \end{bmatrix}.$$

Using $\mathbf{P}^{-1}$ and (2.1-28), we compute the row matrix

$$\boldsymbol{\rho}^T = [0 \quad 1] \, \mathbf{P}^{-1} = [0.5 \quad 0],$$

which is just the last row of $\mathbf{P}^{-1}$, and then from (2.1-27) we obtain the transformation matrix

$$\mathbf{M}^{-1} = \begin{bmatrix} \boldsymbol{\rho}^T \\ \boldsymbol{\rho}^T \mathbf{A} \end{bmatrix} = \begin{bmatrix} 0.5 & 0 \\ -0.5 & 0.5 \end{bmatrix} \Rightarrow \mathbf{M} = \begin{bmatrix} 2 & 0 \\ 2 & 2 \end{bmatrix}.$$

Thus, we can transform the **V**-space system

$$\dot{\mathbf{V}} = \mathbf{AV} + \mathbf{B}u$$

into an **x**-space system in companion form using the transformation

$$\mathbf{x} = \mathbf{M}^{-1}\mathbf{V} = \tfrac{1}{2}\begin{bmatrix} V_1 \\ V_2 - V_1 \end{bmatrix}.$$

In particular,

$$\dot{\mathbf{x}} = \mathbf{M}^{-1}\mathbf{AMx} + \mathbf{M}^{-1}\mathbf{B}u,$$

where

$$\mathbf{M}^{-1}\mathbf{AM} = \begin{bmatrix} 0.5 & 0 \\ -0.5 & 0.5 \end{bmatrix}\begin{bmatrix} -1 & 1 \\ 2 & -2 \end{bmatrix}\begin{bmatrix} 2 & 0 \\ 2 & 2 \end{bmatrix} = \begin{bmatrix} 0 & 1 \\ 0 & -3 \end{bmatrix}$$

$$\mathbf{M}^{-1}\mathbf{B} = \begin{bmatrix} 0.5 & 0 \\ -0.5 & 0.5 \end{bmatrix}\begin{bmatrix} 0 \\ 2 \end{bmatrix} = \begin{bmatrix} 0 \\ 1 \end{bmatrix}$$

and

$$\mathbf{CM} = [-1 \quad 1]\begin{bmatrix} 2 & 0 \\ 2 & 2 \end{bmatrix} = [0 \quad 2],$$

which yield

$$\dot{x}_1 = x_2$$

$$\dot{x}_2 = -3x_2 + u$$

$$y = 2x_2.$$

Input–Output Form

We can transform a state-space SISO system to a unique equivalent N_x-order IO form if, and only if, the **observability matrix**

$$Q = \begin{bmatrix} C \\ CA \\ CA^2 \\ \vdots \\ CA^{N_x-1} \end{bmatrix} \qquad (2.1\text{-}30)$$

is of maximum rank ($|Q| \neq 0$ for single-output systems). If this **observability condition** is satisfied, then the system is said to be **observable**. Like the controllability condition previously discussed, the observability condition should also be satisfied in any properly designed control system since, as we will see in Chapter 8, it determines whether or not the state $x(t)$ can be reconstructed from a record of the output $y(t)$ and the control input $u(t)$ over some finite time interval.

To perform the transformation from SISO form to IO form, we differentiate the output $y(t)$ N_x times. Using the state equations (2.1-5) to substitute for $\dot{x}$ at each step yields the following system of equations:

$$y = Cx + Du$$

$$\dot{y} = CAx + D\dot{u} + CBu$$

$$\ddot{y} = CA^2x + D\ddot{u} + C\{B\dot{u} + ABu\}$$

$$y^{(3)} = CA^3x + Du^{(3)} + C\{B\ddot{u} + AB\dot{u} + A^2Bu\} \qquad (2.1\text{-}31)$$

$$\vdots \qquad \vdots$$

$$y^{(N_x-1)} = CA^{N_x-1}x + Du^{(N_x-1)} + C\{Bu^{(N_x-2)} + ABu^{(N_x-3)} + \cdots$$
$$+ A^{N_x-2}Bu\}$$

$$y^{(N_x)} = CA^{N_x}x + Du^{(N_x)} + C\{Bu^{(N_x-1)} + ABu^{(N_x-2)} + \cdots + A^{N_x-1}Bu\}.$$

The first N_x of these equations must be solved for x in terms of y, u, and their derivatives. For SISO systems the observability matrix Q is square and the observability condition assures the existence of an inverse. Thus, $\mathbf{x}$ will be a unique function of y, u, and their derivatives. Substituting this result into the $y^{(N_x)}$ equation in (2.1-31) and collecting terms yields an IO system of the form of (2.1-10).

EXAMPLE 2.1-4   **Two-Cart Problem Is Not Observable**

Consider the SISO system previously discussed in Example 2.1-3. The companion form of this system is

$$\begin{bmatrix} \dot{x}_1 \\ \dot{x}_2 \end{bmatrix} = \begin{bmatrix} 0 & 1 \\ 0 & -3 \end{bmatrix} \begin{bmatrix} x_1 \\ x_2 \end{bmatrix} + \begin{bmatrix} 0 \\ 1 \end{bmatrix} u$$

$$y = \begin{bmatrix} 0 & 2 \end{bmatrix} \begin{bmatrix} x_1 \\ x_2 \end{bmatrix}.$$

It follows that

$$\mathbf{Q} = \begin{bmatrix} 0 & 2 \\ 0 & -6 \end{bmatrix}.$$

Since $|\mathbf{Q}| = 0$, $\mathbf{Q}$ does not have an inverse and the observability condition is not satisfied. Let us see what happens if we attempt to convert this system to an IO system anyway. The output equation is given by

$$y = 2x_2.$$

Differentiating the output and using the second state equation to eliminate $\dot{x}_2$ yields

$$\dot{y} = -6x_2 + 2u$$

Repeating this process yields

$$\ddot{y} = 18x_2 - 6u + 2\dot{u}.$$

Eliminating x_2 between the first and last equation yields what appears to be a proper IO system

$$\ddot{y} - 9y = -6u + 2\dot{u}.$$

However, we could also eliminate x_2 between the first and second equation to obtain

$$\dot{y} + 3y = 2u.$$

Thus, we see that the output for the original system (the speed difference) is actually equivalent to a first-order IO system. There is no way from a record of the output from this first-order system to be able to reconstruct the state (the individual speeds) for the second-order system. Note that if we differentiate this latter equation, we obtain

$$\ddot{y} + 3\dot{y} = 2\dot{u}.$$

Substituting for $\dot{y}$ from above yields

$$\ddot{y} - 9y = -6u + 2\dot{u},$$

which is the same as the previous result. Thus, although Example 2.1-3 is controllable, it is not observable. It does not have a unique equivalent second-order IO form.

A Common Matrix System

Consider a particular class of SISO systems in which the **A** matrix is a companion matrix, but the **B** matrix is of the general form $\mathbf{B} = [b_1 \quad b_2 \dots b_{N_x}]^T$, so that the system is not necessarily in companion form. Let the output be given by $y = x_1 + du$. The state-space representation of such an SISO system is given by

$$\begin{aligned}
\dot{x}_1 &= x_2 + b_1 u \\
\dot{x}_2 &= x_3 + b_2 u \\
&\ \vdots
\end{aligned} \tag{2.1-32}$$

$$\dot{x}_{N_x} = -a_1 x_1 - a_2 x_2 - \cdots - a_{N_x} x_{N_x} + b_{N_x} u$$

with output

$$y = x_1 + du. \tag{2.1-33}$$

Such a system often occurs in control applications and it will always satisfy the observability condition since $\mathbf{Q} = \mathbf{I}$. However, it may or may not satisfy the controllability condition.

To transform to the IO representation, we differentiate y N_x times, yielding

$$\dot{y} = x_2 + b_1 u + d\dot{u}$$

$$\ddot{y} = x_3 + b_2 u + b_1 \dot{u} + d\ddot{u}$$

$$\vdots$$

$$y^{(N_x)} = -a_1 x_1 - a_2 x_2 - \cdots - a_{N_x} x_{N_x} + b_{N_x} u + b_{N_x-1}\dot{u} + \cdots$$
$$+ b_1 u^{(N_x-1)} + du^{(N_x)}.$$

The last equation can be expressed in terms of y, u, and their derivatives by using the previous equations to eliminate the x's, yielding

$$y^{(N_x)} = -a_1\{y - du\} - a_2\{\dot{y} - b_1 u - d\dot{u}\} - a_3\{\ddot{y} - b_2 u - b_1 \dot{u} - d\ddot{u}\}$$

$$\vdots$$

$$- a_{N_x}\{y^{(N_x-1)} - b_{N_x-1}u - \cdots - b_1 u^{(N_x-2)} - du^{(N_x-1)}\}$$

$$+ b_{N_x}u + b_{N_x-1}\dot{u} + \cdots + b_1 u^{(N_x-1)} + du^{(N_x)}.$$

$$(2.1\text{-}34)$$

By defining

$$p_0 = a_1 \qquad q_0 = b_{N_x} + da_1 + b_1 a_2 + b_2 a_3 + b_3 a_4 + \cdots + b_{N_x-1}a_{N_x}$$

$$p_1 = a_2 \qquad q_1 = b_{N_x-1} + da_2 + b_1 a_3 + b_2 a_4 + \cdots + b_{N_x-2}a_{N_x}$$

$$\vdots \qquad\qquad \vdots$$

$$p_{N_x-2} = a_{N_x-1} \quad q_{N_x-2} = b_2 + da_{N_x-1} + b_1 a_{N_x}$$

$$p_{N_x-1} = a_{N_x} \qquad q_{N_x-1} = b_1 + da_{N_x}$$

$$q_{N_x} = d$$

$$(2.1\text{-}35)$$

and moving all y terms to the left side of (2.1-34), we obtain an IO system of the form of (2.1-10), provided the coefficients $q_0, \ldots, q_{N_x}$ are not all zero. If all of the q_i coefficients in (2.1-35) were zero, then the right-hand side of these equations implies that $b_i = 0$, $i = 1, \ldots, N_x$, so that the controllability matrix (2.1-25) is a zero matrix and **P** would not have maximum rank.

Equations (2.1-35) can also be used to convert from an IO formulation to the specific SISO formulation given by (2.1-32)–(2.1-33). The conversion is accomplished by solving first for b_1, then b_2, and so on, working the set of equations in (2.1-35) in reverse order. This will produce an observable SISO system, with the **A** matrix in companion form, a nonzero $N_x \times 1$ matrix **B**, and an output $y = x_1 + du$. If no derivatives of u are contained in the IO system (2.1-10), then provided that $q_0 = 1$, the resulting SISO system will not only be observable but also controllable since it will be in companion form. A more general procedure for transforming from IO form to state-space form will be presented after we develop the concept of a Laplace transform "transfer function."

EXAMPLE 2.1-5 **Converting from IO Format to State-Space Format**

Equations (2.1-35) can be used to convert from many IO representations to an equivalent state-space representation. For example, consider the system

$$y^{(3)} - 6\ddot{y} + 11\dot{y} - 5y = -4u + \dot{u}.$$

From (2.1-35) we have $d = q_3 = 0$ and

$$a_1 = -5 \qquad -4 = b_3 + 11b_1 - 6b_2$$
$$a_2 = 11 \qquad 1 = b_2 - 6b_1$$
$$a_3 = -6 \qquad 0 = b_1.$$

Thus, $b_1 = 0$, $b_2 = 1$, $b_3 = 2$, and a state variable representation is given by

$$\dot{x}_1 = x_2$$
$$\dot{x}_2 = x_3 + u$$
$$\dot{x}_3 = 5x_1 - 11x_2 + 6x_3 + 2u$$
$$y = x_1.$$

A check will show that this system is both controllable ($|\mathbf{P}| \neq 0$) and observable ($|\mathbf{Q}| \neq 0$).

EXAMPLE 2.1-6 **Two-Cart Problem Misrepresented**

In Example 2.1-4 we found that, since the observability condition was not satisfied, we could not obtain an equivalent second-order IO representation for this problem. Let us now consider what happens if we transform back the previously obtained second-order equation

$$\ddot{y} - 9y = -6u + 2\dot{u}$$

to a state variable form using conditions (2.1-35). These conditions yield the coefficients $d = 0$ and

$$a_1 = -9 \qquad b_1 = 2$$
$$a_2 = 0 \qquad b_2 = -6$$

corresponding to the observable SISO state-space system

$$\dot{x}_1 = x_2 + 2u$$
$$\dot{x}_2 = 9x_1 - 6u$$
$$y = x_1.$$

Can we transform this system to companion form and recover the original state variable representation? The answer is no! This new state variable system is not controllable since

$$|\mathbf{P}| = \begin{vmatrix} 2 & -6 \\ -6 & 18 \end{vmatrix} = 0.$$

We conclude from this example that the state space representation of an IO system may not be controllable.

Controllability

The original state-space system (2.1-3) and (2.1-4), specifically the pair of matrices $(\mathbf{A},\mathbf{B})$, is said to be **controllable** if there exists an unconstrained control $\mathbf{u}$ that can transfer any initial state $\mathbf{x}$ to any other desired location. As we will see later in Chapter 8, the controllability condition (2.1-25) must be satisfied for a system to satisfy this property. The diagonal form for the state-space system provides us with another way of examining controllability. In particular, if the ith row of $\hat{\mathbf{B}}$ is zero ($\hat{\mathbf{b}}_i^T = \mathbf{0}^T$), then $\mathbf{u}$ does not affect z_i and, hence, some linear combination of the state variables will be unaffected by the control $\mathbf{u}$. In this case, the system (2.1-3) is not controllable. Conversely, if $\hat{\mathbf{B}}$ has no zero rows, then (2.1-3) is controllable.

One consequence of controllability is that by the use of state variable feedback control, it is possible to arbitrarily specify the eigenvalues of the controlled system. If an SISO system is in companion form, then it is also controllable. Eigenvalue placement for such a system can be demonstrated by first noting that the characteristic equation corresponding to the companion matrix (2.1-8) is

$$\lambda^{N_x} + a_{N_x}\lambda^{N_x - 1} + \cdots + a_2\lambda + a_1 = 0.$$

Since the control enters only in the last state equation,

$$\dot{x}_{N_x} = -a_1 x_1 - a_2 x_2 - \cdots - a_{N_x} x_{N_x} + u,$$

which also contains all of the constants in the characteristic equation, it follows that if we let

$$u = -\mathbf{k}^T\mathbf{x},$$

where $\mathbf{k}^T = [k_1 \ldots k_{N_x}]$ is a vector of constants, then through an appropriate choice for $\mathbf{k}$ all of the constants in the characteristic equation can be arbitrarily adjusted. In other words, for an SISO system in companion form, a control system containing state variable feedback of the form $\mathbf{u} = -\mathbf{k}^T\mathbf{x}$ will produce a new system $\overline{\mathbf{A}}$ matrix

$$\overline{\mathbf{A}} = [\mathbf{A} - \mathbf{B}\mathbf{k}^T],$$

for which the eigenvalues can be made arbitrary. As we will see later, if all the eigenvalues in the $\overline{\mathbf{A}}$ matrix have negative real parts, then the system will be asymptotically stable to the origin from any point in state space. Note that if a system is in companion form or it can be transformed to companion form, it is a completely controllable system.

EXAMPLE 2.1-7   Incomplete Controllability

As an example of a system that is not completely controllable, consider the system with state equations

$$\dot{x}_1 = -x_1$$

$$\dot{x}_2 = x_1 + x_2 + u.$$

The control input $u(t)$ clearly has no effect at all on the evolution of the state variable $x_1(t)$. If our objective is to transfer $\mathbf{x}(t)$ to $\mathbf{0}$, we can still design a controller to do so, since $x_1(t)$ approaches zero automatically (albeit at its own pace). Thus, complete controllability may not be required for all control situations.

We will usually require that a system be controllable before designing a controller for it. This may require some alteration of the basic system, such as additional control inputs or additional coupling between state variables, so that the control does affect all state variables. The concept of controllability plays a key role in modern control theory. Later we will illustrate the use of the alternate test (2.1-25) for controllability, which does not require that we first find the eigenvalues and eigenvectors of the system.

As should be clear from the previous developments, a system in companion form is a special, restricted case of a single-input single-output state-space system, and an IO system is an even more restricted representation. A general state-space system may have multiple inputs, multiple outputs, and may or may not be controllable or observable.

2.2 TRANSFER FUNCTIONS AND BLOCK DIAGRAMS

Laplace Transform

In terms of block diagrams, the IO formulation of a linear system, given by (2.1-10), can be described and analyzed succinctly by using the **Laplace transform** to convert the differential equation into an algebraic equation. We will generally designate the Laplace transform of variables such as $y(t)$ by a capitalized function $Y(s)$. In this notation, t represents the time domain associated with the differential equation (2.1-10) and s (a complex variable) an s domain associated with the Laplace transform. By definition

$$Y(s) = \mathscr{L}\{y(t)\} \triangleq \int_0^\infty y(t)e^{-st}\, dt, \qquad \textbf{(2.2-1)}$$

where $\mathscr{L}\{\cdot\}$ stands for the Laplace transform operator. We will denote the **inverse Laplace transform** by

$$\mathscr{L}^{-1}\{Y(s)\} = y(t). \qquad \textbf{(2.2-2)}$$

Note that although $y(t)$ is a real-valued function, its Laplace transform $Y(s)$ is a complex function of the complex variable s. This will require a certain amount of complex arithmetic, but not much, since we will not be concerned with computing the solution to the differential equation by computing the inverse Laplace transform. Instead, we will simply use some results based on the Laplace transform to generate the differential equation for a dynamical system represented in block diagram form.

The Laplace transform is a linear operator:

$$\mathcal{L}\{ay_1(t) + by_2(t)\} = a\mathcal{L}\{y_1(t)\} + b\mathcal{L}\{y_2(t)\}.$$

Thus it is well-suited for use with linear dynamical systems.

The Laplace transforms of derivatives are given by

$$\mathcal{L}\{\dot{y}(t)\} = sY(s) - y(0)$$

$$\mathcal{L}\{\ddot{y}(t)\} = s^2 Y(s) - sy(0) - \dot{y}(0) \qquad \text{(2.2-3)}$$

$$\vdots$$

$$\mathcal{L}\{y^{(N_x)}(t)\} = s^{N_x} Y(s) - s^{N_x - 1}y(0) - s^{N_x - 2}\dot{y}(0) - \cdots - y^{(N_x - 1)}(0),$$

where $y^{(i)}(0)$ is the initial condition on the ith derivative. The Laplace transforms of integrals are given by

$$\mathcal{L}\left\{ \int y(t)\, dt \right\} = \frac{1}{s} Y(s)$$

$$\mathcal{L}\left\{ \int \int y(t)\, dt\, dt \right\} = \frac{1}{s^2} Y(s) \qquad \text{(2.2-4)}$$

$$\vdots$$

$$\mathcal{L}\left\{ \underbrace{\int \cdots \int}_{n} y(t)\, dt \ldots dt \right\} = \frac{1}{s^n} Y(s).$$

There is one other property of the Laplace transform that is often useful when properly applied. The property is given by the **final value theorem**

$$\lim_{t \to \infty} y(t) = \lim_{s \to 0} sY(s), \qquad \text{(2.2-5)}$$

provided that both limits exist.

EXAMPLE 2.2-1   **Final Value Theorem**

One use of the final value theorem is to determine the ultimate value of a solution to a differential equation without solving the differential equation. To illustrate this, consider the first-order system

$$\dot{x} - ax = u,$$

where $a < 0$ and the input is a constant $u(t) \equiv k$. Taking the Laplace transform of both sides, we get

$$sX(s) - x(0) - aX(s) = U(s),$$

where $U(s) = k/s$ for the constant input. Solving for $X(s)$ yields

$$X(s) = \frac{x(0) + \dfrac{k}{s}}{s - a}.$$

Now, since $x(t)$ approaches a limit as $t \to \infty$ for $a < 0$, the final value theorem gives the result

$$\lim_{t \to \infty} x(t) = \lim_{s \to 0} \left[\frac{sx(0) + k}{s - a} \right] = -\frac{k}{a},$$

which agrees with the equilibrium result obtained directly from the differential equation by setting $\dot{x} = 0$ with $u = k$.

In a more general setting, the final value theorem allows us to determine the ultimate value of a function $y(t)$, given its Laplace transform $Y(s)$, without having to evaluate the inverse Laplace transform of $Y(s)$. Since the final value theorem does not require that we actually determine the inverse Laplace transform, it can be a mistake to assume too much about the ultimate behavior of some function of time when the function itself is not known. Another caution about this theorem is that the word "limit" has a precise mathematical meaning. Consider the function of time $y(t) = \sin t$ whose Laplace transform is

$$Y(s) = \frac{1}{s^2 + 1}.$$

From the final value theorem we have

$$\lim_{s \to 0} s Y(s) = 0,$$

which could lead one to believe that $y(t) = \sin t \rightarrow 0$ as $t \rightarrow \infty$; clearly, this is a false conclusion. In fact, the "limit" as $t \rightarrow \infty$ does not exist for $\sin t$, since it continues to oscillate without converging to some specific value. In practice, the final value theorem should be applied only to asymptotically stable systems, where the assumption of some ultimate limit is usually justified.

Transfer Function

Consider the derivative and integral results in (2.2-3) and (2.2-4). We observe that, except for the initial condition terms, we could think of s and $1/s$ as derivative and integral operators, respectively. In fact, this is the primary property of the Laplace transform that we will use. Specifically, even though Laplace transform methods represent a useful way to solve constant coefficient linear differential equations (provided the inverse transform can be determined), we will not use them in this way but, instead, we will employ the Laplace transform as a means of modeling the dynamical system.

Taking the Laplace transform of the input–output system (2.1-10) yields

$$(s^{N_x} + p_{N_x-1}s^{N_x-1} + \cdots + p_1 s + p_0)Y(s) + I_y(s)$$
$$= (q_0 + q_1 s + \cdots + q_{N_x-1}s^{N_x-1} + q_{N_x}s^{N_x})U(s) + I_u(s), \quad \textbf{(2.2-6)}$$

where $I_y(s)$ and $I_u(s)$ are the initial condition terms obtained from (2.2-3). Note that the transformed output $Y(s)$ is multiplied by an N_x-order polynomial

$$P(s) = s^{N_x} + p_{N_x-1}s^{N_x-1} + \cdots + p_1 s + p_0, \quad \textbf{(2.2-7)}$$

and the input transform $U(s)$ is multiplied by an N_x- (or lower-) order polynomial

$$Q(s) = q_0 + q_1 s + \cdots + q_{N_x-1}s^{N_x-1} + q_{N_x}s^{N_x}. \quad \textbf{(2.2-8)}$$

The order of $Q(s)$ will depend on which, if any, of the coefficients are zero.

If we set all initial conditions equal to zero in (2.2-6), then the ratio of the Laplace transform of the output divided by the Laplace transform of the input is given by the ratio of the two polynomials $Q(s)$ and $P(s)$

$$\frac{Y(s)}{U(s)} = \frac{Q(s)}{P(s)}.$$

By definition this ratio is called the **transfer function** $G(s)$

$$G(s) \triangleq \frac{Q(s)}{P(s)} \qquad \text{(2.2-9)}$$

and the output is related to the input by

$$Y(s) = G(s)U(s). \qquad \text{(2.2-10)}$$

The transfer function is a compact representation of the differential equation describing the system and must be written as the ratio of two polynomials. If these polynomials contain a common factor, for example, if $Q(s)$ and $P(s)$ both contain the factor $(s\text{-}a)$, then this factor *cannot* be canceled. To do so would reduce the order of the system.

EXAMPLE 2.2-2   **Cancellation**

Consider the system (see Example 2.1-6) given by

$$G(s) = \frac{-6 + 2s}{s^2 - 9} = \frac{2(s - 3)}{(s + 3)(s - 3)},$$

with the corresponding differential equation

$$\ddot{y} - 9y = -6u + 2\dot{u}.$$

A solution to this second-order differential equation, say, for $u(t) \equiv 0$, involves two constants of integration, yielding

$$y(t) = C_1 e^{-3t} + C_2 e^{3t}.$$

If we were to cancel the $(s - 3)$ term in $G(s)$, we would reduce the order of the system, yielding

$$\dot{y} + 3y = 2u,$$

with a solution, again for $u(t) \equiv 0$, involving only one integration constant

$$y(t) = c_1 e^{-3t}.$$

Clearly, the two systems are not equivalent. Canceling a term like $(s - a)$, with $a > 0$, is particularly bad because it hides an unstable term in the solution.

As a general rule, any algebraic operation on a transfer function, for example, multiplication, cancellation, and so on, that may occur in block diagram reduction is valid so long as the final result does not change the order of the original system.

Under the transfer function notation, if we replace s^n with the derivative operator $D^n \triangleq d^n()/dt^n$, then the transfer function implies the system differential equation (in operator form)

$$P(D)y = Q(D)u.$$

Setting the initial conditions equal to zero in defining the transfer function is a device that allows us to recover the differential equation from the two polynomials that form the transfer function. One must use the actual initial conditions if the Laplace transformation is to be employed to obtain a solution to the differential equation. That is, the solution $y(t)$ to the differential equation is not given by the inverse Laplace transform

$$y(t) = \mathcal{L}^{-1}\{G(s)U(s)\}$$

unless all the initial conditions on $y(t)$ and $u(t)$ are, in fact, equal to zero.

Our concern will be with systems that ultimately can be made asymptotically stable. A characteristic of such systems is that any transient phenomena caused by initial conditions on $y(t)$ or $u(t)$ will damp out in time. Thus, in the long run, the transfer function also provides a proper description of the ultimate dynamic behavior of an asymptotically stable system.

The use of the Laplace transform reduces our IO description of the system to an algebraic expression. We now have three ways to describe a linear SISO system. A state-space representation, an IO representation, and a transfer function representation all can be used to describe the same system.

2.3   BLOCK DIAGRAM ALGEBRA

For an IO system as illustrated in Figure 2.3-1, the output is the product of the input and transfer function

$$Y(s) = G(s)U(s).$$

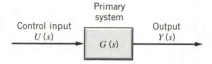

Figure 2.3-1 Block diagram for an input–output system.

The transfer function concept allows us to examine a complicated dynamical system in terms of the transfer functions for the various elements in the system. We connect these elements in a block diagram to show the interaction of the elements in the system. If we use the algebraic properties of transfer functions, the block diagram always can be reduced to a single block of the form shown in Figure 2.3-1. From this we obtain the overall transfer function of the system, which then yields the governing differential equation for the system.

EXAMPLE 2.3-1   **Blocks in Cascade**

As an example, consider the Line-of-Sight (LOS) missile in Section 1.3, where we obtained the state equations

$$\ddot{\alpha} + \beta_2\dot{\alpha} + \beta_1\alpha = \beta_3\delta$$

$$\dot{\gamma} = \beta_2\alpha$$

$$\dot{z} = V\gamma,$$

with δ as the input and z as the output.

Each of these differential equations may be thought of as an IO device, with corresponding transfer functions

$$G_1(s) = \frac{\beta_3}{s^2 + \beta_2 s + \beta_1}$$

$$G_2(s) = \frac{\beta_2}{s}$$

$$G_3(s) = \frac{V}{s}$$

Figure 2.3-2a shows the block diagram for the overall system, which has a cascade structure in which the output from each block is the input to the next block.

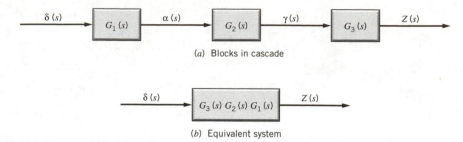

(a) Blocks in cascade

(b) Equivalent system

Figure 2.3-2 Block diagram reduction for blocks in cascade.

For blocks in cascade, as illustrated in Figure 2.3-2b, the equivalent transfer function is given by

$$G(s) = G_3(s)G_2(s)G_1(s). \qquad (2.3\text{-}1)$$

To verify the result in (2.3-1), note that

$$Z(s) = G_3(s)\gamma(s) = G_3(s)G_2(s)\alpha(s) = G_3(s)G_2(s)G_1(s)\delta(s).$$

Hence, the LOS missile has the transfer function

$$G(s) = \frac{Z(s)}{\delta(s)} = \frac{V\beta_2\beta_3}{s^2(s^2 + \beta_2 s + \beta_1)}. \qquad (2.3\text{-}2)$$

This result can also be obtained directly from the state equations by converting to IO formulation in terms of the output $y = z$

$$\dot{y} = \dot{z} = V\gamma$$
$$\ddot{y} = V\dot{\gamma} = \beta_2 V\alpha$$
$$y^{(3)} = \beta_2 V\dot{\alpha}$$
$$y^{(4)} = \beta_2 V\ddot{\alpha} = \beta_2 V(-\beta_2\dot{\alpha} - \beta_1\alpha + \beta_3\delta).$$

Therefore,

$$y^{(4)} + \beta_2 y^{(3)} + \beta_1\ddot{y} = \beta_3\beta_2 V\delta,$$

which yields the same overall transfer function as in (2.3-2).

EXAMPLE 2.3-2   **Feedback Loops**

The rotational speed of a low-inductance armature-controlled dc motor is governed by a first-order equation of the form [see (1.3-48) with $p_0 = k_B k_\Gamma / R_A J$, $q_0 = k_\Gamma / R_A J$, $y = \omega$, $\Gamma_{ex} = 0$]

$$\dot{y} + p_0 y = q_0 u,$$

which has a block diagram representation given by Figure 2.3-1 with

$$G(s) = \frac{q_0}{s + p_0}.$$

However, if the Laplace transform of each term is considered separately, then this system has an equivalent block diagram representation with a feedback loop, as illustrated in Figure 2.3-3, with

$$G_1(s) = \frac{q_0}{s}$$

$$G_2(s) = \frac{p_0}{q_0}$$

and with the feedback adding negatively ($-\Rightarrow$ negative feedback) at the summing junction.

For any G_1 and G_2 in an SISO feedback loop of the form shown in Figure 2.3-3, the overall transfer function is given by

$$G(s) = \frac{Y(s)}{U(s)} = \frac{G_1(s)}{1 + G_1(s)G_2(s)}. \tag{2.3-3}$$

To establish this result, let $E(s)$ denote the output of the summing junction

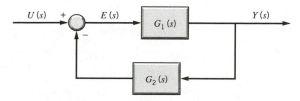

Figure 2.3-3 Block diagram of a feedback loop.

in Figure 2.3-3. Then

$$Y(s) = G_1(s)E(s) = G_1(s)[U(s) - G_2(s)Y(s)].$$

Solving for $Y(s)$ and dividing by $U(s)$ yields the transfer function (2.3-3). For our example of a general first-order system, applying (2.3-3) yields

$$G(s) = \frac{\dfrac{q_0}{s}}{1 + \dfrac{p_0}{s}} = \frac{q_0}{s + p_0}, \qquad \text{(2.3-4)}$$

which is the result previously obtained. Note that in order for the transfer function in (2.3-4) to be interpreted properly, it *must* be reduced to a ratio of two polynomials. This also has the added benefit of simplifying any subsequent block diagram algebra.

EXAMPLE 2.3-3   **Block Diagram Reduction**

As a more complex example, consider the inverted pendulum in Section 1.3. If $\theta(t)$ is small, so that $\sin\theta \approx \theta$, then Equation (1.3-49) for the rod becomes

$$J\ddot{\theta} = \Gamma + mg\ell\theta. \qquad \text{(2.3-5)}$$

The dynamics of the armature-controlled dc motor are given by (1.3-42) and (1.3-44), which can be written as

$$\Gamma = k_\Gamma I_A \qquad \text{(2.3-6)}$$

$$R_A I_A + L_A \dot{I}_A = E - k_B \dot{\theta}. \qquad \text{(2.3-7)}$$

To obtain a block diagram that will display all the interactions, we define

$$\psi = \dot{\theta} \qquad \text{(2.3-8)}$$

$$\Sigma_\Gamma = \Gamma + mg\ell\theta \qquad \text{(2.3-9)}$$

$$\Sigma_E = E - k_B \psi. \qquad \text{(2.3-10)}$$

These definitions allow (2.3-5) and (2.3-7) to be written as

$$J\dot{\psi} = \Sigma_{\Gamma} \tag{2.3-11}$$

$$R_A I_A + L_A \dot{I}_A = \Sigma_E. \tag{2.3-12}$$

Taking the Laplace transform of (2.3-6) and (2.3-8)–(2.3-12), we obtain after some rearrangement

$$\theta(s) = \frac{\psi(s)}{s} \tag{2.3-13}$$

$$\psi(s) = \frac{\Sigma_{\Gamma}(s)}{Js} \tag{2.3-14}$$

$$\Sigma_{\Gamma}(s) = \Gamma(s) + mg\ell\theta(s) \tag{2.3-15}$$

$$\Gamma(s) = k_{\Gamma} I_A(s) \tag{2.3-16}$$

$$I_A(s) = \frac{\Sigma_E(s)}{(R_A + L_A s)} \tag{2.3-17}$$

$$\Sigma_E(s) = E(s) - k_B \psi(s). \tag{2.3-18}$$

From these equations we obtain the block diagram shown in Figure 2.3-4, where

$$G_1(s) = \frac{1}{s} \tag{2.3-19}$$

$$G_2(s) = \frac{1}{Js} \tag{2.3-20}$$

$$G_3(s) = \frac{1}{(R_A + L_A s)}. \tag{2.3-21}$$

This figure clearly illustrates the relationships between the various components of the motor and rod system. This block diagram also incorporates the dynamics as given by the differential equations.

Just as the differential equations can be combined and reduced to a more compact form, so can the block diagram representation. The following algebraic procedure, for example, can be used to obtain an overall

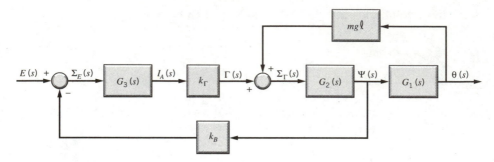

Figure 2.3-4 Block diagram of a rod and motor inverted pendulum.

transfer function for the system. From the block diagram IO relations we have

$$\Gamma(s) = k_\Gamma G_3(s)[E(s) - k_B\psi(s)] \qquad (2.3\text{-}22)$$

$$\psi(s) = G_2(s)[\Gamma(s) + mg\ell\theta(s)] \qquad (2.3\text{-}23)$$

$$\theta(s) = G_1(s)\psi(s). \qquad (2.3\text{-}24)$$

From these equations we obtain

$$\theta(s) = \frac{k_\Gamma G_1 G_2 G_3}{1 + k_\Gamma k_B G_2 G_3 - mg\ell G_1 G_2} E(s). \qquad (2.3\text{-}25)$$

In terms of the transfer functions $G_1(s)$, $G_2(s)$, and $G_3(s)$, (2.2-35) becomes

$$\theta(s) = \frac{k_\Gamma E(s)}{L_A Js^3 + R_A Js^2 + (k_\Gamma k_B - L_A mg\ell)s - R_A mg\ell}, \qquad (2.3\text{-}26)$$

which is the same transfer function relationship obtained if one were to take the Laplace transform of (1.3-50) with $\sin\theta \approx \theta$ and $\cos\theta \approx 1$.

Controller Sensitivity and Disturbance Rejection

Figure 2.3-5 illustrates one typical feedback control system, with command and uncertain inputs, designed to control an output $y(t)$ of a dynamical system (called the "primary system" to distinguish it from the overall system).

Consider first the case of no uncertain input and let $y_1(t)$ denote the corresponding output response to a command input $r(t)$ with all initial

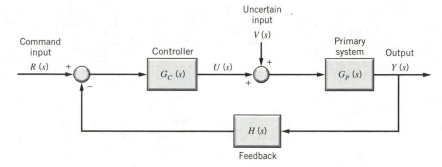

Figure 2.3-5 Block diagram of a feedback control system.

conditions equal to zero. The transfer function in this case is

$$G_1(s) = \frac{Y_1(s)}{R(s)} = \frac{G_C(s)G_P(s)}{1 + G_C(s)G_P(s)H(s)},$$

as can be verified using the block diagram reduction methods depicted in Figure 2.3-6. Similarly, let $y_2(t)$ be the zero initial conditions response for the case of an uncertain input $V(s)$ with no command input. The transfer function as obtained in Figure 2.3-7 is given by

$$G_2(s) = \frac{Y_2(s)}{V(s)} = \frac{G_P(s)}{1 + G_C(s)G_P(s)H(s)}.$$

Since the system is linear, the combined response of the system to both the command and uncertain inputs, with all initial conditions at zero, can be determined by **superposition**

$$Y(s) = Y_1(s) + Y_2(s) = G_1(s)R(s) + G_2(s)V(s)$$

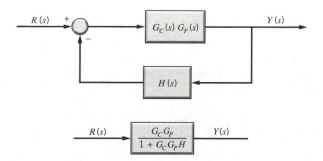

Figure 2.3-6 No uncertain input.

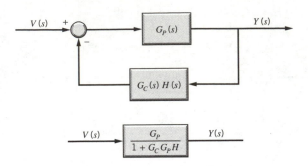

Figure 2.3-7 No command input.

to yield

$$Y(s) = \frac{G_C(s)G_P(s)R(s) + G_P(s)V(s)}{1 + G_C(s)G_P(s)H(s)}.$$

To investigate the effects of feedback control, consider each of the block transfer functions $G_C(s)$, $G_P(s)$, and $H(s)$ as having some magnitude factor, called the **gain** or amplification of the element. In particular, since the inputs to the controller are usually small quantities, such as voltages, displacements, and so on, the transfer function $G_C(s)$ almost always contains some power amplification device. For example, suppose that a particular block corresponds to a first-order system of the form

$$\epsilon\dot{y} + y = Ku(t),$$

which for a step input $u(t) \equiv \bar{u}$ with $\epsilon > 0$ has the steady-state solution $y(t) \to K\bar{u}$. In the steady state ($\dot{y} = 0$), we could say that the system has a gain of K. We obtain this same result from the transfer function

$$Y(s) = \frac{K}{\epsilon s + 1} U(s)$$

by applying the final value theorem (2.2-5) with $U(s) = \bar{u}/s$ for a step input.

Suppose that $G_C(s)$ and $H(s)$, which are free to be selected by the control system designer, have been chosen so that, loosely speaking, the system has a **high loop gain**

$$|G_C(s)G_P(s)H(s)| \gg 1.$$

Then

$$Y(s) \approx \frac{R(s)}{H(s)} + \frac{V(s)}{G_C(s)H(s)}$$

and we see that the response will be essentially independent of the system to be controlled! That is, it depends more on $H(s)$ and $G_C(s)$ than $G_P(s)$. Furthermore, if the system also has a **high gain controller**

$$|G_C(s)| \gg 1,$$

then

$$Y(s) \approx \frac{R(s)}{H(s)}$$

and the response will essentially be independent of the disturbance. Finally, if, in addition, the system has **unity feedback**

$$|H(s)| = 1,$$

then

$$Y(s) \approx R(s)$$

and the response will tend to track the input despite the disturbance.

To investigate the sensitivity of the system to changes in its parameters, let dG_C, dG_P, and dH denote small perturbations in the element transfer functions $G_C(s)$, $G_P(s)$, and $H(s)$, respectively. These may represent uncertainties in the transfer functions or actual changes due to wear or other processes.

For a given transfer function $G(s)$, the **sensitivity** $S_K(G)$ of $G(s)$ with respect to a change in some quantity K is defined as the fractional change in $G(s)$ divided by the fractional change in K

$$S_K(G) \triangleq \frac{dG/G}{dK/K} = \frac{K}{G}\frac{dG}{dK}. \qquad (2.3\text{-}27)$$

For the command response transfer function $G_1(s)$ with a high loop gain,

$$S_{G_P}(G_1) = \frac{G_P}{G_1}\frac{dG_1}{dG_P} = \frac{1}{1 + G_C(s)G_P(s)H(s)} \approx 0$$

and

$$S_H(G_1) = \frac{H}{G_1}\frac{dG_1}{dH} = \frac{-G_C(s)G_P(s)H(s)}{1 + G_C(s)G_P(s)H(s)} \approx -1.$$

Thus for a high loop gain, the command response transfer function is relatively insensitive to changes in the transfer function of the system to be controlled (as previously indicated), but it is highly sensitive to changes in the feedback elements. This means that the feedback elements must be precise and the cost of this precision is one of the factors that makes closed-loop control systems more expensive than open-loop systems.

A similar development can be performed for the disturbance transfer function, but we leave this as an exercise for the reader.

This discussion on sensitivity and disturbance rejection is only approximate at this stage, since the dynamic response was discussed in a loose sense, and the results should be viewed only as general trends. In particular, a high controller gain may not always be as desirable as this discussion makes it seem. As we shall see later, in certain types of systems a high controller gain can actually make a stable system become unstable. This type of behavior is common. One example is provided by the "gain" knob on some x-y analog plotters; turn it too high and the plotter arm may begin to chatter.

2.4 TRANSFER MATRIX

Use of the Laplace transform to develop a transfer function description of a dynamical system is not restricted to IO systems. It can be applied to any linear constant-coefficient state-space system. Such systems generally have multiple inputs and outputs, so that we may end up with a matrix of transfer functions, with each element in the matrix being the transfer function from an input to an output.

For a multiple-input multiple-output state-space system of the form

$$\dot{\mathbf{x}} = \mathbf{A}\mathbf{x} + \mathbf{B}\mathbf{u} \tag{2.4-1}$$

$$\mathbf{y} = \mathbf{C}\mathbf{x} + \mathbf{D}\mathbf{u}, \tag{2.4-2}$$

taking the Laplace transform of both equations and setting the initial condition terms to zero yields

$$\mathbf{Y}(s) = \mathbf{G}(s)\mathbf{U}(s), \tag{2.4-3}$$

where

$$\mathbf{G}(s) = \mathbf{C}[s\mathbf{I} - \mathbf{A}]^{-1}\mathbf{B} + \mathbf{D} \tag{2.4-4}$$

is the $N_y \times N_u$ **transfer matrix**. In matrix form, the output is obtained by premultiplying the input by the transfer matrix.

In scalar form the relationship between the inputs $u_j(t), j = 1, \ldots, N_u$, and a particular output $y_i(t)$, $i = 1, \ldots, N_y$, is given by

$$Y_i(s) = \sum_{j=1}^{N_u} G_{ij}(s)U_j(s). \tag{2.4-5}$$

Now suppose that all the initial conditions and inputs are zero, except that $u_j(t)$ is a unit impulse $U_j(s) = 1$. Then $Y_i(s) = G_{ij}(s)$. This result provides an experimental way of determining the IO transfer functions for a device; apply a unit impulse to one of the inputs u_j and record the time response of the outputs $y_i(t)$ $i = 1, \ldots, N_y$. Then numerically compute the transfer functions $G_{ij}(s)$ from the definition (2.2-1) as the Laplace transforms of the outputs and repeat the process for each input. Some related approaches will be discussed in Chapter 5.

Developing State Equations from the Transfer Matrix

If the transfer function or transfer matrix is known for a system, perhaps by experimental determination, then state-space models for the system can be readily developed.

SISO Systems

For SISO systems the transfer matrix reduces to a scalar transfer function and we can convert from IO form to state-space form directly from (2.4-4), yielding a system in companion form.

EXAMPLE 2.4-1   **Companion Form**

Consider a system with transfer function

$$G(s) = \frac{10 + 2s}{s^2 + 5s + 6}.$$

This system will be converted to state space form in Exercise 2.5-5(c) using Equations (2.1-35). This will result in a system in which the **A** matrix is a companion matrix, but the system is not in companion form because $\mathbf{B} \neq [0 \quad 1]^T$. To put the system directly into companion form, we can use (2.4-4)

$$\frac{10 + 2s}{s^2 + 5s + 6} = [c_1 \quad c_2] \begin{bmatrix} s & -1 \\ a_1 & s + a_2 \end{bmatrix}^{-1} \begin{bmatrix} 0 \\ 1 \end{bmatrix} + D$$

$$= \frac{[c_1 \quad c_2] \begin{bmatrix} s + a_2 & 1 \\ -a_1 & s \end{bmatrix} \begin{bmatrix} 0 \\ 1 \end{bmatrix}}{s^2 + a_2 s + a_1} + D$$

$$= \frac{c_1 + c_2 s}{s^2 + a_2 s + a_1} + D.$$

By equating both sides of this expression, it follows that

$$a_1 = 6, \qquad a_2 = 5, \qquad c_1 = 10, \qquad c_2 = 2, \qquad D = 0.$$

The resultant state-space system is given by

$$\dot{x}_1 = x_2$$
$$\dot{x}_2 = -6x_1 - 5x_2 + u$$
$$y = 10x_1 + 2x_2,$$

which may be checked by converting back to IO form.

An alternate approach based on block diagrams is shown in Figure 2.4-1. The output $Y(s)$ is related to the input $U(s)$ by $Y(s) = G(s)U(s)$, where $G(s) = Q(s)/P(s)$ and $P(s)$ and $Q(s)$ are polynomials. We separate $P(s)$ and $Q(s)$ by introducing an intermediate step, yielding

$$Z(s) = \frac{U(s)}{P(s)}$$

$$Y(s) = Q(s)Z(s).$$

By appropriate choices of the state variables, we can use these relations to convert directly from IO form to state-space companion form.

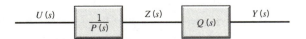

Figure 2.4-1 Using block diagrams to obtain a state-space companion form.

EXAMPLE 2.4-2   **Companion Form from Block Diagram**

Consider again Example 2.4-1, where

$$P(s) = s^2 + 5s + 6$$
$$Q(s) = 10 + 2s.$$

Thus,

$$\ddot{z} + 5\dot{z} + 6z = u$$
$$y = 10z + 2\dot{z}.$$

Letting

$$x_1 = z$$
$$x_2 = \dot{z},$$

we immediately obtain

$$\dot{x}_1 = x_2$$
$$\dot{x}_2 = -6x_1 - 5x_2 + u$$
$$y = 10x_1 + 2x_2,$$

which is the same result obtained previously.

___

Multiple-Input Multiple-Output Systems

We now present a general procedure for constructing a state-space model from a transfer matrix model. Suppose that a multi-input multi-output linear system is described in terms of a transfer matrix $\mathbf{G}(s)$ by

$$\mathbf{Y}(s) = \mathbf{G}(s)\mathbf{U}(s),$$

where

$$\mathbf{G}(s) = \frac{1}{P(s)} \begin{bmatrix} Q_{11}(s) & \cdots & Q_{1N_u}(s) \\ \vdots & & \vdots \\ Q_{N_y1}(s) & \cdots & Q_{N_yN_u}(s) \end{bmatrix}, \qquad \textbf{(2.4-6)}$$

and each $Q_{ij}(s)$ is a polynomial of order equal to or less than the polynomial $P(s)$. From this representation we wish to develop a state-space model of the form (2.1-3) and (2.1-4) that is both controllable and observable.

___

EXAMPLE 2.4-3   **Two-Input Single-Output System**

The following IO system:

$$\ddot{y} + 3\dot{y} + 2y = u_1 + 3u_2 + \dot{u}_2$$

has two inputs and one output. Taking the Laplace transform of each side with all initial conditions set equal to zero yields

$$\{s^2 + 3s + 2\}Y(s) = U_1(s) + 3U_2(s) + sU_2(s).$$

This yields the transfer functions

$$Y(s) = \frac{1}{s^2 + 3s + 2}\{U_1(s) + [s + 3]U_2(s)\},$$

which is equivalent to the transfer matrix representation

$$Y(s) = \left[\frac{1}{s^2 + 3s + 2} \quad \frac{s + 3}{s^2 + 3s + 2}\right]\begin{bmatrix} U_1(s) \\ U_2(s) \end{bmatrix}.$$

In order to develop a general procedure for determining state equations from a transfer matrix, we first obtain a decoupled model in which the **A** matrix is diagonal. Then a suitable nonsingular coordinate transformation is applied to yield a state-space representation having specified properties, such as a set of desired eigenvectors or an **A** matrix that is in companion form.

We compute the eigenvalues $\lambda_1, \ldots, \lambda_{N_x}$ from the **characteristic equation**

$$P(\lambda) = 0, \tag{2.4-7}$$

where $P(s)$ is the common denominator in the transfer matrix (2.4-6) and N_x the order of the polynomial $P(s)$. This is the same characteristic equation as in (2.1-21); it is characteristic of the system, independent of the number of inputs or outputs or the representation employed to describe the system. We assume that the resulting eigenvalues are distinct and we define

$$\mathbf{A} = \begin{bmatrix} \lambda_1 & & & \\ & \lambda_2 & & \mathbf{0} \\ & & \ddots & \\ & \mathbf{0} & & \lambda_{N_x} \end{bmatrix}. \tag{2.4-8}$$

The next step is to determine an $N_x \times N_u$ matrix **B**, an $N_y \times N_x$ matrix **C**, and an $N_y \times N_u$ matrix **D** such that the following conditions hold:

i. The transfer matrix corresponds to a state-space system, that is,

$$\mathbf{G}(s) = \mathbf{C}[s\mathbf{I} - \mathbf{A}]^{-1}\mathbf{B} + \mathbf{D}.$$

ii. B has no zero rows (controllability satisfied).
iii. C has no zero columns (observability satisfied).

Usually, the system of equations that result from equating like powers of s in condition i involves fewer equations than unknowns in **B**, **C**, and **D**, so there is some degree of freedom in choosing the elements of **B**, **C**, and **D**. The resulting decoupled state-space system is

$$\dot{z} = Az + Bu$$

$$y = Cz + Du.$$

From the diagonalized representation, which may have complex-valued matrices, we can change to a final set of state variables $x = Mz$ by choosing a set of desired linearly independent eigenvectors as the columns of the transformation matrix $M = [\xi_1, \ldots, \xi_{N_x}]$. This transformation yields

$$\dot{x} = \tilde{A}x + \tilde{B}u$$

$$y = \tilde{C}x + Du,$$

where $\tilde{A} = MAM^{-1}$, $\tilde{B} = MB$, and $\tilde{C} = CM^{-1}$. The matrix $\tilde{A}$ will have eigenvalues λ_i and the chosen eigenvectors. For real-valued matrices, complex conjugate eigenvectors should be chosen for any corresponding conjugate eigenvalues.

EXAMPLE 2.4-4   **Specified Eigenvectors**

Suppose we want to develop a state-space model, with specified eigenvectors, for the system of Example 2.4-3 that has the transfer matrix

$$G(s) = \left[\frac{1}{s^2 + 3s + 2} \quad \frac{s + 3}{s^2 + 3s + 2} \right].$$

The characteristic equation

$$\lambda^2 + 3\lambda + 2 = 0$$

yields the eigenvalues $\lambda_1 = -1$, $\lambda_2 = -2$. Thus, the diagonalized state equations will have the **A** matrix given by

$$A = \begin{bmatrix} -1 & 0 \\ 0 & -2 \end{bmatrix}.$$

The condition $G(s) = C[sI - A]^{-1}B + D$ yields

$$\frac{[1 \quad s + 3]}{s^2 + 3s + 2} = [c_1 \quad c_2] \begin{bmatrix} \dfrac{1}{s + 1} & 0 \\ 0 & \dfrac{1}{s + 2} \end{bmatrix} \begin{bmatrix} b_{11} & b_{12} \\ b_{21} & b_{22} \end{bmatrix} + [d_1 \quad d_2].$$

Thus, the elements of **B**, **C**, and **D** must satisfy

$$1 \quad = d_1 s^2 + (c_1 b_{11} + c_2 b_{21} + 3d_1)s + (2c_1 b_{11} + c_2 b_{21} + 2d_1)$$

$$s + 3 = d_2 s^2 + (c_1 b_{12} + c_2 b_{22} + 3d_2)s + (2c_1 b_{12} + c_2 b_{22} + 2d_2).$$

Equating like powers of s and solving the resulting equations yield

$$d_1 = 0, \qquad d_2 = 0, \quad c_1 b_{11} = 1, \quad c_2 b_{21} = -1, \quad c_1 b_{12} = 2, \quad c_2 b_{22} = -1.$$

The observability condition iii requires $c_1 \neq 0$ and $c_2 \neq 0$. Therefore, these results can be solved for the b_{ij} in terms of c_1 and c_2. The resulting b_{ij} are all nonzero, so the controllability condition is also satisfied. The choice of c_1 and c_2 is arbitrary, as long as they are nonzero. For this example we will choose $c_1 = c_2 = 1$, yielding

$$\dot{z} = Az + Bu$$

$$y = Cz,$$

where $\mathbf{u} = [u_1 \quad u_2]^T$, $\mathbf{z} = [z_1 \quad z_2]^T$, and

$$A = \begin{bmatrix} -1 & 0 \\ 0 & -2 \end{bmatrix}, \qquad B = \begin{bmatrix} 1 & 2 \\ -1 & -1 \end{bmatrix}, \qquad C = [1 \quad 1].$$

Now, suppose that we want to change from this diagonal form to a state-space system having a set of specified eigenvectors, such as $\xi_1 = [1 \quad -1]^T$ and $\xi_2 = [1 \quad -2]^T$. As a final step, we transform the diagonalized system by using a coordinate transformation $\mathbf{x} = \mathbf{Mz}$, where $\mathbf{M} = [\xi_1, \xi_2]$ and ξ_1 and ξ_2 are the desired eigenvectors. The coordinate transformation yields

$$\dot{x} = \bar{A}x + \bar{B}u$$

$$y = \bar{C}x,$$

where

$$\bar{A} = \begin{bmatrix} 0 & 1 \\ -2 & -3 \end{bmatrix}, \qquad \bar{B} = \begin{bmatrix} 0 & 1 \\ 1 & 0 \end{bmatrix}, \qquad \bar{C} = [1 \quad 0].$$

Note that our choice of eigenvectors happened to produce an $\bar{\mathbf{A}}$ matrix in companion form with the same eigenvalues as $\mathbf{A}$. Of course, the eigenvectors of $\bar{\mathbf{A}}$ and $\mathbf{A}$ differ as this was our design objective. If our objective were instead simply to produce an $\bar{\mathbf{A}}$ matrix in companion form, we could have chosen $\boldsymbol{\xi}_1 = [1 \quad \alpha]^T$ and $\boldsymbol{\xi}_2 = [1 \quad \beta]^T$ as variables, computed $\bar{\mathbf{A}} = \mathbf{MAM}^{-1}$, and then solved for the parameters α and β to satisfy the two companion matrix conditions $\bar{a}_{11} = 0$ and $\bar{a}_{12} = 1$.

2.5 EXERCISES

2.5-1 In many applications a matrix raised to some power must be evaluated. Show how diagonalization can be used to simplify this process. In particular, use diagonalization to find $\mathbf{A}^{10}$, where

$$\mathbf{A} = \begin{bmatrix} 0 & 1 \\ -2 & -3 \end{bmatrix}.$$

Do not compute $\mathbf{A}^{10}$ as $\mathbf{AA} \ldots \mathbf{A}$ or $\mathbf{A}^2\mathbf{A}^2 \ldots \mathbf{A}^2$, and so on; use diagonalization.

Hint: Apply a coordinate transformation to $\mathbf{x}_{k+1} = \mathbf{A}\mathbf{x}_k$.

2.5-2 Transform the following systems to state-space decoupled form:

(a) $\begin{bmatrix} \dot{x}_1 \\ \dot{x}_2 \end{bmatrix} = \begin{bmatrix} 16 & 63 \\ -4 & -16 \end{bmatrix} \begin{bmatrix} x_1 \\ x_2 \end{bmatrix} + \begin{bmatrix} 0 \\ 1 \end{bmatrix} u, \qquad y = [1 \quad 0] \begin{bmatrix} x_1 \\ x_2 \end{bmatrix}$

(b) $\begin{bmatrix} \dot{x}_1 \\ \dot{x}_2 \end{bmatrix} = \begin{bmatrix} 0 & 2 \\ -8 & 0 \end{bmatrix} \begin{bmatrix} x_1 \\ x_2 \end{bmatrix} + \begin{bmatrix} 0 & 1 \\ 1 & 2 \end{bmatrix} \begin{bmatrix} u_1 \\ u_2 \end{bmatrix},$

$\begin{bmatrix} y_1 \\ y_2 \end{bmatrix} = \begin{bmatrix} 1 & 0 \\ 0 & 1 \end{bmatrix} \begin{bmatrix} x_1 \\ x_2 \end{bmatrix}$

2.5-3 The following second-order SISO systems appear to be very close to being in companion form:

$$\text{System 1:} \quad \dot{x}_1 = x_2 + u$$
$$\dot{x}_2 = -x_1 + u$$

$$\text{System 2:} \quad \dot{x}_1 = x_2$$
$$\dot{x}_2 = -x_1 + 2u$$

(a) Show from the definition of companion form that neither of these systems is in companion form.

(b) Show that each of these systems satisfies the controllability condition.

(c) Convert each system to companion form using (2.1-26)– (2.1-28).

2.5-4 Obtain an equivalent IO representation of the following systems, if such a representation exists:

(a) $\dot{x}_1 = x_2 + 2u$
$\quad \dot{x}_2 = -3x_1 - 4x_2 + 5u$
$\quad y = x_1$

(b) $\dot{x}_1 = -x_1 + 2u$
$\quad \dot{x}_2 = x_1$
$\quad y = x_2$

(c) $\dot{x}_1 = x_1$
$\quad \dot{x}_2 = 2x_1 + 3x_2 + u$
$\quad y = x_1$

(d) $\dot{x}_1 = x_1$
$\quad \dot{x}_2 = 2x_1 + 3x_2 + u$
$\quad y = x_2$

2.5-5 Express each of the following IO systems in SISO state-space form by using (2.1-35). That is, find **A**, **B**, and d so that

$$\dot{x} = Ax + Bu$$

$$y = x_1 + du,$$

where

(a) $\dot{y} + 3y = 2u$
(b) $\ddot{y} + 5\dot{y} + 6y = 10u$
(c) $\ddot{y} + 5\dot{y} + 6y = 10u + 2\dot{u}$
(d) $y^{(3)} + 2\ddot{y} + 3\dot{y} + 2y = u + 3\dot{u} + 2\ddot{u}$.

2.5-6 Show that $Q = I$ for the system given by (2.1-32)–(2.1-33).

2.5-7 For the two carts in Example 2.1-3, is the companion form state-space system controllable? Is it observable? If not, given the stated control objective $y \to 0$, how could the state-space representation be changed to achieve a controllable and observable system?

2.5-8 Is the linearized toy train of Exercise 1.5-8 controllable?

2.5-9 Consider the problem of balancing the double inverted pendulum shown in Figure 2.5-1. Defining $x = [\theta, \dot{\theta}, \varphi, \dot{\varphi}]^T$, $u = \Gamma/m\ell^2$, and $\omega^2 = g/\ell$, we obtain the linearized state equations below. Determine whether the system is controllable or not.

$$\begin{bmatrix} \dot{x}_1 \\ \dot{x}_2 \\ \dot{x}_3 \\ \dot{x}_4 \end{bmatrix} = \begin{bmatrix} 0 & 1 & 0 & 0 \\ 2\omega^2 & 0 & -\omega^2 & 0 \\ 0 & 0 & 0 & 1 \\ -2\omega^2 & 0 & 2\omega^2 & 0 \end{bmatrix} \begin{bmatrix} x_1 \\ x_2 \\ x_3 \\ x_4 \end{bmatrix} + \begin{bmatrix} 0 \\ 1 \\ 0 \\ -1 \end{bmatrix} u$$

2.5-10 Show that Example 2.1-5 is both controllable and observable.

2.5-11 Obtain the transfer function for each system in Exercise 2.5-5.

2.5-12 An IO system is given by

$$\ddot{y} + 2\dot{y} + 3y = 2u + \dot{u}.$$

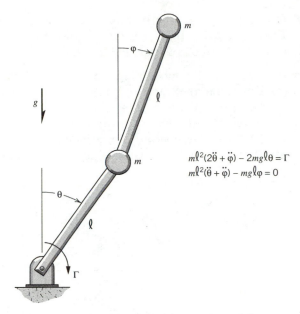

$$ml^2(2\ddot{\theta} + \ddot{\varphi}) - 2mgl\theta = \Gamma$$
$$ml^2(\ddot{\theta} + \ddot{\varphi}) - mgl\varphi = 0$$

Figure 2.5-1 Double-pinned inverted pendulum system for Exercise 2.5-9.

(a) Find the transfer function $G_P(s)$ for this system.

(b) Determine a state-space representation for the IO system in (a) using (2.1-35).

(c) Consider now feeding back the output as in Figure 2.3-3. Specifically, $G_1 = G_P$ and $G_2 = 1$. Determine the overall transfer function for this modified system. What difference exists between the transfer functions of the original IO system and the modified system?

2.5-13 For the feedback control system in Figure 2.3-5, discuss qualitatively the sensitivity of the disturbance transfer function $G_2(s)$ with respect to changes in $G_P(s)$, $G_C(s)$, and $H(s)$, as was done for the command transfer function $G_1(s)$ at the end of Section 2.3.

2.5-14 Use the transfer matrix (2.4-4) to convert the general first-order IO system of the form

$$\dot{y} + p_0 y = q_0 u + q_1 \dot{u}$$

to a state-space system of the form

$$\dot{x} = -p_0 x + q_0 u$$
$$y = \left(\frac{1 - p_0 q_1}{q_0}\right) x + q_1 u.$$

2.5-15 Use the transfer matrix (2.4-4) to convert the IO systems in Exercise 2.5-5 to state-space companion form. Repeat this process using the method of Example 2.4-2.

2.5-16 A system has the block diagram shown in Figure 2.5-2.
  (a) Determine the transfer function $G_1(s) = Y_1(s)/U_1(s)$.
  (b) Determine the transfer function $G_2(s) = Y_2(s)/U_2(s)$.
  (c) Determine the overall matrix transfer function from Figure 2.5-2 and the superposition principle that $Y(s) = Y_1(s) + Y_2(s)$.
  (d) Develop a diagonal state-space representation of the system for $T = 0.1$.

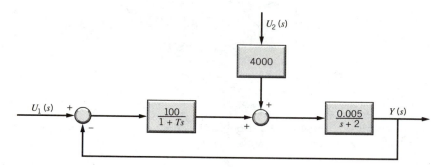

Figure 2.5-2 Block diagram for Exercise 2.5-16.

Chapter **3**

Free Response

3.1 EQUILIBRIUM AND STABILITY

A fundamental concept useful in characterizing the behavior of a
dynamical system is associated with the ability of a system with no
uncertain inputs to remain near or return to a constant solution called an
equilibrium solution (or **equilibrium point**). A solution is an equilibrium
solution if it satisfies the system differential equations (in IO or state-
space form) with a constant input and with all time derivatives equal to
zero. In the case of a nonzero constant input, a resulting equilibrium
solution is said to be "forced," a situation we will discuss in Chapter 4.
For a zero input, the equilibrium solution is said to be "free."

An equilibrium solution is said to be **stable** if any solution starting
sufficiently near the equilibrium point stays near it. An equilibrium
solution is said to be **asymptotically stable** if every solution starting
sufficiently near the equilibrium point not only stays near the equilibrium
point but also approaches it as $t \rightarrow \infty$. For a linear system, stability
properties are global ("sufficiently near" can be arbitrarily far away), and
we may then refer to the system itself as being either stable or unstable.

Frequently, an equilibrium point represents some desired nominal
operating condition for the dynamical system, and the task of a feedback
controller is to automatically return the dynamical system to this
equilibrium point. This is known as the **regulator problem**, and it will be
one of our principle areas of study in the design of feedback control
systems in Chapters 6 to 8.

With all inputs to a dynamical system set to zero, a dynamic response is
still possible by having the system start at a value of **x** (or y and its
derivatives, for an IO system) other than zero. The corresponding motion
is called the **free response**. We can think of the free response as
characterizing the motion of the system subject to an impulsive type of
disturbance from equilibrium. That is, a large disturbance of a very short
duration, which will instantaneously move the system from the equilib-

rium point to some other point, can be considered simply as a change in initial conditions. This approach allows us to examine the effect of impulsive type of disturbances without having to model them using "impulse functions."

The stability of a dynamical system is associated with the free response of the system. In a state-space setting the stability of a linear system is completely determined by the **A** matrix. In a control system application, we add control elements to a system, in part, to modify the primary system's dynamical stability characteristics. The resulting overall system is a new dynamical system, whose stability properties are associated with the free response of the overall system. For example, for a linear system of the form

$$\dot{\mathbf{x}} = \mathbf{Ax} + \mathbf{Bu},$$

we might choose a linear **state feedback control** according to

$$\mathbf{u} = -\mathbf{Kx},$$

where **K** is an $N_u \times N_x$ **feedback gain matrix**. This yields a new linear system

$$\dot{\mathbf{x}} = \tilde{\mathbf{A}}\mathbf{x},$$

where

$$\tilde{\mathbf{A}} = \mathbf{A} - \mathbf{BK}.$$

The stability of this new system is determined by the free response of the new system, specifically, by the matrix $\tilde{\mathbf{A}}$.

3.2   FREE RESPONSE OF INPUT-OUTPUT SYSTEMS

The free response of a single-input single-output IO system is easily obtained by using classical methods to integrate the system. The IO system with scalar input $u(t) \equiv 0$ is given by

$$y^{(N_x)} + p_{N_x-1}y^{(N_x-1)} + \cdots + p_1\dot{y} + p_0 y = 0. \tag{3.2-1}$$

If we assume a solution, with γ and λ constant, of the form

$$y(t) = \gamma e^{\lambda t}, \tag{3.2-2}$$

then by differentiating and substituting into (3.2-1), we obtain

$$(\lambda^{N_x} + p_{N_x-1}\lambda^{N_x-1} + \cdots + p_1\lambda + p_0)\gamma e^{\lambda t} = 0.$$

Let $P(\lambda)$ denote the polynomial multiplying $\gamma e^{\lambda t}$. In order for $\gamma e^{\lambda t}$ to be a nonzero solution, the polynomial $P(\lambda)$ must be zero:

$$P(\lambda) \triangleq \lambda^{N_x} + p_{N_x-1}\lambda^{N_x-1} + \cdots + p_1\lambda + p_0 = 0. \qquad (3.2\text{-}3)$$

Equation (3.2-3) is called the **characteristic equation** of the dynamical system. Note that the polynomial $P(\lambda)$ is the same one defined in connection with the transfer function. Since $P(\lambda)$ is an N_x-order polynomial, it will always have exactly N_x roots, called the **characteristic roots** or **eigenvalues** of the system. The eigenvalues may be real or complex, with complex roots occurring in conjugate pairs since the coefficients in $P(\lambda)$ are all real. Every root corresponds to a solution of the form (3.2-2). Since the IO system is a linear differential equation, the sum of solutions of the form (3.2-2),

$$y(t) = \gamma_1 e^{\lambda_1 t} + \cdots + \gamma_{N_x} e^{\lambda_{N_x} t}, \qquad (3.2\text{-}4)$$

is also a solution, as one can verify by direct substitution. Furthermore, every solution of (3.2-1) can be expressed in the form of (3.2-4) by proper choice of the coefficients $\gamma_1 \ldots \gamma_{N_x}$. If some of the eigenvalues are complex conjugate pairs, then the corresponding coefficients will also form complex conjugate pairs, since $y(t)$ is real-valued. Finally, if some of the eigenvalues are repeated, the corresponding coefficients will be explicit functions of time (Hildebrand, 1962, p. 9). Specifically, if $\lambda = \lambda_i = \cdots = \lambda_{i+m}$ is a root of multiplicity m, then the corresponding coefficients $\gamma_i \ldots \gamma_{i+m}$ will be of the form

$$\gamma_i = \alpha_i$$

$$\gamma_{i+1} = \alpha_{i+1}t$$

$$\gamma_{i+2} = \alpha_{i+2}t^2 \qquad (3.2\text{-}5)$$

$$\vdots$$

$$\gamma_{i+m} = \alpha_{i+m}t^m.$$

To obtain the free response, the coefficients in (3.2-4) are determined by requiring that (3.2-4) satisfy not only the differential equation (3.2-1) but also the initial conditions on $y(0)$, $\dot{y}(0)$, $\ldots$, $y^{(N_x-1)}(0)$. For example, if the roots to (3.2-3) are not repeated, then these requirements yield constant γ's that are determined from the following **Van der Monde equation**:

$$
\begin{bmatrix} y(0) \\ \dot{y}(0) \\ \vdots \\ y^{(N_x-1)}(0) \end{bmatrix}
=
\begin{bmatrix} 1 & \cdots & 1 \\ \lambda_1 & \cdots & \lambda_{N_x} \\ \vdots & & \vdots \\ \lambda_1^{N_x-1} & \cdots & \lambda_{N_x}^{N_x-1} \end{bmatrix}
\begin{bmatrix} \gamma_1 \\ \gamma_2 \\ \vdots \\ \gamma_{N_x} \end{bmatrix}. \qquad (3.2\text{-}6)
$$

Because of their fundamental nature, we will examine in some detail both a general first-order IO system and a general second-order IO system. The root of the characteristic equation for a first-order system is real. The response of a second-order system may differ considerably from a first-order system, since the roots to the characteristic equation may be complex conjugates. Thorough knowledge of the free response of these two systems is important because the total response of any higher-order system will be a combination of first-order and second-order responses. This follows from the fact that all roots of the characteristic equation are either real or complex conjugates.

First-Order Systems

A general first-order IO system with $u(t) \equiv 0$ is given by

$$\dot{y} + p_0 y = 0. \tag{3.2-7}$$

The corresponding characteristic equation

$$\lambda + p_0 = 0$$

has the solution $\lambda = -p_0$, which is real. It follows from (3.2-4) and (3.2-6) that the solution for the free response is given by

$$y(t) = y(0)e^{-p_0 t}. \tag{3.2-8}$$

The behavior of this solution is completely characterized in Figure 3.2-1. Note that if $\lambda < 0$, the solution will asymptotically approach the equilibrium solution $y(t) \equiv 0$ as $t \to \infty$ no matter what finite initial condition is imposed. If $\lambda = 0$, then the solution is $y(t) \equiv y(0) = $ constant. We thus conclude that the system will be stable whenever $\lambda \leq 0$ and asymptotically stable if $\lambda < 0$. Conversely, it follows from (3.2-8) that the system is unstable if $\lambda > 0$. This completely exhausts the possibilities for a first-order IO system.

For an asymptotically stable first-order system, the rate at which the system returns to the equilibrium solution is determined by the magnitude of the characteristic root λ. The initial displacement is reduced by a factor of $1 - 1/e$ ($\approx 63\%$) when

$$t = T \triangleq \frac{1}{|\lambda|}. \tag{3.2-9}$$

The time T defined by (3.2-9), $e^{\lambda T} = e^{-1}$, is called the **return time** or **system time constant**. The return time gives a measure of how quickly an asymptotically stable system will return to the equilibrium state after

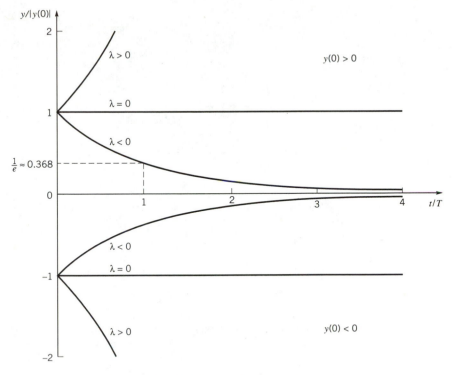

Figure 3.2-1 Response of a first-order system.

being disturbed from it. A large return time corresponds to a small value of $|\lambda|$ and vice versa.

Second-Order Systems

A general second-order IO system with $u(t) \equiv 0$ is given by

$$\ddot{y} + p_1\dot{y} + p_0 y = 0, \qquad (3.2\text{-}10)$$

with the characteristic equation

$$\lambda^2 + p_1\lambda + p_0 = 0. \qquad (3.2\text{-}11)$$

The most frequently quoted example of a second-order system is the so-called **damped harmonic oscillator**. An example of such a device is given by the spring, mass, and shock absorber system depicted in Figure 3.2-2.

We consider a linear spring, with force proportional to displacement, and a linear shock absorber, with force proportional to velocity. If we denote the spring constant by k, the shock absorber constant by β, and the

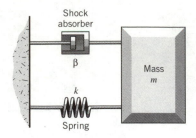

Figure 3.2-2 Damped harmonic oscillator.

mass by $m > 0$, then it follows from Newton's second law that

$$p_0 = \frac{k}{m}$$

$$p_1 = \frac{\beta}{m}.$$

A shock absorber is an energy dissipating device that is also called a **damper**. If $\beta = 0$ (no damping), the spring-mass system will oscillate at its undamped **natural frequency** defined by

$$\omega_n = \sqrt{\frac{k}{m}}.$$

For the case where $\omega_n \neq 0$, we also define a dimensionless **damping ratio**

$$\zeta = \frac{\beta}{2m\omega_n}.$$

In terms of the natural frequency and damping ratio, the coefficients p_0 and p_1 can be written as

$$p_0 = \omega_n^2$$
$$p_1 = 2\zeta\omega_n.$$

In what follows we will study a general second-order system that will include the damped harmonic oscillator as a special case. For those situations in which we particularly want to think of the system in terms of the damped oscillator, we will replace the constants p_0 and p_1 with the definitions given above. The usefulness of this substitution will become apparent as we proceed.

There are three possibilities for the two roots to the characteristic equation (3.2-11). They can be real and distinct, real and equal, or complex conjugates. The character of the free response is different for each case.

Real Distinct Eigenvalues

Consider first the real and distinct case. Let λ_1 and λ_2 represent the two roots. It follows from (3.2-4) that the solution is given by

$$y(t) = \gamma_1 e^{\lambda_1 t} + \gamma_2 e^{\lambda_2 t}. \tag{3.2-12}$$

Using initial conditions to evaluate γ_1 and γ_2 from (3.2-6), we obtain

$$\gamma_1 = \frac{\lambda_2 y(0) - \dot{y}(0)}{\lambda_2 - \lambda_1} \quad \text{and} \quad \gamma_2 = \frac{\dot{y}(0) - \lambda_1 y(0)}{\lambda_2 - \lambda_1}. \tag{3.2-13}$$

That is, the solution is a sum of two first-order solutions of the type shown in Figure 3.2-1. Since the solution is a sum, the system will be stable if and only if both $\lambda_1 \leq 0$ and $\lambda_2 \leq 0$, and asymptotically stable if and only if λ_1 and λ_2 are negative. If both λ_1 and λ_2 are negative, then it is possible to define a return time by

$$T = \frac{1}{\min |\lambda_i|}, \quad i = 1, 2. \tag{3.2-14}$$

If the system is asymptotically stable and λ_1 and λ_2 differ somewhat, then one of the exponential decay terms in (3.2-12) will decrease noticeably more slowly than the other. The time it takes for the system to return to a small neighborhood of the equilibrium state will be determined by the more slowly decaying exponential. The return time gives a proper measure of this time.

Real Repeated Eigenvalues

If the characteristic roots for a second-order IO system are real and equal, a modification of the above procedure must be used. Let λ be the repeated root. We can use (3.2-5), in which case we let $\gamma_1 = \alpha_1$, $\gamma_2 = \alpha_2 t$ and proceed as before, evaluating α_1 and α_2 from initial conditions, to obtain

$$y(t) = y(0)e^{\lambda t} + [\dot{y}(0) - \lambda y(0)]t e^{\lambda t}. \tag{3.2-15}$$

Another method is to temporarily consider the roots to be distinct by letting

$$\lambda_1 = \lambda \tag{3.2-16}$$

$$\lambda_2 = \lambda + \epsilon \tag{3.2-17}$$

and then proceed as with distinct eigenvalues, using (3.2-6), until the following solution for the output is obtained:

$$y(t) = \frac{[(\lambda + \epsilon)y(0) - \dot{y}(0)]e^{\lambda t} + [\dot{y}(0) - \lambda y(0)]e^{(\lambda + \epsilon)t}}{\epsilon}. \tag{3.2-18}$$

If we take the limit as $\epsilon \to 0$, we obtain the right-hand side as a ratio of 0/0. Applying L'Hôpital's rule allows the evaluation of this limit, which yields (3.2-15). Only the first term on the right-hand side of (3.2-15) yields the same response as for a first-order system illustrated in Figure 3.2-1. Figure 3.2-3 shows the general nature of the second term, provided that $\dot{y}(0) \neq \lambda y(0)$. Since the solution is the sum of these two terms, the system will be asymptotically stable when $\lambda < 0$ and unstable if $\lambda \geq 0$. In particular, for repeated roots with $\lambda = 0$ the system is unstable, because to the second term in (3.2-15).

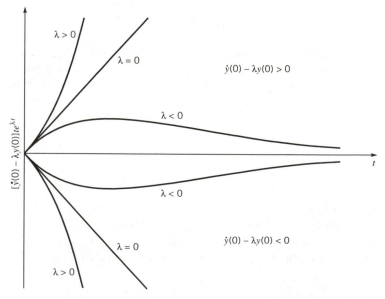

Figure 3.2-3 A plot of the second term in (3.2-15).

Complex Conjugate Eigenvalues

The final possibility is that the characteristic roots for a second-order system are complex conjugates. In this case, it is convenient to think of the system in terms of the damped oscillator and replace p_1 and p_0 by

$$p_1 = \omega_n^2$$
$$p_1 = 2\zeta\omega_n.$$

The characteristic equation is then written as

$$\lambda^2 + 2\zeta\omega_n\lambda + \omega_n^2 = 0 \tag{3.2-19}$$

and its roots are given by

$$\lambda = -\zeta\omega_n \pm i\omega_n\sqrt{1 - \zeta^2}. \tag{3.2-20}$$

We obtain complex roots, provided that $\omega_n \neq 0$ and $|\zeta| < 1$. Since ω_n appears in each term, without loss of generality we can assume that $\omega_n > 0$. If we define a new quantity called the **damped frequency**

$$\omega_d = \omega_n\sqrt{1 - \zeta^2}, \tag{3.2-21}$$

the solution (3.2-4) can be written as

$$y(t) = e^{-\zeta\omega_n t}\{\gamma_1 e^{i\omega_d t} + \gamma_2 e^{-i\omega_d t}\}. \tag{3.2-22}$$

Using Euler's formula

$$e^{i\theta} = \cos\theta + i\sin\theta,$$

we obtain

$$y(t) = e^{-\zeta\omega_n t}[(\gamma_1 + \gamma_2)\cos(\omega_d t) + i(\gamma_1 - \gamma_2)\sin(\omega_d t)]. \tag{3.2-23}$$

Note that for $y(t)$ to be real, the coefficients of the sine and cosine terms must be real, Hence, γ_1 and γ_2 must be a complex conjugate pair. Using the initial conditions to evaluate these coefficients, we obtain

$$y(t) = e^{-\zeta\omega_n t}\left[y(0)\cos(\omega_d t) + \frac{\dot{y}(0) + y(0)\zeta\omega_n}{\omega_d}\sin(\omega_d t)\right]. \tag{3.2-24}$$

The resulting motions, illustrated in Figure 3.2-4 for positive and negative ζ, are called damped harmonic motion.

Figure 3.2-4a illustrates a typical case of **underdamped** harmonic motion, which occurs when $0 < \zeta < 1$ and $\omega_n > 0$. This would correspond

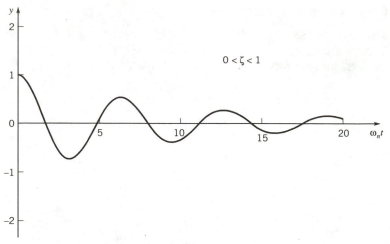

(a) Underdamped harmonic oscillator

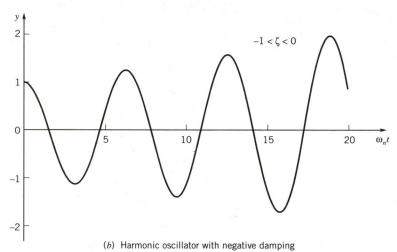

(b) Harmonic oscillator with negative damping

Figure 3.2-4 Output response of a second-order system.

to the physical system illustrated in Figure 3.2-2 with sufficiently small damping to satisfy $0 < \zeta < 1$. In this case, the equilibrium solution $y(t) \equiv 0$ is asymptotically stable because of the dissipation of energy by the shock absorber. However, the approach to the equilibrium state is oscillatory. The return time for damped harmonic motion is defined by

$$T = \frac{1}{|\mathrm{Re}(\lambda)|} = \frac{1}{\zeta\omega_n},$$ (3.2-25)

where $\mathrm{Re}(\lambda)$ denotes the real part of λ. Figure 3.2-4b illustrates unstable oscillatory motion with negative damping, obtained when $-1 < \zeta < 0$. This would not correspond to the physical situation depicted in Figure 3.2-2, but could correspond to some system in which energy is fed into the system.

The quantity ω_n is called the **undamped natural frequency**, and this is the frequency the system would oscillate at if $\zeta = 0$. When the damping ratio $\zeta = 1$, the system is said to be **critically damped**, since this is the point at which the system has real repeated eigenvalues with no oscillation. If $\zeta > 1$, the system is said to be **overdamped**. The larger ζ is for this case, the slower the system will be to return to equilibrium. For a given ω_n, the critically damped system will have the shortest return time but, as we shall see, the underdamped system will generally yield better overall performance. Note that with $0 < \zeta < 1$, the system oscillates at a **damped frequency** given by (3.2-21) that is always less than the natural frequency ω_n.

These quantities are given a geometric interpretation in Figure 3.2-5 in terms of the location of the eigenvalues (3.2-20) in the complex plane. Since complex roots occur in conjugate pairs, Figure 3.2-5 is symmetric about the real axis, and we only display the upper half-plane. The modulus (absolute value) of λ and the angle from the positive real axis to λ are given by

$$r = |\lambda| \triangleq \sqrt{\lambda\lambda^*} = \sqrt{\mathrm{Re}(\lambda)^2 + \mathrm{Im}(\lambda)^2} = \omega_n$$

$$\theta = \angle \lambda \triangleq \tan^{-1}\left[\frac{\mathrm{Im}(\lambda)}{\mathrm{Re}(\lambda)}\right] = \tan^{-1}\left(\frac{\sqrt{1-\zeta^2}}{-\zeta}\right),$$

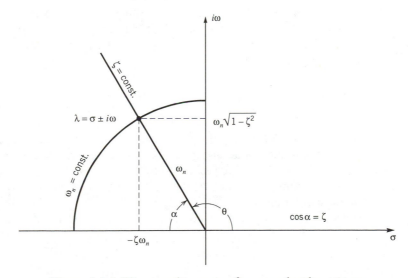

Figure 3.2-5 The complex roots of a second-order system.

where $\text{Re}(\cdot)$ and $\text{Im}(\cdot)$ denote real and imaginary parts, respectively, $(\cdot)^*$ denotes the conjugate of a complex quantity (obtained by replacing i by $-i$), and $\tan^{-1}(\cdot,\cdot)$ is interpreted as the two-argument arc tangent function [as in the Fortran $\text{ATAN2}(\cdot,\cdot)$ function] in order to determine the proper quadrant.

The eigenvalues are located a distance ω_n from the origin. In particular, for a given value of ω_n, the eigenvalues (as functions of ζ) lie on a circle of radius ω_n centered at the origin. Also note in Figure 3.2-5 that the eigenvalues are located at an angle α to the negative real axis given by $\cos\alpha = \zeta$. Therefore for a given value of ζ, the eigenvalues (as functions of $\omega_n \geq 0$) all lie along a straight line through the origin at an angle α with the negative real axis. Note that the eigenvalues are in the left half-plane [$\text{Re}(\lambda) \leq 0$] for $\zeta \geq 0$. As the damping ratio is increased from $\zeta = 0$ to $\zeta = 1$, the angle will decrease from $\alpha = \pi/2$ (no damping, imaginary roots) to $\alpha = 0$ (critical damping, repeated real roots).

Based on these geometric results, a feedback control specification such as "damping ratio $\geq \bar{\zeta}$" means that, in the complex plane, the eigenvalues must lie in a cone with half-angle $\bar{\alpha} = \cos^{-1}(\bar{\zeta})$, centered at the origin and symmetric about the negative real axis. An additional specification of "natural frequency $\leq \bar{\omega}$" would truncate this cone at radius $\bar{\omega}$. Frequently, performance specifications on the design of a feedback control system can be reduced to constraints such as these on the location of the eigenvalues of the overall system.

Higher-Order Systems

Since an IO system is linear, the free response of any higher-order IO system will be a sum of the first- and second-order responses already noted. This follows from the fact that the roots of the characteristic equation are either real or complex, with the complex roots appearing as conjugate pairs. For higher-order asymptotically stable systems, the return time is defined by

$$T = \frac{1}{\min|\text{Re}(\lambda_i)|}, \qquad i = 1, 2, \ldots, N_x, \qquad \text{(3.2-26)}$$

and the eigenvalue corresponding to T is called the **dominant eigenvalue**, or the dominant eigenvalue pair if it is one of a complex conjugate pair. The notion of dominant eigenvalues can often be used to approximate a high-order dynamical system by a lower-order system with the same dominant eigenvalues.

As an example, consider a third-order system with transfer function

$$G(s) = \frac{K}{(\tau s + 1)(s^2 + 2\zeta\omega_n s + \omega_n^2)}.$$

In Figure 3.2-6 the free response of this system is illustrated for two cases. In both cases, the roots are "separated" as shown and consist of a real root and a complex conjugate root pair. In case a the return time is determined by the complex root pair, and in case b the return time is determined by the real root. Note that because of the separation in the real part of the roots, in case a the third-order system behaves like a second-order system after a short time, and in case b the third-order system behaves like a first-order system after a short time.

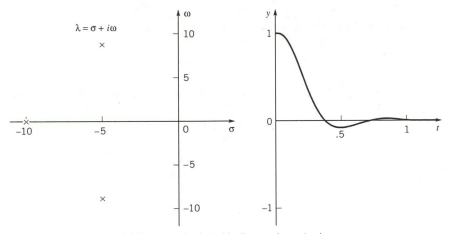

(a) Response dominated by the complex root pair

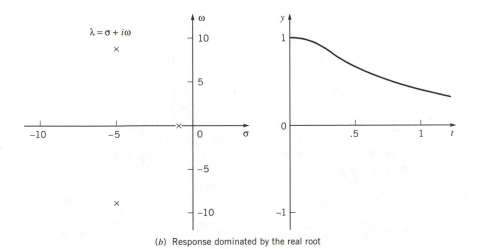

(b) Response dominated by the real root

Figure 3.2-6 Response of a third-order system.

3.3   FREE RESPONSE OF STATE SPACE SYSTEMS

We now return to the general state variable representation of a linear dynamical system, to investigate the free response of a system of the form

$$\dot{\mathbf{x}} = \mathbf{A}\mathbf{x}. \tag{3.3-1}$$

By analogy with an IO system, we seek a solution of the form

$$\mathbf{x}(t) = e^{\lambda t}\boldsymbol{\xi}.$$

Substituting this solution into (3.3-1) yields

$$\lambda e^{\lambda t}\boldsymbol{\xi} = \mathbf{A}e^{\lambda t}\boldsymbol{\xi}.$$

Canceling the nonzero factor $e^{\lambda t}$, we obtain the **eigenvector equation**

$$\mathbf{A}\boldsymbol{\xi} = \lambda\boldsymbol{\xi},$$

or

$$[\lambda\mathbf{I} - \mathbf{A}]\boldsymbol{\xi} = \mathbf{0},$$

which has a nonzero solution $\boldsymbol{\xi}$ if and only if the following **characteristic equation** is satisfied:

$$|\lambda\mathbf{I} - \mathbf{A}| = 0, \tag{3.3-3}$$

where $|\cdot|$ denotes the determinant. Equation (3.3-3) is an N_x-order polynomial equation, and the corresponding polynomial is the same as $P(\lambda)$ in (3.2-3). Indeed, this polynomial is a characteristic of the dynamical system, regardless of whether the system is expressed in state-space, IO, or transfer function form.

   Thus, we see that a nonzero solution to (3.3-1) is obtained only if λ is an eigenvalue of $\mathbf{A}$ and $\boldsymbol{\xi}$ is the corresponding eigenvector. Assume, for the moment, that the eigenvalues of $\mathbf{A}$ are distinct. Then the eigenvectors of $\mathbf{A}$ form a set of N_x linearly independent vectors, and the general solution (called the "homogeneous" solution) is obtained by forming the linear combination

$$\mathbf{x}(t) = \gamma_1 e^{\lambda_1 t}\boldsymbol{\xi}_1 + \cdots + \gamma_{N_x} e^{\lambda_{N_x} t}\boldsymbol{\xi}_{N_x}. \tag{3.3-4}$$

This solution can also be written as

$$\mathbf{x}(t) = \mathbf{M}e^{\mathbf{\Lambda} t}\boldsymbol{\gamma}, \tag{3.3-5}$$

where

$$\gamma = [\gamma_1 \dots \gamma_{N_x}]^T \tag{3.3-6}$$

$$\mathbf{M} = [\xi_1, \dots, \xi_{N_x}] \tag{3.3-7}$$

and

$$e^{\Lambda t} \triangleq \begin{bmatrix} e^{\lambda_1 t} & & & \\ & e^{\lambda_2 t} & & \mathbf{0} \\ & & \ddots & \\ \mathbf{0} & & & e^{\lambda_{N_x} t} \end{bmatrix}. \tag{3.3-8}$$

The coefficients $\gamma_1, \dots, \gamma_{N_x}$ are constants that can be determined from the initial conditions by evaluating (3.3-5) at $t = 0$, yielding a system of N_x equations

$$\mathbf{x}(0) = \mathbf{M}\gamma.$$

Since the eigenvectors are linearly independent, this sytem of equations will yield a unique result for the coefficients as given by

$$\gamma = \mathbf{M}^{-1}\mathbf{x}(0).$$

The solution in terms of initial conditions is then given by

$$\mathbf{x}(t) = \mathbf{M}e^{\Lambda t}\mathbf{M}^{-1}\mathbf{x}(0). \tag{3.3-9}$$

If some of the eigenvalues are repeated, then the eigenvectors may or may not be linearly independent. In any case, a general solution can still be specified in the form of (3.3-9), through the use of a **matrix perturbation** technique (Luenberger, 1979, p. 89). In particular, it is always possible to perturb by small amounts, less than ϵ in magnitude, those elements of **A** that cause repeated eigenvalues. This will make the resulting eigenvalues distinct and will yield N_x linearly independent eigenvectors. The general solution then follows from (3.3-9) in the limit as $\epsilon \to 0$.

EXAMPLE 3.3-1	**Unstable Second-Order System**

For any real number a, including $a = 0$, consider a second-order system given by

$$\begin{bmatrix} \dot{x}_1 \\ \dot{x}_2 \end{bmatrix} = \begin{bmatrix} 0 & 1 \\ a^2 & 0 \end{bmatrix} \begin{bmatrix} x_1 \\ x_2 \end{bmatrix}.$$

From (3.3-3) the characteristic equation

$$\lambda^2 - a^2 = 0$$

yields the eigenvalues $\lambda_1 = a$ and $\lambda_2 = -a$. The system is unstable, since with $a \neq 0$, one root is positive, and when $a = 0$, both roots are equal to zero. From (3.3-2) the corresponding eigenvectors can be written as

$$\xi_1 = [1 \quad a]^T$$
$$\xi_2 = [1 \quad -a]^T.$$

Then for $a \neq 0$ from (3.3-9), we obtain

$$\begin{bmatrix} x_1(t) \\ x_2(t) \end{bmatrix} = \begin{bmatrix} 1 & 1 \\ a & -a \end{bmatrix} \begin{bmatrix} e^{at} & 0 \\ 0 & e^{-at} \end{bmatrix} \begin{bmatrix} \dfrac{1}{2} & \dfrac{1}{2a} \\ \dfrac{1}{2} & -\dfrac{1}{2a} \end{bmatrix} \begin{bmatrix} x_1(0) \\ x_2(0) \end{bmatrix},$$

which yields

$$\begin{bmatrix} x_1(t) \\ x_2(t) \end{bmatrix} = \frac{1}{2} \begin{bmatrix} e^{at} + e^{-at} & \dfrac{e^{at} - e^{-at}}{a} \\ ae^{at} - ae^{-at} & e^{at} + e^{-at} \end{bmatrix} \begin{bmatrix} x_1(0) \\ x_2(0) \end{bmatrix}.$$

This can be written more compactly, in terms of hyperbolic functions, as

$$\begin{bmatrix} x_1(t) \\ x_2(t) \end{bmatrix} = \begin{bmatrix} \cosh at & \dfrac{1}{a} \sinh at \\ a \sinh at & \cosh at \end{bmatrix} \begin{bmatrix} x_1(0) \\ x_2(0) \end{bmatrix}.$$

For the special case where $a = 0$, we have $\lambda_1 = \lambda_2 = 0$ and $\xi_1 = \xi_2 = [1 \quad 0]^T$, and the matrix **M** in this case does not have an inverse. But if we let a be a small positive quantity, then the solution $\mathbf{x}(t)$ is obtained from the above result in the limit as $a \to 0$. Appyling L'Hôpital's rule yields

$$\begin{bmatrix} x_1(t) \\ x_2(t) \end{bmatrix} = \begin{bmatrix} 1 & t \\ 0 & 1 \end{bmatrix} \begin{bmatrix} x_1(0) \\ x_2(0) \end{bmatrix}.$$

Henceforth, we will generally assume that the eigenvalues of **A** are distinct. If not, the above perturbation technique can be applied.

Decoupled State Equations

We have previously observed that if $\mathbf{A}$ were a diagonal matrix of the form

$$\mathbf{\Lambda} = \text{diag}[\lambda_1, \ldots, \lambda_{N_x}] \triangleq \begin{bmatrix} \lambda_1 & & & \\ & \lambda_2 & & \mathbf{0} \\ & & \cdot & \\ \mathbf{0} & & & \cdot \\ & & & & \lambda_{N_x} \end{bmatrix}$$

for some numbers $\lambda_1, \ldots, \lambda_{N_x}$, then the state equations would be "decoupled." If this were the case, then each resulting equation of the form

$$\dot{x}_i = \lambda_i x_i, \qquad i = 1, \ldots, N_x$$

would have the corresponding solution

$$x_i(t) = e^{\lambda_i t} x_i(0).$$

In general, the $\mathbf{A}$ matrix is not in diagonal form. However, the above observation suggests a different procedure for obtaining a solution to the state equations (3.3-1). From our previous discussion of diagonalization, we can decouple the state equations by applying a coordinate transformation

$$\mathbf{z} = \mathbf{M}^{-1}\mathbf{x}, \tag{3.3-10}$$

where $\mathbf{M}$ is the **eigenvector (modal) matrix** (3.3-7) with the eigenvectors of $\mathbf{A}$ as its columns. We assume that the eigenvectors are linearly independent so that $\mathbf{M}^{-1}$ exists. As previously noted, a sufficient condition for this is that the eigenvalues of $\mathbf{A}$ be distinct. Applying this transformation to

$$\dot{\mathbf{x}} = \mathbf{A}\mathbf{x}$$

yields

$$\mathbf{M}\dot{\mathbf{z}} = \mathbf{A}\mathbf{M}\mathbf{z},$$

so that

$$\dot{\mathbf{z}} = \mathbf{M}^{-1}\mathbf{A}\mathbf{M}\mathbf{z}.$$

By definition

$$\mathbf{\Lambda} = \mathbf{M}^{-1}\mathbf{A}\mathbf{M},$$

so that

$$\dot{\mathbf{z}} = \mathbf{\Lambda}\mathbf{z}.$$

Using the definition (3.3-8), we obtain the solution in the form

$$\mathbf{z}(t) = e^{\Lambda t}\mathbf{z}(0).$$

The corresponding $\mathbf{x}(t)$ solution then follows directly from $\mathbf{x} = \mathbf{Mz}$. Applying this transformation, we obtain

$$\mathbf{M}^{-1}\mathbf{x}(t) = e^{\Lambda t}\mathbf{M}^{-1}\mathbf{x}(0)$$

or

$$\mathbf{x}(t) = \mathbf{M}e^{\Lambda t}\mathbf{M}^{-1}\mathbf{x}(0), \tag{3.3-11}$$

which is the same result given by (3.3-9).

Note that the eigenvector transformation matrix $\mathbf{M}$ and the decoupled state $\mathbf{z}$ generally have complex elements, since the eigenvalues may be complex. However, since any complex eigenvalues (and their corresponding eigenvectors) occur in conjugate pairs, the final resulting solution $\mathbf{x}(t)$ will always be real.

From the homogeneous solution (3.3-4), or from (3.3-11) and (3.3-8) in the case of distinct eigenvalues, we can see clearly the **fundamental stability criteria** for linear systems: in order for an equilibrium state to be **stable**, the real part of all eigenvalues must be ≤ 0. If all eigenvalues have real parts < 0, then the equilibrium is **asymptotically stable**. The equilibrium is **unstable** if even one eigenvalue has a real part > 0.

A special case occurs if any repeated eigenvalues have zero real parts. As we have previously noted, it is possible to have repeated roots and still have a set of linearly independent eigenvectors. In this special case, the equilibrium will be stable, but not asymptotically stable. Otherwise, with repeated eigenvalues the matrix perturbation technique must be used, and we will obtain terms involving polynomials in t. For repeated eigenvalues with zero real parts, there will be no counteracting exponentially decaying multiplier for these terms, so that the system would be unstable.

The State Transition Matrix

The solution (3.3-9) may be thought of in terms of a transformation matrix. That is, the state at time t is obtained from the state at time $t = 0$ through a linear transformation of the form

$$\Phi(t) = \mathbf{M}e^{\Lambda t}\mathbf{M}^{-1}.$$

More generally, for any linear constant coefficient system

$$\dot{\mathbf{x}} = \mathbf{Ax},$$

the state at any time t_2 can be related to the state at any other time t_1 by the linear transformation

$$\mathbf{x}(t_2) = \mathbf{\Phi}(t_2 - t_1)\mathbf{x}(t_1) \tag{3.3-12}$$

that depends only on the length of time interval $t_2 - t_1$. The notation is simplified by letting $t = t_2 - t_1$, where we usually have $t_1 = 0$. We will use this simplified notation unless we specifically want to denote two particular times and not just the length of a time interval. Because of the transition feature provided by (3.3-12), the $N_x \times N_x$ matrix $\mathbf{\Phi}(t)$ defined by this equation is called the **state transition matrix**. It has some important properties that follow directly from (3.3-12).

If $\mathbf{\Phi}(t)$ is known, then the initial state $\mathbf{x}(0)$ can be propagated to any time t (forward or backward) by a simple matrix multiplication

$$\mathbf{x}(t) = \mathbf{\Phi}(t)\mathbf{x}(0), \tag{3.3-13}$$

rather than by having to integrate the differential equations of motion. If the state transition matrix has only been computed for one time interval Δt, the state can still be propagated using

$$\mathbf{x}(\Delta t) = \mathbf{\Phi}(\Delta t)\mathbf{x}(0)$$

$$\mathbf{x}(2\Delta t) = \mathbf{\Phi}(\Delta t)\mathbf{x}(\Delta t) = \mathbf{\Phi}^2(\Delta t)\mathbf{x}(0)$$

$$\vdots$$

$$\mathbf{x}(k\Delta t) = \mathbf{\Phi}(\Delta t)\mathbf{x}(\{k - 1\}\Delta t) = \mathbf{\Phi}^k(\Delta t)\mathbf{x}(0),$$

with the advantages that these results are exact (except for possible round-off error involved in performing the multiplication numerically), and $\mathbf{\Phi}(t)$ does not need to be recomputed for each new time t.

Another important property of $\mathbf{\Phi}(t)$ that also follows from the definition is that an inverse for $\mathbf{\Phi}(t)$ exists for all times t. To see this, we note from (3.3-12) that with $t_1 = 0$, $t_2 = t$

$$\mathbf{x}(t) = \mathbf{\Phi}(t)\mathbf{x}(0)$$

and with $t_2 = 0$, $t_1 = t$

$$\mathbf{x}(0) = \mathbf{\Phi}(-t)\mathbf{x}(t)$$

$$= \mathbf{\Phi}(-t)\mathbf{\Phi}(t)\mathbf{x}(0).$$

Hence, the inverse transformation

$$\mathbf{\Phi}^{-1}(t) = \mathbf{\Phi}(-t) \tag{3.3-14}$$

is obtained from $\mathbf{\Phi}(t)$ just by replacing t by $-t$. This result implies that there is a unique trajectory $\mathbf{x}(t)$ that passes through a particular initial point $\mathbf{x}(0)$ in state space. At any time $t > 0$ (or $t < 0$), it is possible to uniquely recover the initial state $\mathbf{x}(0)$ by moving backward (or forward) in time along the trajectory.

Another property is that for any times t_1, t_2, and t_3

$$\mathbf{\Phi}(t_3 - t_1) = \mathbf{\Phi}(t_3 - t_2)\mathbf{\Phi}(t_2 - t_1). \tag{3.3-15}$$

In particular,

$$\mathbf{\Phi}(t)\mathbf{\Phi}(-\tau) = \mathbf{\Phi}(t - \tau). \tag{3.3-16}$$

Also note that $\mathbf{x}(t) = \mathbf{\Phi}(t)\mathbf{x}(0)$ implies that $\mathbf{\Phi}(t)$ must satisfy the matrix differential equation and boundary condition

$$\dot{\mathbf{\Phi}} = \mathbf{A}\mathbf{\Phi}, \qquad \mathbf{\Phi}(0) = \mathbf{I}. \tag{3.3-17}$$

These two results in (3.3-17) could be used to compute $\mathbf{\Phi}(t)$, but they serve better as an algebraic check on the correctness of $\mathbf{\Phi}(t)$ after it has been determined by other means. Based on (3.3-17), we note that $\mathbf{x}(t) = \mathbf{\Phi}(t)\mathbf{x}(0)$ satisfies both the differential equation (3.3-1) and the initial conditions. This confirms that the solution $\mathbf{x}(t)$, which is unique, can in fact be written in terms of a linear transformation.

The Retro-Time System

It follows from either (3.3-9) or (3.3-11) that the state transition matrix for the system

$$\dot{\mathbf{x}} = \mathbf{A}\mathbf{x} \tag{3.3-18}$$

is given by

$$\mathbf{\Phi}(t) = \mathbf{M}e^{\mathbf{\Lambda}t}\mathbf{M}^{-1}. \tag{3.3-19}$$

From (3.3-14) the inverse of $\mathbf{\Phi}(t)$ is obtained as

$$\mathbf{\Phi}^{-1}(t) = \mathbf{M}e^{-\mathbf{\Lambda}t}\mathbf{M}^{-1}. \tag{3.3-20}$$

Since the eigenvalues of $-\mathbf{A}$ are just the negative of the eigenvalues of $\mathbf{A}$, it follows that the solution to

$$\dot{\mathbf{x}} = -\mathbf{A}\mathbf{x} \tag{3.3-21}$$

is given by

$$\mathbf{x}(t) = \mathbf{M}e^{-\mathbf{\Lambda}t}\mathbf{M}^{-1}\mathbf{x}(0). \tag{3.3-22}$$

Since $\Phi(t)$ for system (3.3-21) is just $\Phi^{-1}(t)$ for system (3.3-18), these two systems are related in an interesting way. If we think of (3.3-18) as the **forward system**, then (3.3-21) may be thought of as the **retro system**. That is, the solution to (3.3-21) is the same as that obtained from (3.3-18) by **integrating backward** in time (where initial conditions are the final state of the forward system). Note that if the forward system is stable, then the retro system is unstable.

Laplace Transform Representation of $\Phi(t)$

We computed the state transition matrix in Example 3.3-1 by actually constructing $\mathbf{M}$ from the eigenvectors as in (3.3-7), constructing $e^{\Lambda t}$ from the eigenvalues as in (3.3-8), and then multiplying the results according to (3.3-19). However, an alternative method is available for calculating the state transition matrix, through the use of inverse Laplace transforms. Taking the Laplace transform of (3.3-1) yields

$$s\mathbf{X}(s) - \mathbf{x}(0) = \mathbf{A}\mathbf{X}(s).$$

Thus,

$$\mathbf{X}(s) = [s\mathbf{I} - \mathbf{A}]^{-1}\mathbf{x}(0).$$

Taking the inverse Laplace transform yields the solution

$$\mathbf{x}(t) = \mathscr{L}^{-1}\{[s\mathbf{I} - \mathbf{A}]^{-1}\}\mathbf{x}(0), \tag{3.3-23}$$

from which it follows that

$$\Phi(t) = \mathscr{L}^{-1}\{[s\mathbf{I} - \mathbf{A}]^{-1}\}. \tag{3.3-24}$$

This method may be convenient for those readers familiar with Laplace transform techniques. It has the advantage of automatically taking care of repeated roots. However, it has two disadvantages: it requires calculating the inverse Laplace transform, and it does not yield any eigenvector information.

Series Representation of $\Phi(t)$

We should also note one other method for constructing the state transition matrix. For the scalar first-order system

$$\dot{x} = ax,$$

the solution is given by

$$x(t) = e^{at}x(0)$$

and the state transition "matrix" in this case is the scalar e^{at}. Since the exponential function can be written as an infinite series

$$e^{at} = 1 + at + a^2 \frac{t^2}{2!} + a^3 \frac{t^3}{3!} + \cdots,$$

it suggests defining the $N_x \times N_x$ **matrix exponential** in an analogous form:

$$e^{\mathbf{A}t} \overset{\Delta}{=} \mathbf{I} + \mathbf{A}t + \mathbf{A}^2 \frac{t^2}{2!} + \mathbf{A}^3 \frac{t^3}{3!} + \cdots. \tag{3.3-25}$$

Thus,

$$\frac{d}{dt}(e^{\mathbf{A}t}) = \mathbf{A} + \mathbf{A}^2 t + \mathbf{A}^3 \frac{t^2}{2!} + \mathbf{A}^4 \frac{t^3}{3!} + \cdots = \mathbf{A}e^{\mathbf{A}t}.$$

Using this definition, we observe that the matrix exponential satisfies the differential equations and initial conditions in (3.3-17). Hence, we conclude that

$$\boldsymbol{\Phi}(t) = e^{\mathbf{A}t}. \tag{3.3-26}$$

The series expansion for the matrix exponential can be used to approximate the state transition matrix $\boldsymbol{\Phi}(\Delta t)$ for small time steps Δt by substituting Δt for t in (3.3-25). However, it can also be used to compute the actual state transition matrix. To do so, we make use of the **Cayley–Hamilton theorem** from linear algebra, which states that every square matrix satisfies its own characteristic equation.

In particular, the characteristic equation

$$0 = |\lambda \mathbf{I} - \mathbf{A}|$$

yields an N_x-order polynomial equation (3.2-3)

$$0 = P(\lambda) \overset{\Delta}{=} \lambda^{N_x} + p_{N_x - 1}\lambda^{N_x - 1} + \cdots + p_1 \lambda + p_0.$$

From the Cayley-Hamilton theorem, $\mathbf{A}$ satisfies

$$0 = P(\mathbf{A}) = \mathbf{A}^{N_x} + p_{N_x - 1}\mathbf{A}^{N_x - 1} + \cdots + p_1 \mathbf{A} + p_0 \mathbf{I}.$$

By solving this equation for $\mathbf{A}^{N_x}$, we can develop expressions for it and all higher powers of $\mathbf{A}$. These can then be substituted into (3.3-25) to yield a finite series for the matrix exponential

$$e^{\mathbf{A}t} = \alpha_0(t)\mathbf{I} + \alpha_1(t)\mathbf{A} + \alpha_2(t)\mathbf{A}^2 + \cdots + \alpha_{N_x - 1}(t)\mathbf{A}^{N_x - 1}. \tag{3.3-27}$$

In addition to this result, we also have the fact that the eigenvalues themselves satisfy the characteristic equation. Therefore, if we apply the same Cayley–Hamilton procedure used for $\mathbf{A}$ to the scalar matrices $[\lambda_i]$, we get

$$e^{\lambda_i t} = \alpha_0(t) + \alpha_1(t)\lambda_i + \alpha_2(t)\lambda_i^2 + \cdots + \alpha_{N_x-1}(t)\lambda_i^{N_x-1}, \qquad \textbf{(3.3-28)}$$

for each eigenvalue λ_i, $i = 1, \ldots, N_x$. Assuming the eigenvalues are distinct, we see that Equations (3.3-28) constitute a set of N_x linearly independent equations, which we can solve for the coefficients $\alpha_j(t)$, $j = 0, \ldots, N_x - 1$. Then the state transition matrix follows from (3.3-27).

EXAMPLE 3.3-2   **Cayley–Hamilton Computation of $\Phi(t)$**

Consider the same system as in Example 3.1-1, given by

$$\begin{bmatrix} \dot{x}_1 \\ \dot{x}_2 \end{bmatrix} = \begin{bmatrix} 0 & 1 \\ a^2 & 0 \end{bmatrix} \begin{bmatrix} x_1 \\ x_2 \end{bmatrix}.$$

The characteristic equation

$$P(\lambda) = \lambda^2 - a^2 = 0$$

yields the eigenvalues $\lambda_1 = a$, $\lambda_2 = -a$.
    Applying (3.3-28) to each eigenvalue, we obtain

$$e^{at} = \alpha_0(t) + \alpha_1(t)\lambda_1 = \alpha_0(t) + a\alpha_1(t)$$
$$e^{-at} = \alpha_0(t) + \alpha_1(t)\lambda_2 = \alpha_0(t) - a\alpha_1(t),$$

where the right-hand sides stop after terms of order $N_x - 1 = 1$ in λ, since the characteristic polynomial is of order $N_x = 2$. Solving these equations, we have

$$\alpha_0(t) = \frac{e^{at} + e^{-at}}{2} = \cosh at$$

$$\alpha_1(t) = \frac{e^{at} - e^{-at}}{2a} = \frac{1}{a}\sinh at$$

Then (3.3-27) yields the state transition matrix

$$\Phi(t) = e^{\mathbf{A}t} = \alpha_0(t)\mathbf{I} + \alpha_1(t)\mathbf{A} = \begin{bmatrix} \cosh at & \dfrac{1}{a}\sinh at \\ a\sinh at & \cosh at \end{bmatrix},$$

which agrees with the results of Example 3.3-1.

It should be clear, from this procedure and from our previous procedures for computing $\Phi(t)$, that $\exp[\mathbf{A}t]$ *cannot* be computed simply by taking the exponent of each element of the matrix $\mathbf{A}$ (unless, of course, $\mathbf{A}$ is a diagonal matrix). To avoid any possible confusion, we will always use the notation $\Phi(t)$ for the state transition matrix.

3.4   STATE SPACE TRAJECTORIES

The IO formulation and the state-space formulation of a dynamical system present more than just two mathematical ways of looking at the system. They also offer two geometric ways of looking at the system. Under the IO format the output of the system represents the dominant feature, and the output is usually plotted as a function of time for a reference set of initial conditions. Examples are given in Figures 3.2-4 and 3.2-6. In the state-space formulation the state variables represent the dominant feature. As in the IO formulation, the output and/or each state variable can be plotted as a function of time. However, it is also informative to plot the state variables against each other in a state space where each axis corresponds to one of the state variables. The corresponding solution curve plots are called **trajectories**. Since (3.3-1) uniquely defines a velocity $\dot{\mathbf{x}}$ for every point in state space, no two trajectories in state space can intersect. Although this process can be thought of in any number of dimensions, obviously it is practical for at most three-dimensional systems and is most useful for two-dimensional systems. Note that there is no real difference between the IO format and the state representation for one-dimensional systems, so first-order systems would always be plotted as output versus time.

In order to illustrate a variety of possible state-space plots, consider the second-order system

$$\ddot{y} + 2\zeta\omega_n\dot{y} + \omega_n^2 y = 0, \tag{3.4-1}$$

in which ζ and ω_n can be any real or complex numbers, so long as the resulting coefficients in (3.4-1) are real. With $x_1 = y$, an equivalent state-space representation is given by

$$\dot{x}_1 = x_2 \tag{3.4-2}$$

$$\dot{x}_2 = -\omega_n^2 x_1 - 2\zeta\omega_n x_2. \tag{3.4-3}$$

The state-space variables are position and velocity. Figure 3.4-1 illustrates the solution to (3.4-2) and (3.4-3) in state space for different

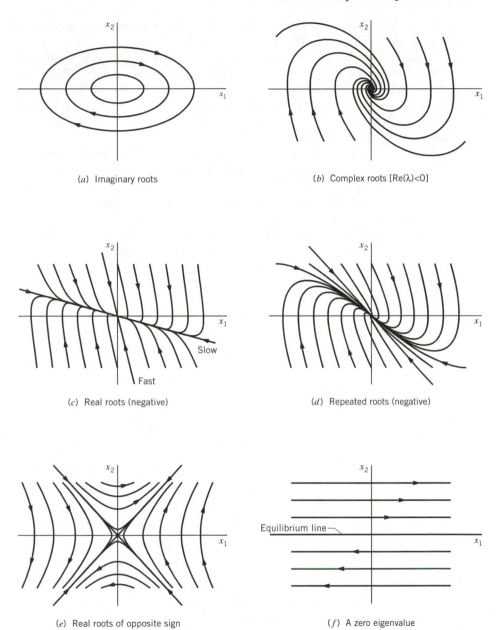

Figure 3.4-1 State-space trajectories for second-order systems.

cases of the parameter values ζ and ω_n. The trajectories were generated by starting from various initial states.

Case a, called a **center**, occurs in an undamped harmonic oscillator, with $\zeta = 0$ and $\omega_n > 0$. The corresponding sustained oscillation is characterized by closed trajectories in state space. Each trajectory curve, in this case, represents initial conditions that have the same total kinetic plus potential energy. Since there is no damping, the energy level remains constant along a trajectory.

Case b, termed a **focus**, corresponds to an underdamped harmonic oscillator, with $\omega_n > 0$ and $0 < \zeta < 1$. Notice how the trajectory spirals into the origin. Figure 3.3-1b clearly illustrates the fact that the total energy of a spring-mass-shock absorber system is continuously decreasing owing to viscous damping in the shock absorber.

For an overdamped system, as in case c with $\zeta > 1$ and $\omega_n > 0$, referred to as a **node**, the sytem has two real negative eigenvalues and two corresponding real eigenvectors. The motion associated with the most negative eigenvalue decays very quickly, more or less in the direction of the corresponding eigenvector that is termed the "fast" eigenvector. The other eigenvalue corresponds to the return time T. The motion corresponding to the associated eigenvector decays slowly more or less in the direction of the eigenvector and is referred to as the "slow" eigenvector.

Case d, also a **node**, corresponds to critical damping ($\zeta = 1$, $\omega_n^2 > 0$), and the system has only one eigenvector with the motion ultimately dominated by the direction of this eigenvector.

Cases b through d are asymptotically stable systems with $\zeta > 0$. They can be made unstable, with trajectories diverging from the origin, by setting $\zeta < 0$. The direction of motion indicated by the arrows in cases b through d reverses itself.

Case e, called a **saddle**, illustrates an unstable system ($\omega_n^2 < 0$, $\zeta = 0$), with one positive and one negative real eigenvalue, analogous to a linearized undamped inverted pendulum.

Case f, termed a **shear**, corresponds to a spring-mass-damper system with no spring and no damper ($\zeta = 0$ and $\omega_n^2 = 0$), that is, just a moving mass with no applied force. The origin is only one of an infinity of possible equilibrium points, all of which are unstable. The mass can be in equilibrium at any position x_1, provided that the speed x_2 is zero. Otherwise, the mass moves at constant speed.

When the eigenvalues are real, as in cases c through f, the corresponding eigenvectors have a geometric significance in the state space; they are the only straight-line trajectories. This result can be seen by considering the case where, at some instant of time t, $\mathbf{x}(t)$ is a point on an eigenvector ξ. That is, the "position" vector from the origin to $\mathbf{x}(t)$ lies along the eigenvector. Hence, for some k

$$\mathbf{x}(t) = k\xi.$$

Then the "velocity" vector

$$\dot{\mathbf{x}} = \mathbf{A}\mathbf{x}(t) = k\mathbf{A}\boldsymbol{\xi} = k\lambda\boldsymbol{\xi}$$

also is directed along the eigenvector since $\mathbf{A}\boldsymbol{\xi} = \lambda\boldsymbol{\xi}$. Thus, the eigenvector is an **invariant set**; any trajectory starting on it remains on it.

One other feature of state-space plots should be noted. For linear systems the state-space trajectories are independent of the scale at which they are viewed. In particular, for any solution $\mathbf{x}(t)$ to the state equations

$$\dot{\mathbf{x}} = \mathbf{A}\mathbf{x}$$

and any scale factor k, the function $k\mathbf{x}(t)$ is also a solution. Thus no matter how far we zoom in or out, the state-space plots will look identical except for the scale on the axes.

3.5 ROUTH–HURWITZ STABILITY CRITERIA

In the design of feedback controllers, one of the basic objectives will be to stabilize the system about some commanded operating point. In this section we present some techniques for analyzing the stability of closed-loop control systems in terms of the parameters in the feedback control system.

We will be concerned with a general nth-order linear constant-coefficient system having the **characteristic equation**

$$0 = P(\lambda) = \lambda^n + p_{n-1}\lambda^{n-1} + p_{n-2}\lambda^{n-2} + \cdots + p_1\lambda + p_0. \quad \text{(3.5-1)}$$

The roots of this equation are the **eigenvalues** of the system. For a feedback control system, the coefficients in the characteristic equation may contain many parameters, both from the primary system to be controlled and from the controller. For a particular set of coefficient values, there are various numerical algorithms available for computing the eigenvalues. Thus, the effect of different parameter values could be studied by repeatedly solving the characteristic equation numerically. Our objective here, however, is to study some stability analysis techniques that do not require repetitive solution of the characteristic equation.

Descartes' Rules and Newton's Identities

A considerable amount of valuable information can be learned without actually solving the characteristic equation. For example, the **rules of Descartes** tell us several things:

1. There are exactly n roots, since the characteristic equation is an nth-order polynomial equation.

2. If complex roots exist, they occur in conjugate pairs $\lambda = \sigma \pm i\omega$, since the coefficients in the characteristic equation are assumed to be real.

3. The number of positive real roots is less than or equal to the number of sign changes in the coefficients of $P(\lambda)$.

4. The number of negative real roots is less than or equal to the number of sign changes in $P(-\lambda)$.

5. In order for all of the roots to have negative (nonpositive) real parts, it is necessary but not sufficient that all of the coefficients be positive (nonnegative) in the characteristic equation.

These results, and others, follow from some basic algebraic relations, called **Newton's identities**, between the coefficients $p_0, p_1, \ldots, p_{n-1}$ of the characteristic polynomial and the eigenvalues $\lambda_1, \ldots, \lambda_n$:

$$p_{n-1} = -\sum_i \lambda_i$$

$$p_{n-2} = \sum_{i \neq j} \lambda_i \lambda_j \qquad\qquad (3.5\text{-}2)$$

$$p_{n-3} = -\sum_{i \neq j \neq k} \lambda_i \lambda_j \lambda_k$$

$$\vdots$$

$$p_0 = (-1)^n \lambda_1 \lambda_2 \ldots \lambda_n,$$

where the sums are over all possible (single, pairwise, three-way, and so on) distinct indices $1, \ldots, n$.

EXAMPLE 3.5-1   Harmonic Oscillator

Consider a harmonic oscillator, such as a spring-mass system with no damping. The characteristic equation can be written as

$$0 = P(\lambda) = \lambda^2 + \omega_n^2.$$

This equation is easily solved and the roots, $\lambda = \pm i\omega_n$, are purely imaginary, so that the underlying dynamical system is stable, but not asymptotically stable. However, for illustrative purposes, let us consider what can be learned without solving the characteristic equation.

From the last of Descartes' rules we know that the system is not asymptotically stable, since the coefficient of λ is zero. If any of the coefficients had been negative, we could have concluded that the system was unstable. There are no sign changes in the coefficients of $P(\lambda)$, since a zero coefficient does not constitute a sign change. Thus, there are no positive real roots. Similarly, since no sign changes exist in the coefficients of $P(-\lambda) = (-\lambda)^2 + \omega_n^2$, there are no negative real roots. Thus, if the roots are real, they must both be zero.

From the first of Newton's identities, we have $p_1 = 0 = -(\lambda_1 + \lambda_2) = -2\sigma$, since complex roots can occur only in conjugate pairs, $\lambda = \sigma \pm i\omega$. Thus, if the roots are complex, they must have zero real parts. Therefore, either the roots are purely imaginary or they are both zero. From the last of Newton's identities, ω_n^2 is equal to the product of the roots. Therefore, if $\omega_n \neq 0$, the roots must be nonzero (hence, imaginary). Otherwise, the roots must both be zero.

Routh's Stability Procedure

Notice that Descartes' third and fourth rules deal exclusively with real roots and only provide upper bounds on the possible number of positive or negative real roots. They do not say directly how many actual positive or negative real roots there are for the characteristic equation (except when the number is zero).

Fortunately, there exists a fairly simple method, valid for real and complex roots, that will tell us exactly how many roots have positive real parts. The method is termed the **Routh–Hurwitz criteria**. The procedure that we will present, due to Routh, involves constructing a table or array of numbers based on the coefficients in the characteristic polynomial. An equivalent procedure, due to Hurwitz, involves n determinants of the coefficients, of order $1, 2, \ldots, n$.

The Routh array for the characteristic equation

$$0 = P(\lambda) = \lambda^n + p_{n-1}\lambda^{n-1} + p_{n-2}\lambda^{n-2} + \cdots + p_1\lambda + p_0$$

has the following triangular form:

$$
\begin{array}{c|ccccccc}
\lambda^n: & 1 & + \cdot\, p_{n-2} & p_{n-4} & p_{n-6} & p_{n-8} & \cdots & 0 \\
\lambda^{n-1}: & p_{n-1} & - \cdot\, p_{n-3} & p_{n-5} & p_{n-7} & \cdots & 0 \\
\lambda^{n-2}: & c_1 & c_2 & c_3 & \cdots & 0 \\
\lambda^{n-3}: & d_1 & d_2 & \cdots & 0 \\
\vdots & \vdots & \vdots & \\
\lambda^1: & e_1 & 0 \\
\lambda^0: & f_1 &
\end{array}
$$

where

$$c_1 = \frac{p_{n-1}p_{n-2} - 1 \times p_{n-3}}{p_{n-1}}, \qquad c_2 = \frac{p_{n-1}p_{n-4} - 1 \times p_{n-5}}{p_{n-1}}, \ldots$$

$$d_1 = \frac{c_1 p_{n-3} - p_{n-1}c_2}{c_1}, \qquad d_2 = \frac{c_1 p_{n-5} - p_{n-1}c_3}{c_1}, \ldots$$

$$\vdots$$

The first row consists of every other coefficient in $P(\lambda)$, descending in order by powers of 2 and starting with the coefficient of λ^n. The second row is constructed in the same manner, starting with the coefficient of λ^{n-1} and descending by powers of 2. All coefficients must be used, including any zero coefficients. For the above procedure, trailing positions in all rows of the Routh array consist of zeros, as needed.

Each element in the third and subsequent rows consists of the negative of a 2×2 determinant (from the two previous rows) divided by a "pivot element." The pivot element is the first element in the preceding row. The first column of the determinant consists of the two elements in column one (boxed) of the preceding two rows. The second column consists of the corresponding two elements (in the previous two rows) in the column just to the right of the element being computed.

The result of this procedure is that *the number of roots with a positive real part is equal to the number of sign changes in column one* of the Routh array. Note that any row derived from the first two rows can be multiplied by any convenient positive number without changing the conclusion.

EXAMPLE 3.5-2   **Routh Array**

For the characteristic equation

$$0 = \lambda^4 + 3\lambda^3 + \lambda^2 + 6\lambda + 2,$$

the corresponding Routh array is

λ^4:	1	1	2
λ^3:	3	6	0
λ^2:	$-1 = \dfrac{3 \times 1 - 1 \times 6}{3}$	$2 = \dfrac{3 \times 2 - 1 \times 0}{3}$	0
λ^1:	$12 = \dfrac{-1 \times 6 - 3 \times 2}{-1}$	0	
λ^0:	2.		

There are two sign changes in column one, so there are exactly two roots with positive real part (either two real roots or a complex conjugate pair).

Special Cases in Routh's Procedure

There are two special cases to consider for Routh's procedure: (1) a zero in column one, but the entire row is not all zero, and (2) a row of zeros.

EXAMPLE 3.5-3   **Zero Pivot Element**

If a pivot element in column one is zero, replace it with a small positive quantity ϵ and proceed as before. After the array is constructed, consider the limiting case as ϵ approaches zero. The same rules apply, with the additional information that if the elements just above and below the zero have the same sign, then the zero corresponds to a pair of imaginary roots. If the last element in column one is zero, then zero is one of the roots of the characteristic equation.

For example, the characteristic equation

$$0 = \lambda^5 + 2\lambda^4 + 4\lambda^3 + 8\lambda^2 + 10\lambda + 6$$

yields the Routh array

λ^5:	1	4	10
λ^4:	2	8	6
λ^3:	ϵ	7	0
λ^2:	$\left(\dfrac{8\epsilon - 14}{\epsilon}\right) \to -\infty$	6	0
λ^1:	$\left(7 - \dfrac{6\epsilon^2}{8\epsilon - 14}\right) \to 7$	0	
λ^0:	$6.$		

Thus, there are two roots with a positive real part.

EXAMPLE 3.5-4   Row of Zeros

A row of zeros (not counting the last row) corresponds to roots located symmetrically about the origin, as indicated for several situations in Figure 3.5-1. The polynomial associated with the row preceding the zero row corresponds to these roots.

The procedure in this case is to construct an auxiliary polynomial $\rho(\lambda)$ by using the coefficients of the previous row and then to replace the zero row with the coefficients of the derivative $\rho'(\lambda)$. The power of λ for the first term in $\rho(\lambda)$ is the power of λ associated with the row preceding the zero row, and for each subsequent term in $\rho(\lambda)$ the power of λ is reduced by 2. The polynomial $\rho(\lambda)$ will be either even $[\rho(-\lambda) = \rho(\lambda)]$ or odd $[\rho(-\lambda) = -\rho(\lambda)]$. The roots of $\rho(\lambda) = 0$ will also be roots of the original characteristic equation, corresponding to symmetrically located roots of the original characteristic equation.

As an example, the characteristic equation

$$0 = \lambda(\lambda^2 - 1)(\lambda^2 + 2\lambda + 4) = \lambda^5 + 2\lambda^4 + 3\lambda^3 - 2\lambda^2 - 4\lambda$$

yields the Routh array

λ^5:	1	3	-4
λ^4:	2	-2	0
λ^3:	4	-4	$0 \Rightarrow \rho(\lambda) = 4\lambda^3 - 4\lambda$
λ^2:	$0 \to 12$	$0 \to -4$	$0 \Leftarrow \rho'(\lambda) = 12\lambda^2 - 4$
λ^1:	$-\frac{8}{3}$	0	
λ^0:	$-4.$		

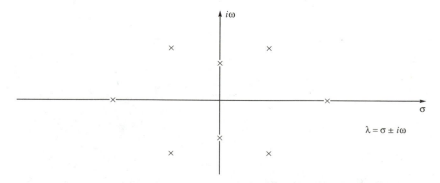

Figure 3.5-1 A possible symmetric root configuration about the origin for a zero row in the Routh array.

Thus, there is one root in the postive right-half side of the complex plane. Also, the roots of $0 = \rho(\lambda) = 4\lambda(\lambda^2 - 1)$ are roots of the characteristic equation, and they are located symmetrically about the origin.

Relative Stability (Shifting the Origin)

In the previous example we applied the Routh procedure to a system that was clearly unstable, because of the negative coefficients in the characteristic equation. However, the Routh procedure can be used for more than determining the number of roots with $\text{Re}(\lambda) > 0$. It can also be used to find the number of roots with $\text{Re}(\lambda) > \sigma$, by defining $s \triangleq \lambda - \sigma$ and examining the shifted characteristic equation $\overline{P}(s) \triangleq P(s + \sigma) = P(\lambda) = 0$ to find the number of roots s of $\overline{P}(s) = 0$ with $\text{Re}(s) > 0$, that is, with $\text{Re}(\lambda) - \sigma > 0$.

EXAMPLE 3.5-5   **Specification on Real Parts of Eigenvalues**

Suppose we want to control the second-order system

$$\ddot{y} + 2\dot{y} + 2y = 3u + \dot{u},$$

under a constant command input $r(t) \equiv \bar{r}$, using a proportional-plus-integral (PI) error feedback controller (see Chapter 6) of the form

$$u = K\left\{ r - y(t) + a \int_0^t [r - y(\tau)]\, d\tau \right\}.$$

For simplicity, suppose that we have chosen $a = 5$ and would like to know what values of K will yield closed-loop eigenvalues that all satisfy $\text{Re}(\lambda) \leq -2$. We could, of course, ask the same question considering both K and a as parameters, but this example will illustrate the concept.

For $a = 5$ the controlled system has the characteristic equation

$$0 = P(\lambda) = \lambda^3 + (2 + K)\lambda^2 + (2 + 8K)\lambda + 15K.$$

Letting $s = \lambda + 2$ (that is, $\lambda = s - 2$), we have

$$0 = \overline{P}(s) = [s - 2]^3 + (2 + K)[s - 2]^2 + (2 + 8K)[s - 2] + 15K$$
$$= s^3 + (K - 4)s^2 + (4K + 6)s + (3K - 4).$$

Applying the Routh procedure to $\overline{P}(s)$ yields the array

$$
\begin{array}{ccc}
s^3: & 1 & 4K + 6 \\[2mm]
s^2: & K - 4 & 3K - 4 \\[2mm]
s^1: & \dfrac{4K^2 - 13K - 20}{K - 4} & 0 \\[2mm]
s^0: & 3K - 4. &
\end{array}
$$

If there are no sign changes in column one, then all roots satisfy $\mathrm{Re}(s) \le 0$, that is, the roots of the original characteristic equation satisfy $\mathrm{Re}(\lambda) \le -2$. Thus, we require that K satisfy all of the following conditions:

$$K \ge 4$$

and

$$K \ge 4.3892 \quad \text{or} \quad K \le -1.1392$$

and

$$K \ge \frac{4}{3}.$$

Therefore, we have $\mathrm{Re}(\lambda) \le -2$ for all closed-loop eigenvalues, provided that $K \ge 4.3892$.

This example also illustrates another aspect of using the Routh array to investigate the effect of varying parameters in the dynamical system. Each of the four limiting values of K corresponds to a zero in column one of the Routh array for the complex s plane. By examining the Routh array for K values slightly above and below these critical values, we can determine the nature of the roots that cause the zero. For example, suppose $K = 4.3892 + \epsilon$, where $\epsilon > 0$ is small. Then column one contains no sign changes and $\overline{P}(s) = 0$ has no roots with $\mathrm{Re}(s) > 0$. On the other hand, for $K = 4.3892 - \epsilon$, the s^1 coefficient in column one becomes negative, yielding two sign changes and thus two roots with $\mathrm{Re}(s) > 0$. Therefore, we can conclude that at $K = 4.3892$ a pair of complex conjugate roots becomes purely imaginary. Similarly, for $K = 4/3 + \epsilon$, the s^2 coefficient in column one is negative, yielding two sign changes, whereas for $K = 4/3 - \epsilon$, the s^2 and s^0 coefficients in column one are negative, yielding three sign changes. Thus, at $K = 4/3$ we conclude that a single s root passes through the imaginary axis. This must correspond to a real root passing through the origin. We will make use of this parametric analysis technique again in subsequent chapters.

3.6 EXERCISES

3.6-1 The magnetic supsension system discussed in Section 1.3 is typical of an unstable system that can be stabilized by using state variable feedback. Equations (1.3-10)–(1.3-11) are of the form

$$\dot{\mathbf{x}} = \mathbf{A}\mathbf{x} + \mathbf{B}u.$$

(a) Show that for $\beta_1 > 0$, one of the eigenvalues of the $\mathbf{A}$ matrix is positive, which makes the system unstable.

(b) Show that there exists constants k_1 and k_2 such that under linear state variable feedback control of the form

$$u = -\mathbf{K}\mathbf{x},$$

where

$$\mathbf{K} = [k_1 \, k_2],$$

the controlled system

$$\dot{\mathbf{x}} = \tilde{\mathbf{A}}\mathbf{x},$$

where

$$\tilde{\mathbf{A}} = \mathbf{A} - \mathbf{B}\mathbf{K},$$

will have both eigenvalues negative and, hence, be a stable system.

3.6-2 Determine the free output response $y(t)$ for each of the systems in Exercise 2.5-5, subject to the initial condition $y(0) = 1$, with all other initial conditions (where applicable) equal to zero.

3.6-3 A state-space system is given by

$$\dot{x}_1 = x_2$$

$$\dot{x}_2 = -a_1 x_1 - a_2 x_2 + u$$

$$y = x_1.$$

(a) Determine the state transition matrix for this system when $a_1 = 2$, $a_2 = 3$, and $u = 0$.

(b) Determine $\Phi^{-1}(t)$ for part (a)

3.6-4 Find the state transition matrix $\Phi(t)$ for the following cases and verify that your result satisfies the conditions in (3.3-17):

(a) $\mathbf{A} = \begin{bmatrix} 0 & 2 \\ -1 & -3 \end{bmatrix}$ (b) $\mathbf{A} = \begin{bmatrix} 16 & 52 \\ -5 & -16 \end{bmatrix}$

(c) $\mathbf{A} = \begin{bmatrix} 0 & 1 \\ -5 & -2 \end{bmatrix}$.

3.6-5 For the systems in Exercise 3.6-4, determine the free response solution $x(t)$ to $\dot{x} = Ax$ with $x(0) = [1 \quad 1]^T$.

3.6-6 Given $\dot{x} = Ax$, with

$$A = \begin{bmatrix} 4 & -2 \\ 1 & 1 \end{bmatrix}, \qquad x(0) = \begin{bmatrix} 1 \\ 0 \end{bmatrix},$$

find $x(t)$ for $t \geq 0$.

3.6-7 The A matrix (1.3-40) and B matrix (1.3-41) for a particular aircraft are given by

$$A = \begin{bmatrix} -0.00657 & 18.01 & -32.17 & 0 \\ -0.000119 & -0.858 & 0 & 1 \\ 0 & 0 & 0 & 1 \\ -0.000025 & -3.314 & 0 & -1.352 \end{bmatrix}$$

and

$$B = [0 \quad 0 \quad 0 \quad -1]^T.$$

(a) Show that the eigenvalues of the A matrix all have negative real parts. The eigenvalues with the longer time period are associated with the so-called phugoid mode and the eigenvalues with the shorter time period are associated with what is termed the short-period mode.

(b) Obtain the solution for all the state variables as a function of time after the aircraft has been subjected to an initial condition disturbance of $x(0) = [10, \quad 0, \quad 0, \quad 0]^T$. Plot the solutions over a time interval of 120 sec and identify the state variables associated with the phugoid mode.

3.6-8 Illustrate the nature of the unstable motion of Example 3.3-1 by making a state-space plot of several solutions starting in the neighborhood of the origin.

3.6-9 For the system defined by (3.4-2)–(3.4-3), with $\zeta = 0$ and $\omega_n > 0$, prove that the resulting trajectories are ellipses centered at the origin. (*Hint*: Integrate dx_2/dx_1.)

3.6-10 For the system in Exercise 3.6-1, use the Routh procedure to determine the ranges of values (possibly negative) of the parameters β_1, β_2, k_1, and k_2 for which the feedback control system will be stable.

3.6-11 A system has the characteristic equation

$$0 = P(\lambda) = \lambda^3 + (2 + K)\lambda^2 + (2 + 8K)\lambda + 15K.$$

Using the Routh procedure, determine the range of values for K so that the system is stable.

3.6-12 A system has the characteristic equation

$$0 = P(\lambda) = \lambda^3 + a\lambda^2 + b\lambda + c.$$

Use the Routh procedure to discuss the range of parameter values for which the system will be stable. In addition, discuss the particular cases where one or more of the parameters (a, b, c) is zero.

Chapter 4

Forced Response

4.1 SEPARATION OF INITIAL CONDITIONS AND INPUTS

In this section we deal with linear systems subject to deterministic inputs. These systems were previously described by (2.1-3) and (2.1-4) which we repeat here as

$$\dot{\mathbf{x}} = \mathbf{A}\mathbf{x} + \mathbf{B}\mathbf{u}(t) \tag{4.1-1}$$

$$\mathbf{y}(t) = \mathbf{C}\mathbf{x}(t) + \mathbf{D}\mathbf{u}(t). \tag{4.1-2}$$

Before presenting a procedure for determining the **forced response** solution $\mathbf{x}(t)$ to (4.1-1) for a specified input $\mathbf{u}(t)$ and initial conditions $\mathbf{x}(0)$, we will first show that the effects caused by initial conditions and the effects caused by the input can be studied separately.

When $\mathbf{u}(t) \equiv 0$, the **free response** solution to (4.1-1) for specified initial conditions $\mathbf{x}(0) = \mathbf{x}_0$ is given by (3.3-13), where the state transition matrix can be computed by using (3.3-19). Since such a solution satisfies (4.1-1) with the input $\mathbf{u}(t) \equiv 0$, it will clearly not satisfy (4.1-1) with $\mathbf{u}(t) \neq 0$.

For a specified input $\mathbf{u}(t)$, let $\mathbf{x}(t)$ be the solution to (4.1-1) satisfying the initial conditions $\mathbf{x}(0) = \mathbf{x}_0$ and let $\mathbf{x}_F(t)$ denote the free response for the same initial conditions $\mathbf{x}_F(0) = \mathbf{x}_0$. We define one more response called the **input response** $\mathbf{x}_I(t)$ as the solution to (4.1-1), for the specified $\mathbf{u}(t)$, with $\mathbf{x}_I(0) = \mathbf{0}$.

From the theory of ordinary differential equations (Coddington and Levinson, 1955, p. 49), we know that for specified initial conditions and a specified input $\mathbf{u}(t)$, the solution $\mathbf{x}(t)$ to (4.1-1) is unique. We will show that this solution can be written as the sum of two terms: one associated with the initial conditions $\mathbf{x}(0)$ and the other associated with the input function $\mathbf{u}(t)$. This makes it possible to study separately the effects of initial conditions and the effects of inputs.

Since (4.1-1) is a linear system, the relationship between the forced response $\mathbf{x}(t)$, the free response $\mathbf{x}_F(t)$, and the input response $\mathbf{x}_I(t)$ is given by

$$\mathbf{x}(t) = \mathbf{x}_F(t) + \mathbf{x}_I(t). \qquad \text{(4.1-3)}$$

To verify this, we first note that $\mathbf{x}(t)$ satisfies the initial conditions:

$$\mathbf{x}(0) = \mathbf{x}_F(0) + \mathbf{x}_I(0) = \mathbf{x}_F(0) = \mathbf{x}_0.$$

Next we note that, from the definitions of the free and input responses,

$$\dot{\mathbf{x}}_F = \mathbf{A}\mathbf{x}_F \qquad \text{(4.1-4)}$$

$$\dot{\mathbf{x}}_I = \mathbf{A}\mathbf{x}_I + \mathbf{B}\mathbf{u}(t). \qquad \text{(4.1-5)}$$

Thus, from (4.1-3),

$$\begin{aligned}
\dot{\mathbf{x}} &= \dot{\mathbf{x}}_F + \dot{\mathbf{x}}_I \\
&= \mathbf{A}\mathbf{x}_F + \mathbf{A}\mathbf{x}_I + \mathbf{B}\mathbf{u}(t) \\
&= \mathbf{A}\mathbf{x} + \mathbf{B}\mathbf{u}(t).
\end{aligned}$$

Therefore, $\mathbf{x}(t)$ satisfies both the differential equation (4.1-1) and the initial conditions.

From our previous study of the free response, if the dynamical system is asymptotically stable, then as $t \to \infty$, the solution to (4.1-1) will approach the input response $\mathbf{x}_I(t)$. Thus, if we are dealing with an asymptotically stable system and are solely interested in the solution as $t \to \infty$, we need only seek the input response $\mathbf{x}_I(t)$.

The separation afforded by (4.1-3) is not used to advantage in the standard classical differential equations technique. There the constants in the **homogeneous solution** (3.3-4) cannot be determined from the initial conditions until after a **particular solution** [any function satisfying (4.1-1), but not necessarily satisfying any specified initial conditions] has also been found. Under that approach, by using an arbitrary "particular" solution, any change in the input function $\mathbf{u}(t)$ requires that the entire process be repeated, to determine a new particular solution and new constants in the homogeneous solution.

4.2 GENERAL SOLUTION OF STATE EQUATIONS

In order to provide a uniform method for analytically determining the forced response of a linear dynamical system, we will use the state transition matrix method previously developed in Section 3.3. Since this method is a state-space method, an IO system will first have to be converted to a state-space system before a solution is determined. The main advantage of the state transition matrix method is that a solution can be formulated for all possible inputs. It also provides a solution even if numerical techniques must be used to obtain eigenvalues and eigenvectors, or to evaluate an integral input term.

To develop the general solution to (4.1-1), we will first diagonalize the system. However, the final result will apply to general linear state-space systems, whether diagonal or not.

The system is decoupled using the eigenvector matrix $\mathbf{M}$ and the coordinate transformation

$$\mathbf{z} = \mathbf{M}^{-1}\mathbf{x}. \tag{4.2-1}$$

Thus, in terms of $\mathbf{z}$, we have

$$\dot{\mathbf{z}} = \mathbf{M}^{-1}\mathbf{A}\mathbf{M}\mathbf{z} + \mathbf{M}^{-1}\mathbf{B}\mathbf{u}. \tag{4.2-2}$$

Let $\hat{\mathbf{B}} = \mathbf{M}^{-1}\mathbf{B} = [\hat{b}_{ij}]$. Then in terms of the diagonalized eigenvalue matrix $\mathbf{\Lambda} = \text{diag}[\lambda_1, \ldots, \lambda_{N_x}]$, we can write (4.2-2) as

$$\dot{\mathbf{z}} = \mathbf{\Lambda}\mathbf{z} + \hat{\mathbf{B}}\mathbf{u}. \tag{4.2-3}$$

This represents a system of N_x decoupled first-order equations, which can be written in terms of the components of $\mathbf{z}$ and $\mathbf{u}$ as

$$\dot{z}_i = \lambda_i z_i + \sum_{j=1}^{N_u} \hat{b}_{ij} u_j, \qquad i = 1, \ldots, N_x. \tag{4.2-4}$$

Integrating Factor for Decoupled System

We integrate (4.2-4) by introducing the **integrating factor** $\exp[-\lambda_i t]$. Subtracting $\lambda_i z_i$ from both sides of (4.2-4) and multiplying by the integrating factor yields

$$e^{-\lambda_i t}\dot{z}_i - \lambda_i e^{-\lambda_i t}z_i = e^{-\lambda_i t}\sum_{j=1}^{N_u} \hat{b}_{ij} u_j,$$

from which we obtain

$$\frac{d}{dt}(e^{-\lambda_i t} z_i) = e^{-\lambda_i t}\sum_{j=1}^{N_u} \hat{b}_{ij} u_j. \tag{4.2-5}$$

Integrating (4.2-5) and multiplying both sides of the result by $\exp[\lambda_i t]$, we get

$$z_i(t) = e^{\lambda_i t}z_i(0) + \int_0^t e^{\lambda_i(t-\tau)} \sum_{j=1}^{N_u} \hat{b}_{ij}u_j(\tau)\, d\tau. \tag{4.2-6}$$

Equations (4.2-6) can now be written more compactly, in terms of vectors and matrices, as

$$\mathbf{z}(t) = e^{\Lambda t}\mathbf{z}(0) + \int_0^t e^{\Lambda(t-\tau)}\mathbf{M}^{-1}\mathbf{B}\mathbf{u}(\tau)\, d\tau. \tag{4.2-7}$$

The solution in terms of the original state variables $\mathbf{x}$ is obtained by again using the transformation (4.2-1), yielding

$$\mathbf{M}^{-1}\mathbf{x}(t) = e^{\Lambda t}\mathbf{M}^{-1}\mathbf{x}(0) + \int_0^t e^{\Lambda(t-\tau)}\mathbf{M}^{-1}\mathbf{B}\mathbf{u}(\tau)\, d\tau. \tag{4.2-8}$$

Multiplying both sides by $\mathbf{M}$ and using (3.3-19) gives the solution

$$\mathbf{x}(t) = \underbrace{\mathbf{\Phi}(t)\mathbf{x}(0)}_{\text{Free response}} + \underbrace{\int_0^t \mathbf{\Phi}(t-\tau)\mathbf{B}\mathbf{u}(\tau)\, d\tau}_{\text{Input response}}. \tag{4.2-9}$$

The state transition matrix $\mathbf{\Phi}(t)$ allows us to express the solution in terms of an arbitrary input $\mathbf{u}(t)$ and separate arbitrary initial conditions $\mathbf{x}(0)$. Note that since

$$\mathbf{\Phi}(t) = \mathbf{M}e^{\Lambda t}\mathbf{M}^{-1}, \tag{4.2-10}$$

if the system is asymptotically stable, the response of the system as $t \to \infty$ will involve only the integral term (input response). If the input response does not die out with time, then, as $t \to \infty$, what remains is termed the **residual response**.

Some Typical Forcing Functions

We are now in a position to examine the response of some basic linear systems subject to various inputs. Of particular interest for many control applications is the relationship between the input $u(t)$ and the output $y(t)$ for SISO systems. The exact relationship between the input and the desired output is very much a function of the application. For example, we may want the system to track a command input, or we may want the system to ignore an uncertain disturbance input. Generally, we will want the control system to satisfy both of these properties.

In order to cope with all the possible design requirements associated with an infinite variety of inputs, it has been standard practice to examine two specific objectives by means of two specific input functions.

The first objective, **tracking**, is associated with how well the output tracks the input. The input function used in this case is a **step function** (or possibly time integrals of a step function). A step function is a good test function for investigating tracking, since it can be used to represent a worst-case type of input, such as a bounded input at one of its limits.

The second objective is the **filtering** capacity of the system with respect to periodic inputs, such as high-frequency "noise." The input function used in this case is a **sine wave**. It represents a good test input, since any periodic function can be constructed as a sum of sine waves through the use of Fourier series (Kreyszig, 1983, p. 462).

Another important system consideration is associated with the response of the system to an **uncertain input**. Since this type of input may vary from a constant input (bias) to a high-frequency input (noise), it is not completely represented by either a step function or a sine wave. The objective in this case will be to determine the maximum deviation of the output from the equilibrium under an uncertain input whose maximum magnitude is known. The input function used in this case is a **piecewise constant function**, corresponding to an uncertain input that may, for example, switch repeatedly between its upper and lower bounds.

In the development that follows we assume that the system of interest is asymptotically stable. Of course, not every system of interest needs to be stable. However, this difficulty can be dealt with readily in the next chapter, so the stability assumption here is not limiting. If the system is asymptotically stable, then any initial condition effect will damp out with time. Since our interest is to focus on the effect of the input, we set all initial conditions equal to zero and consider only the response of the system to an input.

4.3   STEP RESPONSE

Under a constant input $\mathbf{u}(t) \equiv \bar{\mathbf{u}}$, an asymptotically stable system will approach a **forced equilibrium point** as $t \rightarrow \infty$. In state space an equilibrium solution for the linear dynamical system (4.1-1) under a constant input $\bar{\mathbf{u}}$ is of the form $\mathbf{x}(t) \equiv \bar{\mathbf{x}}$, where

$$A\bar{\mathbf{x}} + B\bar{\mathbf{u}} = 0. \tag{4.3-1}$$

An equilibrium solution exists when

$$\text{rank}[\mathbf{A}] = \text{rank}[\mathbf{A}, -\mathbf{B}\bar{\mathbf{u}}]. \qquad (4.3\text{-}2)$$

This condition is implied by the stronger condition that $\mathbf{A}^{-1}$ exists ($|\mathbf{A}| \neq 0$, rank $(\mathbf{A}) = N_x$, all eigenvalues of $\mathbf{A}$ are nonzero). If $\mathbf{A}^{-1}$ exists, then the equilibrium solution $\bar{\mathbf{x}} = -\mathbf{A}^{-1}\mathbf{B}\bar{\mathbf{u}}$ is unique. If these conditions are not satisfied, then an equilibrium solution may or may not exist.

Any time that an equilibrium solution exists, we can always move that solution to the origin by a simple translation of coordinates. In particular, suppose $\bar{\mathbf{z}}$ is an equilibrium of interest for the system

$$\dot{\mathbf{z}} = \mathbf{A}\mathbf{z} + \mathbf{B}\bar{\mathbf{u}}.$$

Since

$$\mathbf{A}\bar{\mathbf{z}} + \mathbf{B}\bar{\mathbf{u}} = 0,$$

we can translate the origin of the coordinate system by defining

$$\mathbf{x} = \mathbf{z} - \bar{\mathbf{z}}.$$

It follows that

$$\dot{\mathbf{x}} = \mathbf{A}\mathbf{z} + \mathbf{B}\bar{\mathbf{u}}$$

and since

$$\mathbf{z} = \mathbf{x} + \bar{\mathbf{z}},$$

we obtain

$$\dot{\mathbf{x}} = \mathbf{A}\mathbf{x} \qquad (4.3\text{-}3)$$

with the equilibrium of interest now being at $\mathbf{x} = \mathbf{0}$.

Note that the stability properties of the forced equilibrium point, as determined by the eigenvalues of the $\mathbf{A}$ matrix, will be the same as the free equilibrium point corresponding to $\mathbf{u} = \mathbf{0}$. If we assume the system is asymptotically stable, all the eigenvalues of the $\mathbf{A}$ matrix will have negative real parts. In particular, $\mathbf{A}$ does not have a zero eigenvalue. Thus, $\mathbf{A}^{-1}$ exists and both the free or forced response equilibriums are unique.

In the following discussion, we will investigate the response of asymptotically stable SISO systems to a scalar step input $u(t) \equiv \bar{u} = $ constant. We will be interested in two features of the response. The first feature is associated with the difference between the input and the new forced steady output. This "steady error" feature is important in applications where the objective is to have the output track the input. The second feature of the response is associated with the manner in which the output approaches the new forced steady output with time. This "transient performance" feature plays a role when the time scale of the application is such that initial tracking performance is important, as well as ultimate tracking accuracy. As with the free response, we will examine general first- and second-order systems in detail, because they are the building blocks for higher-order systems.

First-Order Systems

Consider a general first-order IO system of the form

$$\dot{y} + p_0 y = q_0 u + q_1 \dot{u} \tag{4.3-4}$$

in which $p_0 \neq 0$ and $q_0 \neq 0$. An equivalent (one-dimensional) state-space formulation (see Exercise 2.5-14) is given by

$$\dot{x} = -p_0 x + q_0 u$$

$$y = \left(\frac{1 - p_0 q_1}{q_0} \right) x + q_1 u,$$

as may be verified directly by differentiating the output. In this case, we have a single eigenvalue, $\lambda = -p_0$, and a scalar "eigenvector," $\xi = 1$. Thus, the eigenvector matrix is a scalar, $M = 1$, and the solution, with $x(0) = 0$ and $u(t) \equiv \bar{u} =$ constant, is obtained from (4.2-9) as

$$x(t) = \int_0^t e^{-p_0(t-\tau)} q_0 \bar{u} \, d\tau = \bar{u} \frac{q_0}{p_0} (1 - e^{-p_0 t}). \tag{4.3-5}$$

From the output equation we obtain

$$y(t) = \bar{u} \frac{q_0}{p_0} \left(1 - \frac{p_0 q_1}{q_0} \right) (1 - e^{-p_0 t}) + q_1 \bar{u}. \tag{4.3-6}$$

As $t \to \infty$, the solution $y(t)$ asymptotically approaches the forced **steady residual response**

$$y(t) \to y_\infty = \bar{u} \frac{q_0}{p_0}. \tag{4.3-7}$$

This steady output solution can also be obtained directly from (4.3-4) by setting the input to $u = \bar{u}$ and $\dot{y} = \dot{u} = 0$.

A plot of (4.3-6) with $q_1 = 0$ is shown in Figure 4.3-1. The reader should compare this result with Figure 3.2-1 for the free response of the same system. The stability characteristics associated with the free response and the step response of any linear system are similar, except for a translation in coordinates and a corresponding change in initial conditions.

Steady-State Output Error

The forced steady output in our first-order example will not equal the input unless the steady-state gain q_0/p_0 happens to equal 1. The difference between the forced steady output y_∞ and step input $\bar{u}$ is called the **steady-state output error** e_∞

$$e_\infty \overset{\Delta}{=} \bar{u} - y_\infty. \tag{4.3-8}$$

As a fraction of the step input, we have

$$\frac{e_\infty}{\bar{u}} = \frac{\bar{u} - y_\infty}{\bar{u}}.$$

For a first-order system, the steady-state output error fraction is given by

$$\frac{e_\infty}{\bar{u}} = 1 - \frac{q_0}{p_0}.$$

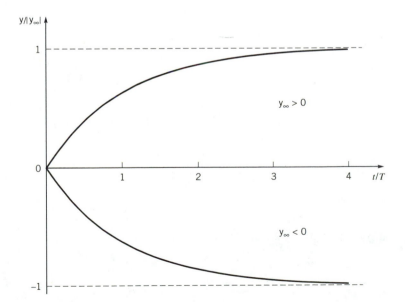

Figure 4.3-1 Response of a stable first-order IO system to a step input (with $q_1 = 0$).

The two parameters p_0 and q_0 are associated with the dynamical system to be controlled and are generally not free to be chosen by the control system designer. If it were possible to choose p_0 or q_0 during the design of the system, then it would be possible to have a system gain $y_\infty/\bar{u} = 1$ and a steady-state output error of zero by choosing $p_0 = q_0$. However, this would not necessarily be a desirable way of obtaining such a result. First, this would fix the return time $T = 1/p_0$ at a value that may not be desirable. Second, such a procedure would not accommodate inaccurate knowledge of p_0 or q_0, or perhaps changes in p_0 or q_0 due to wear, and so on. In the next chapter, using feedback methods, we will learn how to achieve both specified return time requirements and a zero steady output error.

Second-Order Systems

Consider a second-order IO system with $q_1 = q_2 = 0$, of the form

$$\ddot{y} + 2\zeta\omega_n\dot{y} + \omega_n^2 y = q_0 u,$$

subjected to a step input $u(t) \equiv \bar{u} =$ constant, with $y(0) = \dot{y}(0) = 0$ and with $\omega_n > 0$. For $x_1 = y$ and $x_2 = \dot{y}$, the state equations are

$$\begin{bmatrix} \dot{x}_1 \\ \dot{x}_2 \end{bmatrix} = \begin{bmatrix} 0 & 1 \\ -\omega_n^2 & -2\zeta\omega_n \end{bmatrix} \begin{bmatrix} x_1 \\ x_2 \end{bmatrix} + \begin{bmatrix} 0 \\ q_0 \end{bmatrix} u(t). \tag{4.3-9}$$

In terms of the eigenvalues and eigenvectors

$$\lambda_1 = -\zeta\omega_n + \omega_n\sqrt{\zeta^2 - 1}, \qquad \xi_1 = \begin{bmatrix} 1 \\ \lambda_1 \end{bmatrix}$$

$$\lambda_2 = -\zeta\omega_n - \omega_n\sqrt{\zeta^2 - 1}, \qquad \xi_2 = \begin{bmatrix} 1 \\ \lambda_2 \end{bmatrix}, \tag{4.3-10}$$

the state transition matrix $\boldsymbol{\Phi}(t)$ as determined from (4.2-10) can be written as

$$\boldsymbol{\Phi}(t) = [\phi_{ij}(t)] = \frac{1}{\lambda_2 - \lambda_1} \left[\begin{array}{c|c} \lambda_2 e^{\lambda_1 t} - \lambda_1 e^{\lambda_2 t} & -e^{\lambda_1 t} + e^{\lambda_2 t} \\ \hline \lambda_1\lambda_2(e^{\lambda_1 t} - e^{\lambda_2 t}) & -\lambda_1 e^{\lambda_1 t} + \lambda_2 e^{\lambda_2 t} \end{array} \right] \tag{4.3-11}$$

or alternately,

$$\boldsymbol{\Phi}(t) = \begin{bmatrix} \lambda_2 & -1 \\ \lambda_1\lambda_2 & -\lambda_1 \end{bmatrix} \frac{e^{\lambda_1 t}}{\lambda_2 - \lambda_1} + \begin{bmatrix} -\lambda_1 & 1 \\ -\lambda_1\lambda_2 & \lambda_2 \end{bmatrix} \frac{e^{\lambda_2 t}}{\lambda_2 - \lambda_1}.$$

Then for $u(t) \equiv \bar{u}$ and $\mathbf{x}(0) = \mathbf{0}$, the solution to (4.3-9) is

$$\begin{bmatrix} x_1(t) \\ x_2(t) \end{bmatrix} = \bar{u} \int_0^t \begin{bmatrix} \phi_{11}(t - \tau) & \phi_{12}(t - \tau) \\ \phi_{21}(t - \tau) & \phi_{22}(t - \tau) \end{bmatrix} \begin{bmatrix} 0 \\ q_0 \end{bmatrix} d\tau, \qquad (4.3\text{-}12)$$

which yields

$$y(t) = \frac{\bar{u}q_0}{\lambda_2 - \lambda_1} \int_0^t [-e^{\lambda_1(t-\tau)} + e^{\lambda_2(t-\tau)}] d\tau$$

$$= \frac{\bar{u}q_0}{\lambda_2 - \lambda_1} \left[\frac{1 - e^{\lambda_1 t}}{\lambda_1} - \frac{1 - e^{\lambda_2 t}}{\lambda_2} \right]. \qquad (4.3\text{-}13)$$

The solution given by (4.3-13) is valid for any ζ, with $\omega_n > 0$, whether the eigenvalues are real or complex, and even for the case of repeated eigenvalues, where L'Hôpital's rules must be used to evaluate the solution in the limit as $\lambda_1 \to \lambda_2$. In the following, we assume the common situation of $\omega_n > 0$ and $\zeta \geq 0$, as well as continuing with the assumption that $q_1 = q_2 = 0$. Note that if these latter assumptions are not satisfied, the form of the results could differ significantly.

Characterizing the Response

For $\omega_n > 0$ and $\zeta > 0$, we note that the system is asymptotically stable, since $\mathrm{Re}(\lambda_i) = -\zeta\omega_n < 0$. Thus, as $t \to \infty$,

$$y(t) \to y_\infty = \bar{u} \frac{q_0}{\lambda_1 \lambda_2} = \bar{u} \frac{q_0}{\omega_n^2}, \qquad (4.3\text{-}14)$$

and the **steady output error**, as a fraction of the input, is given by

$$\frac{e_\infty}{\bar{u}} = \frac{\bar{u} - y_\infty}{\bar{u}} = 1 - \frac{q_0}{\omega_n^2}. \qquad (4.3\text{-}15)$$

Therefore, even though the system is asymptotically stable, the output will not equal the input unless $q_0 = \omega_n^2$.

The manner in which the step response approaches its steady output value depends on the nature of the eigenvalues. We note that for $\zeta > 1$, the eigenvalues are real, distinct, and negative, and the solution, as given directly by (4.3-13), is said to be **overdamped**. For $\zeta = 1$, the roots are both equal to $-\omega_n$ and this solution, said to be **critically damped**, reduces to

$$y(t) = \bar{u} \frac{q_0}{\omega_n^2} [1 - (1 + \omega_n t)e^{-\omega_n t}]. \qquad (4.3\text{-}16)$$

For $0 \le \zeta < 1$, the eigenvalues are a complex conjugate pair and the corresponding solution, said to be **underdamped**, reduces to

$$y(t) = \bar{u}\frac{q_0}{\omega_n^2}\left[1 - \frac{e^{-\zeta\omega_n t}}{\sqrt{1-\zeta^2}}(\sqrt{1-\zeta^2}\cos\omega_d t + \zeta\sin\omega_d t)\right],$$

which can be written as

$$y(t) = \bar{u}\frac{q_0}{\omega_n^2}\left[1 - \frac{e^{-\zeta\omega_n t}}{\sqrt{1-\zeta^2}}\sin(\omega_d t + \alpha)\right], \qquad \textbf{(4.3-17)}$$

where the **damped frequency** of oscillation ω_d is

$$\omega_d = \omega_n\sqrt{1-\zeta^2} \qquad \textbf{(4.3-18)}$$

and the **phase angle** α is given by

$$\cos\alpha = \zeta, \qquad 0 \le \alpha < \frac{\pi}{2}. \qquad \textbf{(4.3-19)}$$

Performance Measures

For an underdamped system the **return time**, previously discussed in connection with the free response, is given by

$$T = \frac{1}{|\text{Re}(\lambda)|} = \frac{1}{\zeta\omega_n}. \qquad \textbf{(4.3-20)}$$

Figure 4.3-2 shows plots of $y(t)/y_\infty$ versus $\omega_n t$, as obtained from (4.3-13) or (4.3-17) for various values of the damping ratio ζ. Note that for a large amount of damping, $\zeta \ge 1$, the output motion does not oscillate. For a smaller amount of damping, $\zeta < 1$, it takes a longer time for the oscillating system to settle down to the steady state. For many control applications, a value of ζ between $0.5 \le \zeta \le 1$ is used.

In addition to the return time and the steady output error, several other performance criteria, illustrated in Figure 4.3-3, are often used for control system design specifications.

The **integral square error** e_I is defined as the integral of the square of the difference between the error and the steady-state error. Recall that the error under a step input is defined by

$$e(t) = \bar{u} - y(t)$$

with the steady-state output error given by

$$e_\infty = \bar{u} - y_\infty.$$

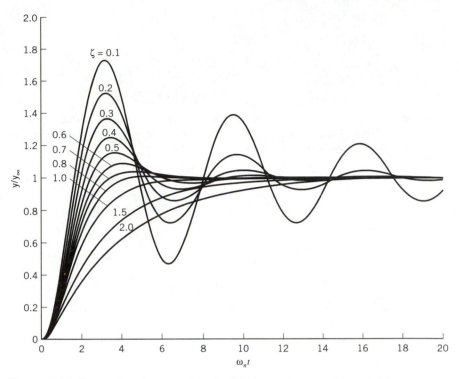

Figure 4.3-2 Response of a second-order IO system to a step input (with $q_1 = q_2 = 0$).

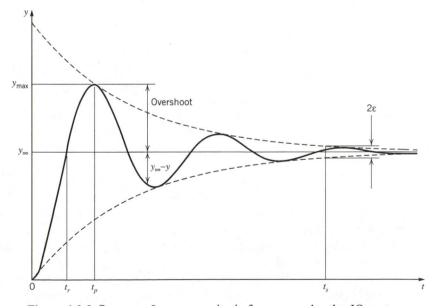

Figure 4.3-3 Some performance criteria for a second-order IO system.

Thus, the integral square error is also given by

$$e_I = \int_0^\infty [y_\infty - y(t)]^2 \, dt.$$

It gives a good overall measure of how well the system is tracking the input with respect to the expected steady state. Using (4.3-17) and evaluating the error integral, we find that the integral square error for the second-order system (4.3-9) is given by

$$e_I = \frac{\bar{u} q_0}{\omega_n^3} \left(\zeta + \frac{1}{4\zeta} \right).$$

It follows that the minimum square error for this case is obtained when $\zeta = 1/2$.

The **rise time** t_r is defined as the first time that the step response passes through its ultimate steady value; $y(t_r) = y_\infty$. The first time that the sine term in (4.3-17) goes to zero is given by

$$\omega_d t_r + \alpha = \pi.$$

Hence,

$$t_r = \frac{\pi - \alpha}{\omega_n \sqrt{1 - \zeta^2}}. \tag{4.3-21}$$

where α is given by (4.3-19). The rise time increases as ω_n decreases or ζ increases. It is infinite for $\zeta \geq 1$, since the response in those cases will never actually reach the ultimate steady value y_∞.

The **settling time** t_s is the time it takes for the step response to settle to within a small neighborhood of its ultimate steady value, that is, the time after which $|1 - y(t)/y_\infty| < \epsilon$. This neighborhood is usually taken to be within 2 or 5% of the final steady-state value ($\epsilon = 0.02$ or 0.05). For underdamped systems with $0 < \zeta < 1$, since the sine term in (4.3-17) ranges between $+1$ and -1, setting its coefficient equal to ϵ yields

$$t_s = -\frac{\ln(\epsilon\sqrt{1 - \zeta^2})}{\zeta\omega_n}, \qquad 0 < \zeta < 1. \tag{4.3-22}$$

The settling time decreases with increases in $\omega_n > 0$ or $\epsilon > 0$. For small ζ the settling time is large (as is the return time), since the oscillatory response decays slowly. On the other hand, for ζ very near 1 or greater than 1, the settling time is again large, since the response approaches its ultimate steady value slowly. For given values of ϵ and ω_n, with $0 < \epsilon < 1$ and $\omega_n > 0$, there exists a value of ζ in the range $0 < \zeta < 1$ that will yield a minimum settling time. For a 2% settling criteria ($\epsilon = 0.02$), the minimum settling time is $\omega_n t_s \approx 5.266$, at a damping ratio $\zeta \approx 0.9096$.

The **peak time**, as illustrated in Figure 4.3-3, is the time at which the step response reaches its maximum overshoot. It is the first time greater than zero at which $\dot{y}(t) = 0$ and is given by

$$t_p = \frac{\pi}{\omega_n \sqrt{1 - \zeta^2}}. \tag{4.3-23}$$

The maximum **overshoot**, as a fraction of the steady output value, is given by

$$\text{OS} = \frac{y_{\max} - y_\infty}{y_\infty} = \exp\left(\frac{-\zeta\pi}{\sqrt{1 - \zeta^2}}\right). \tag{4.3-24}$$

To provide a qualitative discussion of these performance measures for a second-order system, in terms of ζ and ω_n, we first note that the steady output error does not depend on ζ, and the fractional overshoot does not depend on ω_n. Of all the performance measures, only the steady output error and the integral square error depend on the input parameters $\bar{u}$ and q_0; the rest depend on only ζ and ω_n. In particular, these two parameters completely determine the return time, rise time, settling time (for given ϵ), peak time, and overshoot.

The return time, integral square error, rise time, settling time, and peak time all decrease with an increase in ω_n. The rise time and peak time increase with an increase in ζ, whereas the return time, overshoot, and settling time (up to a minimum value) decrease with an increase in ζ. In general, we might like all of these performance measures to be as small as possible. However, as we have learned, there is a direct trade-off of rise time and peak time versus overshoot and settling time. For example, consider the trade-off between rise time and settling time. For most automatic control applications, $0.5 \leq \zeta \leq 1$ is a reasonable range for the damping ratio. Near the low end we get close to minimum integral square error and faster rise times. However, we also obtain considerable oscillation and overshoot and slower settling times. For damping ratios toward the high end, we get faster settling times. Indeed, for a 2% settling criteria ($\epsilon = 0.02$), a value of $\zeta = 0.9$ yields essentially a minimum settling time. However, at high damping ratios we also get slower rise times, but with little overshoot or oscillation. As a result, even though the system settles faster to its ultimate steady output value, initially the response spends less time on the average near its ultimate state. A good compromise might be a damping ratio of about 0.7 (one commercial airplane manufacturer has a fondness for the number $1/\sqrt{2} = 0.707$). On the other hand, applications exist in which any overshoot is to be avoided (such as an automobile cruise control at the posted speed limit?). In such cases, a damping ratio of $\zeta = 1$ will produce the smallest possible settling time without any overshoot, and a damping ratio of $\zeta = 0.9$ will do even better if a small overshoot is tolerable.

4.4   UNCERTAIN INPUTS

For most applications, there will always be some uncertain input entering the system. By and large, these inputs will move the system away from the desired operating conditions. In later chapters we will design controllers for operation with uncertain inputs. In particular, our objective will be to maintain some small neighborhood of a desired operating condition in spite of an uncertain "disturbance" input.

For now, we will be content with understanding the role some of the design parameters play in "buffering" a system from uncertain inputs. We will take as the uncertain input a bounded piecewise constant function $v(t)$ that has a constant value over each time interval of length δ, but with the value of $v(t)$ chosen randomly at the beginning of each time interval, subject to bounds such as $-1 \leq v(t) \leq 1$.

Consider the output of the first-order system

$$\dot{y} + p_0 y = q_0 v, \tag{4.4-1}$$

subject to the uncertain input described above. At the end of the first time interval, $t = \delta$ and the output is

$$y(\delta) = e^{-p_0\delta} y(0) + \frac{q_0}{p_0}(1 - e^{-p_0\delta}) v(0), \tag{4.4-2}$$

where $v(0)$ is the input value used over the first time period. If we let

$$\psi(\delta) = \frac{q_0}{p_0}(1 - e^{-p_0\delta}), \tag{4.4-3}$$

then the output at the end of every δ units of time can be easily expressed by using the previous output as the "initial condition." Thus, for $y(0) = 0$,

$$y(\delta) = \psi(\delta)v(0)$$
$$y(2\delta) = \psi(\delta)v(0)e^{-p_0\delta} + \psi(\delta)v(\delta)$$
$$y(3\delta) = \psi(\delta)v(0)e^{-2p_0\delta} + \psi(\delta)v(\delta)e^{-p_0\delta} + \psi(\delta)v(2\delta)$$
$$\vdots$$
$$y[(n + 1)\delta] = e^{-p_0\delta}y(n\delta) + \psi(\delta)v(n\delta).$$

At the end of any sequence of time intervals, the largest (or smallest) value of y will occur when v is at its maximum (or minimum) value throughout the sequence. In other words, for a first-order system, a constant uncertain input will cause the largest deviation in the output. We have already examined this first-order system subject to a constant input in Section 4.3. We note that if the system (4.4-1) is asymptotically stable ($p_0 > 0$), then the output approaches

$$y_\infty = \frac{q_0}{p_0}\,\bar{v},$$

where $\bar{v}$ is the constant uncertain input.

Thus, we see that for a bounded uncertain input $|v(t)| \leq \bar{v}$, the output of an asymptotically stable first-order system (4.4-1) is also bounded, with the bounds given by $|y(t)| \leq y_\infty$. For large values of p_0, these bounds will be small, and for small values of p_0, these bounds will be large. Recall that this system has an eigenvalue given by $\lambda = -p_0$ and a return time given by $T = 1/p_0$. Thus, the bounds on $y(t)$ can also be examined in terms of these parameters. This system will be well buffered from uncertain inputs if the return time is small or, equivalently, if the eigenvalue is very negative.

The v-Reachable Set

The effect of uncertainty can be examined in a more general setting than the one-dimensional situation discussed above. In particular, let us examine the linear system model given by

$$\dot{\mathbf{x}} = \mathbf{A}\mathbf{x} + \mathbf{B}\mathbf{u} + \mathbf{R}\mathbf{v},$$

in which the uncertain input $\mathbf{v}$ is bounded. As in Exercise 3.6-1, suppose that the control input has been specified in terms of state feedback of the form

$$\mathbf{u} = -\mathbf{K}\mathbf{x},$$

so that the controlled system is of the form

$$\dot{\mathbf{x}} = \tilde{\mathbf{A}}\mathbf{x} + \mathbf{R}\mathbf{v}, \tag{4.4-4}$$

where

$$\tilde{\mathbf{A}} = \mathbf{A} - \mathbf{B}\mathbf{K}.$$

The constants in the matrix $\mathbf{K}$ have been chosen so that (4.4-4) with $\mathbf{v} = \mathbf{0}$ will be stable with some specified eigenvalues for the $\tilde{\mathbf{A}}$ matrix. Consider now the effect of $\mathbf{v}$ on (4.4-4) in terms of motion of the system in state space. The system velocity $\dot{\mathbf{x}}$ is determined by both the system components $\tilde{\mathbf{A}}\mathbf{x}$ and the uncertain inputs $\mathbf{R}\mathbf{v}$. In a small neighborhood of the origin ($\mathbf{x} \approx \mathbf{0}$), the system velocity is almost totally determined by the uncertain input. A great distance away from the origin ($\|\mathbf{x}\| \gg 0$), the system velocity will be mostly influenced by the stabilizing system components, since the magnitude of the term $\mathbf{R}\mathbf{v}$ is limited by the bounds on $\mathbf{v}$. It then follows that there must be some limited region about the origin in which it is possible for the uncertain input to drive the state of the system. We call this region the v-**reachable set.** Clearly, for any given control design choice for $\mathbf{K}$, we would like to have the v-reachable set be as small as possible. For many systems, it is possible to find the v-reachable set. This set may then be used as a qualitative picture of the possible effects of uncertainty on the system.

It can be shown (Gayek and Vincent, 1985) that if all of the eigenvalues of the $\tilde{\mathbf{A}}$ matrix have negative real parts and there exists an input $\mathbf{v}(t)$ for the system (4.4-4) that will generate a trajectory remaining on the boundary of the v-reachable set for all times $t > 0$, then this same control law will drive the system to the boundary of the v-reachable set from any point within the v-reachable set. Furthermore, the existence of such a control law is guaranteed for single-input, controllable, two-dimensional systems of the form of (4.4-4) for which the input bound is of the form $-\infty < v_{\min} < 0 < v_{\max} < \infty$ (Gayek and Vincent, 1988). In this latter situation, the input $v(t)$ is easily determined by applying the **controllability maximum principle** (Grantham and Vincent, 1975). In particular, a single-input two-dimensional system of the form of (4.4-4) can be written as

$$\dot{x}_1 = a_{11}x_1 + a_{12}x_2 + r_1 v$$

$$\dot{x}_2 = a_{21}x_1 + a_{22}x_2 + r_2 v. \tag{4.4-5}$$

Let the bounds on v be given by $|v| \leq \bar{v}$. Using the controllability maximum principle, we can then show that provided (4.4-5) satisfies the conditions stated above, a control law for v that will drive the system along the boundary of the v-reachable set is given by

$$v = \bar{v}\,\text{sgn}[r_2(a_{11}x_1 + a_{12}x_2) - r_1(a_{21}x_1 + a_{22}x_2)], \tag{4.4-6}$$

where $v = \bar{v}$ or $v = -\bar{v}$ when the argument of the sgn function is zero.

EXAMPLE 4.4-1	**The v-Reachable Set for a System with Complex Eigenvalues**

Consider the two-dimensional system (4.4-5) in which $a_{11} = 0$, $a_{12} = 1$, $a_{21} = -1$, $a_{22} = -1$, $r_1 = 0$, $r_2 = 1$, and $\bar{v} = 1$. This is a controllable system that is stable with complex eigenvalues when $v = 0$. A stable two-dimensional system with complex eigenvalues will be driven from any point inside the v-reachable set to the boundary of the v-reachable set under the closed-loop control law (4.4-6). In this case, (4.4-6) can be reduced to

$$v = \begin{cases} \bar{v}\,\mathrm{sgn}(x_2) & \text{if } x_2 \neq 0 \\ \bar{v} & \text{if } x_2 = 0. \end{cases} \tag{4.4-7}$$

Figure 4.4-1 illustrates the trajectory obtained for this system starting at the origin and integrating the system equations with v subject to the control (4.4-6). The trajectory spirals out from the origin and asymptotically approaches the boundary of the v-reachable set.

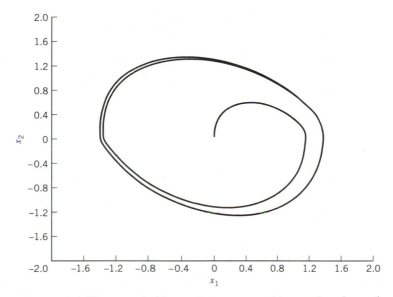

Figure 4.4-1 The v-reachable set for a system with complex eigenvalues.

EXAMPLE 4.4-2 **The v-Reachable Set for a System with Real Eigenvalues**

When the eigenvalues for a two-dimensional stable system are real, equilibrium solutions corresponding to $v(t) = \bar{v}$ or $v(t) = -\bar{v}$ will always lie on the boundary of the v-reachable set. Use of this information allows for an easy determination of the v-reachable set. Figure 4.4-2 illustrates trajectories on the boundary of the v-reachable set for the system (4.4-5) with $a_{11} = 0$, $a_{12} = 1$, $a_{21} = -2$, $a_{22} = -3$, $r_1 = 2$, $r_2 = -3$, and $\bar{v} = 1$. In this case, the eigenvalues for the $\tilde{A}$ matrix are real. A trajectory gen-

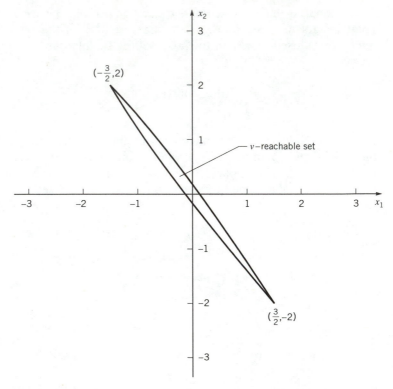

Figure 4.4-2 The v-reachable set for a system with real eigenvalues.

erated using $v(t) = 1$ will move directly from the origin and asymptotically approach the right-hand equilibrium point on the boundary of the v-reachable set. As an alternative, one could simply start there. To then define the boundary of the v-reachable set, the control must be switched from $v = +1$ to $v = -1$ at this point. The resulting trajectory will trace out the lower boundary to the left-hand cusp point that is also an equilibrium point (with $v = -1$). At this point, the control must then be switched to $v = +1$ to yield the upper boundary of the v-reachable set.

An alternate approach for examining uncertain inputs uses stochastic differential equations, in which the inputs are treated as time-varying random variables, usually assumed to have an average value of zero and a Gaussian probability distribution. Noise might be one example of such a random input, but we will approximate noise either by a high-frequency sinusoidal input (if we wish to assume an average value of zero) or, more generally, as a bounded uncertain input, with the bounds corresponding, for example, to two or three standard deviations from the mean. We will not pursue the stochastic approach in this introductory test, but it is an important area of modern control theory, and several complete texts are available on the subject, such as Meditch (1969).

Regardless of whether uncertain inputs are treated stochastically or by reachable sets analysis, one effect of uncertain inputs is that the control system designer generally cannot assure that the state will converge asymptotically to an equilibrium point, even if the $\bar{\mathbf{A}}$ matrix is asymptotically stable. Usually, the best one can do is assure that the state be driven to and kept inside the v-reachable set, where the state may wander around erratically. There are results available (Leitmann, 1981; Corless and Leitmann, 1981; Barmish and Leitmann, 1982) for designing (nonlinear) controllers to accomplish this **ultimate boundedness** objective and to make the v-reachable set small.

4.5 EXERCISES

4.5-1 For systems given by (4.1-1) with **A** matrices given in Exercise 3.6-4 and $\mathbf{B} = [0 \quad 1]^T$, determine the state response $\mathbf{x}(t)$ when $u(t) \equiv 1$ and $\mathbf{x}(0) = [0 \quad 0]^T$.

4.5-2 If $y = x_2$, determine the output response $y(t)$ for each system in Exercise 4.5-1.

4.5-3 Determine the output response $y(t)$ for each system in Exercise 4.5-2 if $\mathbf{x}(0) = [1 \quad 0]^T$.

4.5-4 A servomotor has a response due to a unit step input as shown in Figure 4.5-1. Determine a linear second-order system that would be an appropriate model for this system.

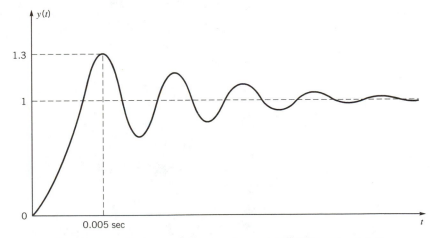

Figure 4.5-1 Time response for the system of Exercise 4.5-4.

4.5-5 With $y(0) = \dot{y}(0) = 0$ and $u(t) \equiv 5$ for the system

$$\ddot{y} + 2\zeta\omega_n\dot{y} + \omega_n^2 y = u(t),$$

determine values for ζ and ω_n for a 25% overshoot and a 4-sec settling time, based on a 2% settling band (that is, $\epsilon = 0.02$). Determine the corresponding peak time and rise time. Based on these results, make an accurate sketch of the response $y(t)$.

4.5-6 For a second-order system as in Exercise 4.5-5, numerically determine the minimum nondimensional settling time $\omega_n t_s$ and the corresponding damping ratio ζ, for a 5% settling band ($\epsilon = 0.05$). Use a numerical root-finding algorithm, not a graphical plot, to find ζ to four significant digits.

4.5-7 For the system in Exercise 4.5-5, plot (on one graph) the following nondimensional quantities for $0 < \zeta < 1$: return time $\omega_n T$, rise time $\omega_n t_r$, peak time $\omega_n t_p$, overshoot OS, and settling time $\omega_n t_s$ (for $\epsilon = 0.02$).

4.5-8 Determine the eigenvalues for Examples 4.4-1 and 4.4-2. Using Figure 4.4-1, show that the equilibrium point corresponding to a constant input $\bar{v} = +1$ in Example 4.4-1 lies inside the v-reachable set. Using Figure 4.4-2, show that the equilibrium point corresponding to a constant input $\bar{v} = +1$ in Example 4.4-2 lies on the boundary of the v-reachable set.

4.5-9 Find the v-reachable set for the system (4.4-5) with $a_{11} = 0$, $a_{12} = 1$, $a_{21} = -1.75$, $a_{22} = -2$, $r_1 = 0$, $r_2 = 1$, and $\bar{v} = 1$. This system is more robust with respect to uncertain inputs than Example 4.4-1, since it has a smaller v-reachable set. Can you explain why it does?

Chapter 5

Frequency Response

5.1 FREQUENCY RESPONSE AND BODE PLOTS

Periodic input functions are the second class of fundamental inputs that we will study. For a particular control application, we may be required not only to move the system to a desired state (as with a step input) but also to isolate the system from vibrations or high-frequency "noise," which may enter as part of the control input. On the other hand, we may be interested in how accurately the controlled system will track a time-varying input, such as a sine function. As we will learn in the following discussion, any asymptotically stable linear system subjected to a sinusoidal input at a forcing frequency ω will oscillate at that forcing frequency. The amplitude of the output will depend on the forcing frequency, and it can be very large, even if the input amplitude is small. Although it may not be possible to completely isolate a linear system from vibrations, we can usually design the control system so that vibration inputs will induce only small amplitude outputs in the system.

Sinusoidal Inputs

Consider a sinusoidal input $u(t) = \bar{u} \sin \omega t$ to the first-order IO system

$$\dot{x} = -p_0 x + q_0 u, \qquad y = x, \tag{5.1-1}$$

with $p_0 \neq 0$ and $q_0 \neq 0$. The solution with $x(0) = 0$ is given by

$$x(t) = \bar{u} \int_0^t e^{-p_0(t-\tau)} q_0 \sin \omega \tau \, d\tau.$$

We obtain the output solution by evaluating this integral and noting that the output y is equal to the state x. Thus,

$$y(t) = \frac{\bar{u} q_0}{p_0^2 + \omega^2} (\omega e^{-p_0 t} + p_0 \sin \omega t - \omega \cos \omega t). \tag{5.1-2}$$

149

If the system is asymptotically stable ($p_0 > 0$), then as $t \rightarrow \infty$, the exponential term approaches zero and we are left with a **residual response** $y_r(t)$, which can be written as

$$y_r(t) = \frac{\bar{u}q_0}{\sqrt{p_0^2 + \omega^2}}\left[\frac{p_0}{\sqrt{p_0^2 + \omega^2}}\sin \omega t - \frac{\omega}{\sqrt{p_0^2 + \omega^2}}\cos \omega t\right],$$

or, more simply, using the trigonometric identity

$$\sin(\omega t + \varphi) = \cos \varphi \sin \omega t + \sin \varphi \cos \omega t,$$

we obtain

$$y_r(t) = Y\sin(\omega t + \varphi), \tag{5.1-3}$$

where the **amplitude** Y, given by

$$Y(\omega) = \bar{u}\frac{q_0/p_0}{\sqrt{1 + [\omega/p_0]^2}}, \tag{5.1-4}$$

and the **phase angle** φ, given by

$$\tan \varphi = \frac{-\omega}{p_0}, \tag{5.1-5}$$

both depend on the input frequency ω.

The input $\bar{u} \sin \omega t$ along with the residual response $y_r(t)$ are illustrated in Figure 5.1-1. In the form of (5.1-3) we observe that the residual response to a sine wave input is also a sine wave at the same frequency ω as the input, but at a different phase. For an asymptotically stable first-order system, the output amplitude is reduced. Also, the phase angle φ is negative, indicating a **phase lag**, as shown in Figure 5.1-1.

Bode Plots

Figure 5.1-2 is a plot of the amplitude ratio $Y(\omega)/\bar{u}$ and the phase angle $\varphi(\omega)$ as functions of frequency ω, when the **steady-state gain** $q_0/p_0 = 1$. Plots such as these are named after a major contributor to "frequency domain" classical control methods and are referred to as **Bode plots**. Note from the phase plot of Figure 5.1-2b that the angle φ always lags after the input. For the first-order system (5.1-1), the phase shift is $-45°$ at $\omega_n = p_0$ and approaches a value of $-90°$ at very high frequencies.

The amplitude plot of Figure 5.1-2a is on a log–log scale. The horizontal scale is proportional to the frequency of the input, and each increase in power of 10 is called a **decade**. The vertical scale is also in powers of 10. However, it is common to designate this scale in terms of **decibels** (db). One power of 10 corresponds to 20 db. For any positive number G, the

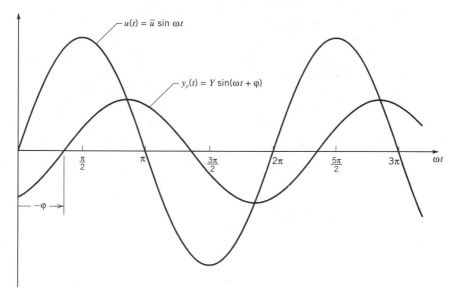

Figure 5.1-1 Phase lag ($\varphi < 0$) residual response of a first-order IO system to a sinusoidal input.

conversion to decibels is given by

$$G \text{ in db} = 20 \log_{10}(G). \tag{5.1-6}$$

The advantage of this logarithmic formulation will become clear shortly, when we investigate some generalizations of these results. Note that if the steady-state gain q_0/p_0 is not 1 (0 db), then the amplitude plot in Figure 5.1-2a will simply be shifted by $20 \log_{10}(q_0/p_0)$ db.

When $\omega/p_0 > 1$, the amplitude ratio in Figure 5.1-2a begins to drop off noticeably from its initial ($\omega = 0$) level. The actual Bode plot, indicated by the dashed curve, is often approximated by the two straight-line asymptotes shown in Figure 5.1-2a, which intersect, as we will show, at the **corner frequency** $\omega_c = p_0$. The **low-frequency asymptote** is the zero db line. The **high-frequency asymptote** is a straight line with a slope of -20 db/decade, typical of any first-order system.

Figure 5.1-2a illustrates the filtering effect of a first-order system; for frequencies above the corner frequency $\omega/p_0 = 1$, the system filters out the input, by reducing the amplitude of the output oscillation. At frequencies below the corner frequency, the amplitude of the output is not appreciably attenuated. For this reason, an asymptotically stable first-order system is said to behave as a **low-pass filter**. The corner frequency, at which filtering begins to occur, is a function of p_0. For large p_0 (short return times, $T = 1/p_0$), ω must be large for this filtering effect to occur,

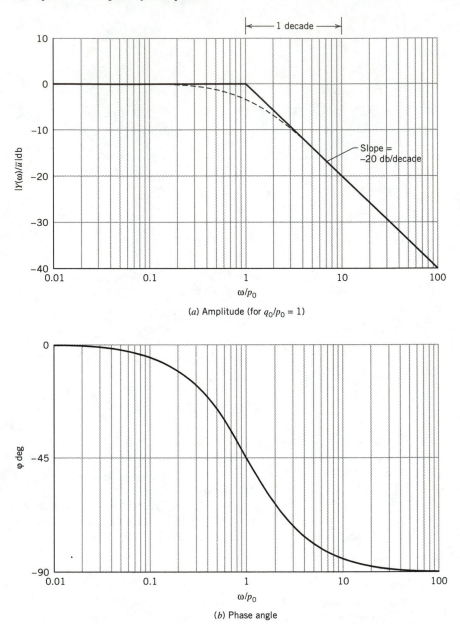

(a) Amplitude (for $q_0/p_0 = 1$)

(b) Phase angle

Figure 5.1-2 Bode plot for a first-order system.

and for small p_0 (large return times), filtering can take place for small ω. The system **bandwidth** ω_b is the frequency range over which the amplitude has been attenuated by less than 3 db compared to the amplitude at $\omega = 0$. For this first-order system, the bandwidth is approximated by the corner frequency ω_c. Thus, a large bandwidth corresponds to a system with a short return time, and vice versa.

Low- and High-Frequency Asymptotes

If we divide both sides of (5.1-4) by $\bar{u}$ and take the log according to (5.1-6), this yields

$$\left.\frac{Y(\omega)}{\bar{u}}\right|_{\text{db}} = 20 \log_{10}\left\{\frac{q_0/p_0}{\sqrt{1 + [\omega/p_0]^2}}\right\},$$

which can also be written as

$$\left.\frac{Y(\omega)}{\bar{u}}\right|_{\text{db}} = 20 \log_{10}\left(\frac{q_0}{p_0}\right) - 10 \log_{10}\left[1 + \left(\frac{\omega}{p_0}\right)^2\right].$$

If $q_0/p_0 = 1$, then

$$\left.\frac{Y(\omega)}{\bar{u}}\right|_{\text{db}} = -10 \log_{10}\left[1 + \left(\frac{\omega}{p_0}\right)^2\right].$$

At low frequencies ($\omega \ll p_0$), this result is approximately given by

$$\left.\frac{Y(\omega)}{\bar{u}}\right|_{\text{db}} \approx -10 \log_{10}(1) = 0,$$

and at high frequencies ($\omega \gg p_0$), it is approximated by

$$\left.\frac{Y(\omega)}{\bar{u}}\right|_{\text{db}} \approx -10 \log_{10}\left(\frac{\omega}{p_0}\right)^2 = -20 \log_{10}\left(\frac{\omega}{p_0}\right).$$

These amplitude approximations are equal at the corner frequency $\omega = p_0$ defined by the intersection of the two asymptotes, which are straight lines on the logarithmic Bode plot.

5.2 RESIDUAL RESPONSE AND THE TRANSFER FUNCTION

For any order system subjected to a sine wave input, the system response and, in particular, the residual output response can be obtained by using the same methods as described for a first-order system. However, if we are interested in just the residual response, then a simpler procedure can be used to obtain this response for IO systems.

Consider any asymptotically stable linear system of the form

$$\dot{\mathbf{x}} = \mathbf{A}\mathbf{x} + \mathbf{B}\mathbf{u}$$

$$\mathbf{y} = \mathbf{C}\mathbf{x} + \mathbf{D}\mathbf{u}.$$

With a specified input $\mathbf{u} = \mathbf{u}(t)$ and any initial condition $\mathbf{x}(0)$, as $t \to \infty$, the forced response approaches the input response

$$\mathbf{x}_I(t) = \int_0^t \mathbf{\Phi}(t - \tau)\mathbf{B}\mathbf{u}(\tau)\,d\tau,$$

where $\mathbf{\Phi}(t)$ is the state transition matrix.

As we have just seen, the input response $\mathbf{x}_I(t)$ may contain terms that die out as $t \to \infty$. Denote these terms by $\boldsymbol{\rho}(t)$, where $\boldsymbol{\rho}(t)$ [and $\dot{\boldsymbol{\rho}}(t)$] approach $\mathbf{0}$ as $t \to \infty$. Then

$$\mathbf{x}_I(t) = \mathbf{x}_r(t) + \boldsymbol{\rho}(t), \tag{5.2-1}$$

where the **residual response** $\mathbf{x}_r(t)$ is the portion of $\mathbf{x}_I(t)$ that does not die out as $t \to \infty$. The corresponding **residual output response** is given by

$$\mathbf{y}_r(t) = \mathbf{C}\mathbf{x}_r(t) + \mathbf{D}\mathbf{u}(t).$$

Since $\mathbf{x}_I(t)$ satisfies the state equations

$$\dot{\mathbf{x}}_I(t) = \mathbf{A}\mathbf{x}_I(t) + \mathbf{B}\mathbf{u}(t),$$

we have

$$\dot{\mathbf{x}}_r(t) + \dot{\boldsymbol{\rho}}(t) = \mathbf{A}\mathbf{x}_r(t) + \mathbf{A}\boldsymbol{\rho}(t) + \mathbf{B}\mathbf{u}(t).$$

Therefore, as $t \to \infty$,

$$\dot{\mathbf{x}}_r(t) = \mathbf{A}\mathbf{x}_r(t) + \mathbf{B}\mathbf{u}(t)$$

and we see that the residual response must also satisfy the state equations, without regard to initial conditions.

For any state-space system, the residual response can be determined by examining the input response as $t \to \infty$, as we have done for the first-order system. This lengthy procedure can be avoided, however, for a single-input single-output system when the input is sinusoidal. In this case, an elegant procedure exists for finding the residual output response in terms of the transfer function.

Consider a sinusoidal input $u(t) = \bar{u} \sin \omega t$ applied to an asymptotically stable IO system of the form

$$y^{(N_x)} + p_{N_x-1}y^{(N_x-1)} + \cdots + p_1\dot{y} + p_0 y = q_0 u + \cdots + q_{N_x}u^{(N_x)}.$$

As $t \to \infty$, the solution $y(t)$ approaches the residual output response $y_r(t)$ that must also satisfy the differential equation (but not necessarily any specified initial conditions). We will now develop a general procedure for obtaining this residual solution, regardless of the order of the system.

For algebraic convenience, let $u(t) = \bar{u}e^{i\omega t} = \bar{u}(\cos \omega t + i \sin \omega t)$. We do this to simplify the resulting development. However, this complex-valued function is a perfectly valid input, which could be used to study both sine and cosine inputs simultaneously. The resulting residual solution will also be complex, with its real part corresponding to the real part of the input and its imaginary part corresponding to the imaginary part of the input.

The residual solution is assumed to be of the form

$$y_r(t) = Ye^{i(\omega t + \varphi)} = Y[\cos(\omega t + \varphi) + i \sin(\omega t + \varphi)].$$

By differentiating both $u(t)$ and $y_r(t)$ as required and substituting into the differential equation, we find that this yields

$$[(i\omega)^{N_x} + p_{N_x-1}(i\omega)^{N_x-1} + \cdots + p_1(i\omega) + p_0]Ye^{i(\omega t + \varphi)}$$
$$= [q_0 + q_1(i\omega) + \cdots + q_{N_x}(i\omega)^{N_x}]\bar{u}e^{i\omega t}.$$

Canceling $e^{i\omega t}$ and solving for $Ye^{i\varphi}$, we obtain

$$Ye^{i\varphi} = \bar{u}G(i\omega) = \bar{u}\frac{Q(i\omega)}{P(i\omega)},$$

where $P(s)$ and $Q(s)$ are the polynomials defined by (2.2-7) and (2.2-8), and $G(s)$ is the transfer function defined by (2.2-9). Since $G(i\omega)$ is a complex number, it can be written in the form

$$G(i\omega) = re^{i\varphi},$$

so that the previous result can be written as

$$\frac{Y}{\bar{u}}e^{i\varphi} = |G(i\omega)|e^{i\tan^{-1}\left[\frac{\text{Im}\{G(i\omega)\}}{\text{Re}\{G(i\omega)\}}\right]}. \tag{5.2-2}$$

Associating the imaginary part of this result with the imaginary part of the input, we conclude that any asymptotically stable linear system subjected to a sinusoidal input $u(t) = \bar{u} \sin \omega t$ will oscillate at the forcing frequency ω (after any transients die out) according to the residual response

$$y_r(t) = Y \sin(\omega t + \varphi), \qquad (5.2\text{-}3)$$

with an amplitude $Y(\omega)$ and a phase angle $\varphi(\omega)$ that depend on the forcing frequency. Both Y and φ can be determined directly from (5.2-2) in terms of the transfer function as

$$\frac{Y}{\bar{u}} = |G(i\omega)| \triangleq \sqrt{[\text{Re}\{G(i\omega)\}]^2 + [\text{Im}\{G(i\omega)\}]^2} \qquad (5.2\text{-}4)$$

and

$$\tan \varphi = \frac{\text{Im}\{G(i\omega)\}}{\text{Re}\{G(i\omega)\}}, \qquad (5.2\text{-}5)$$

where $\text{Re}(\cdot)$ and $\text{Im}(\cdot)$ denote the real and imaginary parts of the transfer function, respectively, and φ is the angle from the positive real axis to the point $G(i\omega)$ in the complex plane.

The transfer function magnitude $|G(i\omega)|$ and phase angle $\varphi(\omega)$ from (5.2-4) and (5.2-5) are called the **frequency response** of the system. They are typically plotted in the form of Bode plots, as illustrated previously for a first-order system.

First-Order Systems

The results given in (5.2-3)–(5.2-5) hold true for any order asymptotically stable IO system. As an example, consider the asymptotically stable first-order system previously discussed:

$$\dot{y} + p_0 y = q_0 u.$$

For a sinusoidal input $u(t) = \bar{u} \sin \omega t$, the residual response is given by (5.2-3)–(5.2-5). For this example, the transfer function is

$$G(s) = \frac{q_0}{s + p_0},$$

from which we have

$$G(i\omega) = \frac{q_0}{p_0 + i\omega}.$$

It follows from (5.2-4) that

$$\frac{Y}{\bar{u}} = \frac{|q_0|}{|p_0 + i\omega|} = \frac{|q_0/p_0|}{\sqrt{1 + (\omega/p_0)^2}}.$$

To obtain φ, we multiply the numerator and denominator of $G(i\omega)$ by the conjugate of the denominator to write $G(i\omega)$ as a complex number in standard form

$$G(i\omega) = \frac{p_0 q_0 - i q_0 \omega}{p_0^2 + \omega^2},$$

and from (5.2-5) we obtain

$$\tan \varphi = -\frac{\omega}{p_0},$$

which are the same results obtained in Section 5.1 as given by Equations (5.1-4) and (5.1-5).

Second-Order Systems

Another important example is the sinusoidally forced oscillation of a second-order system. Consider an asymptotically stable system of the particular form (note the scale factor ω_n^2 on the input)

$$\ddot{y} + 2\zeta\omega_n\dot{y} + \omega_n^2 y = \omega_n^2 u(t), \tag{5.2-6}$$

with $y(0) = \dot{y}(0) = 0$ and $u(t) = \bar{u}\sin\omega t$. Using the transfer function

$$G(s) = \frac{\omega_n^2}{s^2 + 2\zeta\omega_n s + \omega_n^2},$$

we obtain

$$G(i\omega) = \frac{1}{1 - \left(\dfrac{\omega}{\omega_n}\right)^2 + i2\zeta\left(\dfrac{\omega}{\omega_n}\right)}.$$

Using (5.2-3)–(5.2-5), the residual response is given by

$$y_r(t) = Y\sin(\omega t + \varphi), \tag{5.2-7}$$

where

$$\frac{Y(\omega)}{\bar{u}} = |G(i\omega)| = \frac{1}{\sqrt{\left[1 - \left(\dfrac{\omega}{\omega_n}\right)^2\right]^2 + \left(\dfrac{2\zeta\omega}{\omega_n}\right)^2}} \tag{5.2-8}$$

and

$$\tan \varphi = \frac{-2\zeta\omega/\omega_n}{1 - \left(\dfrac{\omega}{\omega_n}\right)^2}. \tag{5.2-9}$$

Figure 5.2-1 shows a Bode plot of the amplitude ratio $Y(\omega)/\bar{u} = |G(i\omega)|$ and the phase angle $\varphi(\omega)$ as functions of the frequency ratio ω/ω_n. When the forcing frequency is near the natural frequency of the system and when ζ is small, the amplitude of the output becomes large and **resonance** is said to occur. For an undamped system ($\zeta = 0$), resonance occurs when $\omega = \omega_n$, and the amplitude becomes infinite. The frequency at which resonance occurs when $\zeta \neq 0$ is obtained by finding the ratio ω/ω_n where $|G(i\omega)|$ is maximized. This process yields the following expression for the **resonant frequency** ω_r:

$$\omega_r = \omega_n\sqrt{1 - 2\zeta^2}, \qquad 0 \le \zeta \le \frac{1}{\sqrt{2}}, \tag{5.2-10}$$

with the corresponding **resonant amplitude** ratio given by

$$|G(i\omega_r)| = \frac{1}{2\zeta\sqrt{1 - \zeta^2}}. \tag{5.2-11}$$

The resonant frequency is always slightly less than the natural frequency of the system, and the resonant amplitude is large for small damping ratios. Resonance does not occur for $\zeta > 1/\sqrt{2} = 0.707$.

For damping ratios greater than about 0.5, the amplitude ratio $|G(i\omega)|$ can be approximated by two straight lines. The **low-frequency asymptote** is the zero db line [because $\omega/\omega_n \to 0$ on the right-hand side of (5.2-8) yields a gain $Y(0)/\bar{u}$ of 0 db]. The **high-frequency asymptote** is a straight line at a slope of -40 db/decade, which is typical of a second-order system. The two asymptotes intersect at the **corner frequency** $\omega/\omega_n = 1$. Also note that the phase angle is $-90°$ at the corner frequency, regardless of the damping ratio, and that the phase angle approaches $-180°$ as $\omega/\omega_n \to \infty$.

From these results, we see that a lightly damped system will oscillate at a very large amplitude when the forcing frequency is near the natural frequency of the system, even if the forcing amplitude is small. To avoid

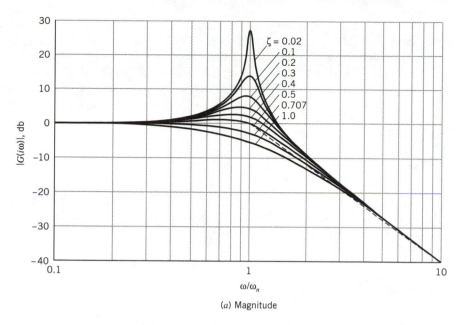

(a) Magnitude

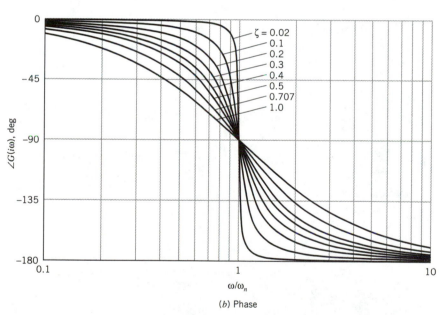

(b) Phase

Figure 5.2-1 Bode plot for $G(s) = \omega_n^2/(s^2 + 2\zeta\omega_n s + \omega_n^2)$.

this resonance phenomenon, armies do not "march" troops across bridges. Sometimes this lesson is learned in unexpected and tragic ways, as evidenced by the collapse of the Tacoma Narrows Bridge. The forcing functions in that instance were Kármán vortex trails that formed from strong winds which produced periodic vortices shed from the bridge. More recently, a new hotel's interior "skywalk" collapsed while people were dancing on it.

Composite Bode Plots

The logarithmic Bode plot representation of the transfer function allows us to develop "templates" for various element transfer functions and to combine them graphically, simply by adding the corresponding amplitude plots and phase plots. Specifically, if

$$G(s) = G_1(s)G_2(s) \ldots = r_1 e^{i\varphi_1} r_2 e^{i\varphi_2} \ldots$$
$$= (r_1 r_2 \ldots)e^{i(\varphi_1 + \varphi_2 + \cdots)}, \qquad \text{(5.2-12)}$$

then this transfer function has a magnitude (in db) of

$$20 \log_{10}[G(i\omega)] = 20 \log_{10}[G_1(i\omega)] + 20 \log_{10}[G_2(i\omega)] + \ldots \quad \text{(5.2-13)}$$

and an angle given by

$$\varphi = \varphi_1 + \varphi_2 + \ldots . \qquad \text{(5.2-14)}$$

To illustrate this process, consider a system with transfer function

$$G(s) = K \frac{1 + 10s}{(1 + s)(1 + 0.1s)} = G_1(s)G_2(s)G_3(s)G_4(s),$$

where the gain is $K = 10$ and

$$G_1(s) = K$$

$$G_2(s) = 1 + 10s$$

$$G_3(s) = \frac{1}{1 + s}$$

$$G_4(s) = \frac{1}{1 + 0.1s}.$$

The gain K is a positive constant. Therefore,

$$|G_1(i\omega)| = K, \qquad \varphi_1 = 0.$$

The linear transfer function $G_2(s)$ is of the form

$$G_2(s) = 1 + Ts$$

with $T = 10$. Thus,

$$|G_2(i\omega)| = \sqrt{1 + (\omega T)^2}, \qquad \varphi_2 = \tan^{-1}(\omega T).$$

Both the transfer functions $G_3(s)$ and $G_4(s)$ are first-order systems, which we have previously discussed, and they have transfer functions of the form

$$G_j(s) = \frac{1}{1 + Ts},$$

with $T = 1$ and $T = 0.1$, respectively. Thus, for $j = 3,4$

$$|G_j(i\omega)| = \frac{1}{\sqrt{1 + (\omega T)^2}}, \qquad \varphi_j = -\tan^{-1}(\omega T).$$

Figure 5.2-2 is a sketch of the approximate amplitude frequency response plots for the four individual transfer functions, as well as for the overall transfer function, obtained simply by adding the amplitudes. The phase angle plots for the individual transfer functions are not shown,

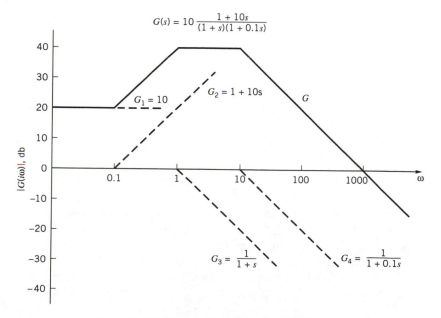

Figure 5.2-2 Composite Bode plot (amplitude).

but they also add to yield the phase angle plot for the overall transfer function.

The amplitude plot for $G_1(i\omega)$ is a horizontal line at $20 \log_{10}(K) = 20$ db, for $K = 10$, and the phase angle plot is a constant $0°$.

The approximate (asymptotic) amplitude plots for $G_3(i\omega)$ and $G_4(i\omega)$ both consist of a 0-db line, followed after their corresponding corner frequencies $\omega_c = 1/T$, by a line with a slope of -20 db/decade. The phase angle plots are asymptotic to $0°$ at $\omega = 0$, pass through $-45°$ at the corner frequency, and are asymptotic to $-90°$ at $\omega = \infty$.

The amplitude and phase angle plots for $G_2(i\omega)$ are mirror-image reflections of the first-order system plots, except for the corner frequency.

Minimum Phase Transfer Functions

For a positive parameter a, consider two transfer functions

$$G_1(s) = s + a$$

$$G_2(s) = s - a$$

which are zero at $s = -a$ and $s = +a$, respectively. Such roots are called **zeros** (see Chapter 7). The magnitudes and phase angles for $G_1(i\omega)$ and $G_2(i\omega)$ are given by

$$|G_1(i\omega)| = |G_2(i\omega)| = \sqrt{a^2 + \omega^2}$$

$$\varphi_1 = \tan^{-1}\left(\frac{\omega}{a}\right)$$

$$\varphi_2 = \tan^{-1}\left(\frac{\omega}{-a}\right) = \pi - \tan^{-1}\left(\frac{\omega}{a}\right).$$

From a Bode magnitude plot (corner frequency $\omega_c = a$, with high-frequency slope $= +20$ db/decade), one could only determine that the corresponding transfer function is $G(s) = s + \alpha$ and that $|\alpha| = a$. However, the Bode magnitude plot does not determine the sign of α. That is, the magnitude plot implies that $G(s) = G_1(s)$ or $G(s) = G_2(s)$, but does not distinguish which of the two is the case, since the magnitudes are the same. The two transfer functions are distinguished by the fact that they have different phase angle plots.

Note that $\varphi_1(\omega) < \varphi_2(\omega)$ for all $\omega > 0$. Also, as ω decreases from $\omega = \infty$ to $\omega = 0$, $\varphi_1(\omega)$ decreases from $\varphi_1(\infty) = \pi/2$ to $\varphi_1(0) = 0$, with a net change in phase angle of $\Delta\varphi_1 \triangleq \varphi_1(0) - \varphi_1(\infty) = -\pi/2$, while $\varphi_2(\omega)$ increases from $\pi/2$ to π, with a net phase change of $\Delta\varphi_2 = +\pi/2$. Thus, $G_1(s)$, representing a zero in the left-half s plane, has minimum phase and minimum net phase change for decreasing ω when compared with $G_2(s)$, which represents a zero in the right-half s plane.

Corresponding magnitude and phase angle results occur for second-order transfer functions representing complex conjugate zeros at $s = \sigma \pm i\omega_n$. The Bode magnitude plots are the same for both $\mathrm{Re}(s) < 0$ and $\mathrm{Re}(s) > 0$ and determine ω_n and $|\sigma|$, but not the sign of σ. Note that for second-order transfer function terms, the accurate (not asymptotic) Bode plot is required to determine $|\sigma|$. The two cases $\mathrm{Re}(s) < 0$ and $\mathrm{Re}(s) > 0$ are distinguished by having differing phase angle plots. $\mathrm{Re}(s) < 0$ produces minimum phase $\varphi(\omega)$ and minimum net phase change ($\Delta\varphi = -\pi$) as ω decreases from infinity to zero. $\mathrm{Re}(s) > 0$ has larger phase and a positive net phase change, $\Delta\varphi = +\pi$.

Similar results hold true for transfer functions of the form

$$G_3(s) = \frac{1}{s + b}$$

$$G_4(s) = \frac{1}{s - b},$$

in which the zeros of the denominator, called **poles** (see Chapter 7), are at $s = -b$ and $s = +b$, respectively, with $b > 0$. The magnitudes

$$|G_3(i\omega)| = |G_4(i\omega)| = \frac{1}{\sqrt{b^2 + \omega^2}}$$

are the same for $G_3(i\omega)$ and $G_4(i\omega)$ and imply that $G(s) = 1/(s + \beta)$, with $|\beta| = b$. But the magnitude plots do not determine the sign of β. The two transfer functions are distinguished by differing phase angles:

$$\varphi_3 = \tan^{-1}\left(\frac{-\omega}{b}\right) = -\tan^{-1}\left(\frac{\omega}{b}\right)$$

$$\varphi_4 = \tan^{-1}\left(\frac{-\omega}{-b}\right) = -\pi + \tan^{-1}\left(\frac{\omega}{b}\right),$$

with $\varphi_3 > \varphi_4$ for $\omega > 0$. As ω decreases from $\omega = \infty$ to $\omega = 0$, $G_3(i\omega)$ has a net phase change $\Delta\varphi_3 = +\pi/2$, whereas $\Delta\varphi_4 = -\pi/2$. Thus, $G_3(s)$, corresponding to a pole in the left-half s plane, has maximum phase $\varphi(\omega)$ and maximum net phase change $\Delta\varphi$ when compared with $G_4(s)$, which corresponds to a pole in the right-half s plane.

Corresponding results occur for complex conjugate poles at $s = \sigma \pm i\omega_n$. The $\mathrm{Re}(s) < 0$ case has maximum phase and maximum net phase change, for decreasing ω, compared with the $\mathrm{Re}(s) > 0$ case.

Historically, an IO system with transfer function $G(s)$ is termed a **minimum phase** system if and only if its transfer function has only poles or zeros with $\text{Re}(s) < 0$ and, possibly, $s = 0$. If any pole or zero, excluding $s = 0$, has $\text{Re}(s) \geq 0$, the system is called **nonminimum phase**. The phrase "minimum phase" is most often applied to the case where a system is asymptotically stable (poles in the left-half s plane). In this case, the phase angle of $G(s)$ is a minimum if all of the zeros have $\text{Re}(s) < 0$.

Minimum phase systems have two important properties, noted by Bode (Takahashi et al., 1970, p. 366). The first is that the Bode magnitude plot alone completely and unambiguously determines the transfer function $G(s)$. The same is also true for the phase angle plot alone, for a minimum phase system. But the fact that the magnitude plots can be easily constructed by using approximate straight-line asymptotes makes the magnitude plots easier to use for determining the corresponding transfer function, except that the actual magnitude plot is required in order to determine $|\sigma|$ for poles or zeros at $s = \sigma \pm i\omega_n$. The second property is that for a minimum phase system with n poles and m zeros, excluding poles or zeros at $s = 0$, the net phase angle change for $G(i\omega)$, as ω decreases from $\omega = \infty$ to $\omega = 0$, is given by $\Delta\varphi = (n - m)\pi/2$.

Experimental Determination of Transfer Functions

As implied by the previous discussion, the process of constructing Bode magnitude plots from composite transfer functions can be reversed, to determine a transfer function from its Bode plot. The amplitude plot for a transfer function can be approximated, for determining corner frequencies, by straight-line segments with slopes that are multiples of ± 20 db/decade. From the slopes and corner frequencies, the individual transfer function terms can be estimated, yielding the overall transfer function, for a minimum phase system. For nonminimum phase systems, the phase-angle plot must also be employed.

We see that Bode plots provide an experimental way to determine a model for an existing system. The procedure consists of applying a sine wave as an input to the system and recording the amplitude and phase angle of the residual output, for a range of input forcing frequencies. The resulting Bode plot is a graphical model of the system. To develop an IO model for the system, one can fit sections of the Bode plot to "templates" of Bode plots for typical elements (first- or second-order poles or zeros) and thus infer the estimated transfer function $G(i\omega)$. For more accurate results, one can perform a numerical least-squares curve fit for the coefficients in the transfer function polynomials.

5.3   UNDAMPED SECOND-ORDER SYSTEMS

If $\zeta = 0$ in (5.2-6), then the method of Section 5.2 for determining the residual response is not applicable, since the system is not asymptotically stable. However, the method of Section 5.1 used to examine a first-order system is still applicable. Since the undamped harmonic oscillator ($\zeta = 0$) is an important special case, we will examine it here in more detail. There are many problems in vibration control where the damping coefficient is sufficiently small to assume $\zeta = 0$. The solution phenomena that we will discover also apply to the case of very small damping, since the ultimate residual response will not be dominant until after a long time interval.

We will consider an undamped second-order system of the form

$$\ddot{y} + \omega_n^2 y = u(t). \tag{5.3-1}$$

This has an equivalent state-space form given by (4.3-9) with $\zeta = 0$ and $q_0 = 1$. It follows from (4.3-10) that the eigenvalues are given by $\lambda_1 = i\omega_n$, $\lambda_2 = -i\omega_n$. The state transition matrix

$$\Phi(t) = \begin{bmatrix} \phi_{11}(t) & \phi_{12}(t) \\ \phi_{21}(t) & \phi_{22}(t) \end{bmatrix} \tag{5.3-2}$$

is given by (4.3-11). The state-space solution (4.2-9) can be written as

$$\mathbf{x}(t) = \Phi(t)\mathbf{x}(0) + \int_0^t \Phi(t - \tau) \begin{bmatrix} 0 \\ 1 \end{bmatrix} u(\tau) \, d\tau. \tag{5.3-3}$$

With all initial conditions zero [$\mathbf{x}(0) = \mathbf{0}$], the state solution is of the form

$$\mathbf{x}(t) = \int_0^t \begin{bmatrix} \phi_{12}(t - \tau)u(\tau) \\ \phi_{22}(t - \tau)u(\tau) \end{bmatrix} d\tau. \tag{5.3-4}$$

Since the output is just the first component of the state, we obtain

$$y(t) = \int_0^t \phi_{12}(t - \tau)u(\tau) \, d\tau. \tag{5.3-5}$$

It follows then from (4.3-11) that the solution for the output is given by

$$y(t) = \frac{1}{\omega_n} \int_0^t \frac{e^{i\omega_n(t-\tau)} - e^{-i\omega_n(t-\tau)}}{2i} u(\tau) \, d\tau \tag{5.3-6}$$

or, equivalently,

$$y(t) = \frac{1}{\omega_n} \int_0^t \sin \omega_n(t - \tau) \, u(\tau) \, d\tau. \tag{5.3-7}$$

With the sinusoidal input $u(t) = \bar{u} \sin \omega t$, we obtain

$$y(t) = \frac{\bar{u}}{\omega_n} \int_0^t \sin \omega_n(t - \tau) \sin \omega\tau \, d\tau. \tag{5.3-8}$$

If we use a table of integrals, or the convolution formula of Laplace transform theory (see Exercise 5.5-8), this result can be integrated to yield

$$y(t) = \frac{\overline{u}}{\omega_n^2 - \omega^2} \left(\sin \omega t - \frac{\omega}{\omega_n} \sin \omega_n t \right). \tag{5.3-9}$$

It follows that in the undamped case, the harmonic oscillator will vibrate not at just the input frequency, but as a sum of both the input frequency and the natural frequency of the system. This allows for some rather strange-looking outputs.

For example, if ω is very close to ω_n (for example, $\omega = 1$ rad/sec, $\omega_n = 1.1$ rad/sec), the output is as shown in Figure 5.3-1. This result is known as the phenomenon of **beats**. This effect is often heard in the ringing of bells. The beat frequencies and lingering sound of Japanese and other oriental bells are well known. This represents an interesting application of the fast Fourier transformation methods, which we will discuss in the next section. Recently, with these methods, the dynamical characteristics of the Korean Yi-dynasty bell were studied by (Chung, Kong, and Yum, 1987).

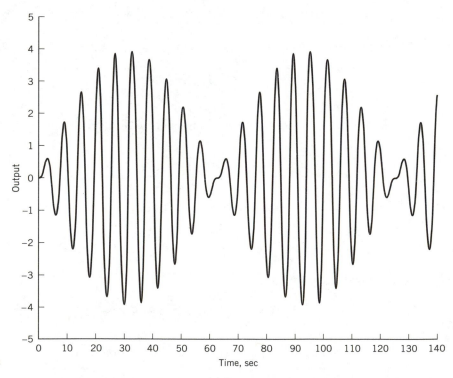

Figure 5.3-1 Solution to (5.3-1) with $\omega_n = 1$ rad/sec and $u(t) = 0.4 \sin(1.1t)$.

Note that as $\omega \to \omega_n$, we can obtain the solution from (5.3-9) by applying L'Hôpital's rules, yielding

$$y(t) = \frac{\bar{u}}{2\omega_n^2} \left(\sin \omega_n t - \omega_n t \cos \omega_n t\right). \qquad \textbf{(5.3-10)}$$

For $\omega = \omega_n = 1$ rad/sec, we obtain the result shown in Figure 5.3-2. This results in unstable motion termed **resonance**, clearly a phenomena to be avoided in most control applications.

If $k\omega + m\omega_n = 0$, where k and m are integers, the motion will repeat itself after a sufficient number of cycles (Nayfeh and Mook, p. 26). The number of cycles may be few or many, depending on the actual values used for ω and ω_n. For example, if $\omega = 6$ rad/sec and $\omega_n = 1$ rad/sec, the output illustrated in Figure 5.3-3 is found to repeat itself after every π cycles. However, with $\omega = 3.35$ rad/sec and $\omega_n = 1$ rad/sec, the output illustrated in Figure 5.3-4 has not repeated itself after 70 sec. In both of these plots, $\bar{u} = 0.4$. The state-space plot of this same system (over a much longer time period) is illustrated in Figure 5.3-5.

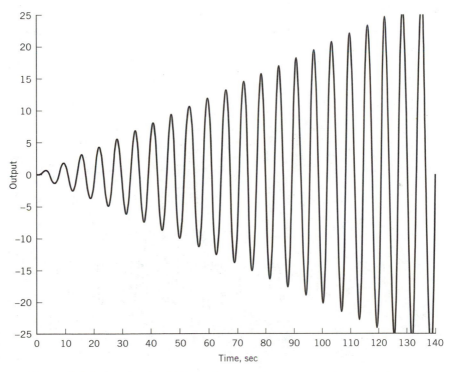

Figure 5.3-2 Solution to (5.3-1) with $\omega_n = 1$ rad/sec and $u(t) = 0.4 \sin(t)$.

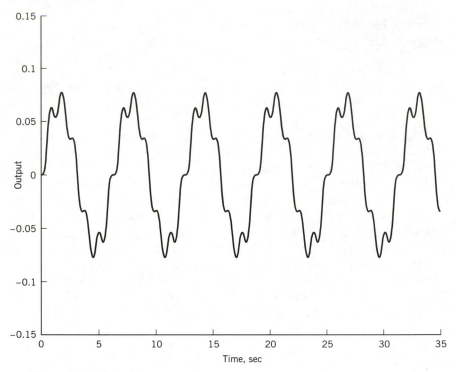

Figure 5.3-3 Solution to (5.3-1) with $\omega_n = 1$ rad/sec and $u(t) = 0.4\sin(6t)$.

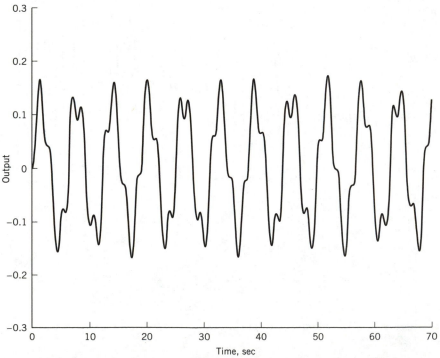

Figure 5.3-4 Solution to (5.3-1) with $\omega_n = 1$ rad/sec and $u(t) = 0.4\sin(3.35t)$.

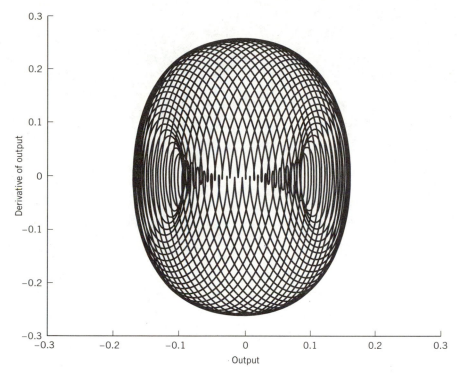

Figure 5.3-5 A state-space plot $(y, \dot{y})$ of the system used to generate Figure 5.3-4.

5.4 FOURIER ANALYSIS

The fact that the sum of a few periodic functions can have a very complicated form would seem to imply that it would be impossible to ever determine the underlying simple structure of an output signal by measuring only its time history. This inverse problem is important in many applications where it is important to know what frequencies of a lightly damped flexible system have been "excited" by (certain or uncertain) periodic inputs. Fortunately, it is possible to solve the inverse problem numerically. Estimates for the underlying sine-cosine structure of any periodic output can be determined using the **fast Fourier transform** (**FFT**) to be discussed at the end of this section.

Fourier Series

Any periodic function $y(t)$ of period T that is piecewise continuous and has a piecewise continuous derivative with respect to time on the interval $0 \le t \le T$ can be represented (Kreyszig, 1983, p. 472) by the **Fourier series**

$$y(t) = \frac{1}{T} \int_0^T y(\tau) \, d\tau + \sum_{n=1}^{\infty} \left[A_n \cos\left(\frac{2\pi n}{T} t\right) + B_n \sin\left(\frac{2\pi n}{T} t\right) \right], \quad (5.4\text{-}1)$$

where

$$A_n = \frac{2}{T} \int_0^T y(\tau) \cos\left(\frac{2\pi n}{T}\tau\right) d\tau$$

$$B_n = \frac{2}{T} \int_0^T y(\tau) \sin\left(\frac{2\pi n}{T}\tau\right) d\tau.$$

(5.4-2)

Using Euler's equations

$$e^{i\theta} = \cos\theta + i\sin\theta$$

$$e^{-i\theta} = \cos\theta - i\sin\theta,$$

(5.4-3)

we can obtain the following equivalent representation of the Fourier series:

$$y(t) = \frac{1}{T} \int_0^T y(\tau) \, d\tau + \frac{1}{2} \sum_{n=1}^{\infty} \left[(A_n - iB_n) \exp\left(i\frac{2\pi n}{T}t\right) \right.$$

$$\left. + (A_n + iB_n) \exp\left(-i\frac{2n\pi}{T}t\right) \right], \quad (5.4\text{-}4)$$

where

$$A_n - iB_n = \frac{2}{T} \int_0^T y(\tau) \exp\left(-i\frac{2\pi n}{T}\tau\right) d\tau$$

$$A_n + iB_n = \frac{2}{T} \int_0^T y(\tau) \exp\left(i\frac{2\pi n}{T}\tau\right) d\tau.$$

(5.4-5)

If we now substitute (5.4-5) into (5.4-4), after some rearrangement, we obtain a more compact equivalent representation of the Fourier series

$$y(t) = \sum_{n=-\infty}^{\infty} C_n \exp\left(i\frac{2\pi n}{T}t\right),$$

(5.4-6)

where

$$C_n = \frac{1}{T} \int_0^T y(\tau) \exp\left(-i\frac{2\pi n}{T}\tau\right) d\tau.$$

(5.4-7)

In this form, Equations (5.4-6)–(5.4-7) are known as the **continuous Fourier transform pair**, with C_n being the **continuous Fourier transform** of $y(t)$. If we take the limit of (5.4-6) and (5.4-7) as $T \to \infty$, the continuous Fourier transform pair is known simply as the **Fourier transform pair**, with C_n being the **Fourier transform** of $y(t)$.

Discrete Fourier Transform

If we are interested in approximating $y(t)$ over a unit interval $[0,1]$, then this can be done using the **discrete Fourier transform** (Press et al., 1988, p. 404), which is of the form

$$y(t) \approx \sum_{n=1}^{N} C_n \exp\left(i\frac{2\pi n}{T}t\right), \qquad (5.4\text{-}8)$$

where

$$C_n = \frac{1}{N}\sum_{j=0}^{N-1} y(t_j)\exp(-i2\pi n t_j) \qquad (5.4\text{-}9)$$

and

$$t_j = \frac{i}{N}, \qquad 0 \le j \le N - 1. \qquad (5.4\text{-}10)$$

That is, t_j forms a uniformly spaced grid on the unit interval.

The discrete Fourier transformation pair is a sum approximation to the continuous Fourier transformation pair. It has the advantage that it can be implemented directly on a digital computer. That is to say, given an output reading $y(t_j)$ at a finite number of points over an interval, the coefficients for a Fourier series approximation of this output can be obtained directly from (5.4-9). As the number N of data points increases, the discrete Fourier transformation should converge to the continuous Fourier transform, provided that $y(t)$ is continuous and there are no roundoff errors in the calculations.

There are numerical methods (Press et al., 1988, p. 407) that will compute the discrete Fourier transformation more efficiently than a direct application of (5.4-9). Such methods are referred to as a **fast Fourier transformation** (FFT) and can be found in several software packages such as MATHCAD. The use of an FFT routine is usually quite simple. For example, consider the output shown in Figure 5.3-3. Suppose we created a data file with a sample of this output taken 512 times over a time interval of $T = 8\pi$ sec. Supplying this data file to the MATHCAD FFT routine provides the information shown in Figure 5.4-1. Each spike on this graph represents the magnitude of the coefficient C_n. Each tick mark is $2\pi/T$ rad/ sec. Thus, the spikes at 4 ticks and 24 ticks represent frequencies of 1 and 6 rad/sec, respectively. We see that in this instance, the FFT routine exactly recovers the structure of the original components of the output signal. Furthermore, the individual components of C_n can be printed out. From this we can determine an analytical expression for the approximate solution. In this case, we obtain a result very close to the actual solution,

Figure 5.4-1 Output from FFT of data used to plot Figure 5.3-3.

which from (5.3-9) is

$$y(t) = -0.01143 \sin(6t) + 0.06857 \sin(t). \tag{5.4-11}$$

5.5 EXERCISES

5.5-1 Using numerical methods, plot the output response $y(t)$ of the systems
(a) $\dot{y} + 2y = u$, $y(0) = 0$
(b) $\ddot{y} + 2\dot{y} + 4y = u$, $y(0) = \dot{y}(0) = 0$
for all the following inputs:
   (i) $u = 1$
   (ii) $u = \sin t$
   (iii) $u = \sin 2t$
   (iv) $u = \sin 10t$.
Compare for each case the results obtained with the analytical residual response $y_r(t)$ given by (5.2-3).

5.5-2 For the following system, using state transition matrix methods with $\mathbf{x}(0) = \mathbf{0}$, determine the residual response $y_r(t)$ when
(a) $u(t) = 3$
(b) $u(t) = 3 \sin 2t$

$$\begin{bmatrix} \dot{x}_1 \\ \dot{x}_2 \end{bmatrix} = \begin{bmatrix} 1 & 1 \\ -4 & -3 \end{bmatrix} \begin{bmatrix} x_1 \\ x_2 \end{bmatrix} + \begin{bmatrix} 0 \\ 1 \end{bmatrix} u$$

$$y = [1 \quad 2] \begin{bmatrix} x_1 \\ x_2 \end{bmatrix}.$$

5.5-3 One one sheet of semilog graph paper, sketch $|G_j(i\omega)|$ and $\angle G_j(i\omega) = \varphi$, for $j = 1, \ldots, 4$, in a Bode plot for an IO system with the transfer function

$$G_4(s) = \frac{10}{s(0.5s + 1)} = G_1(s)G_2(s)G_3(s),$$

where $G_1(s) = 10$, $G_2(s) = 1/s$, and $G_3(s) = 1/(0.5s + 1)$.

5.5-4 For the system given by

$$\dot{\mathbf{x}} = \mathbf{A}\mathbf{x} + \mathbf{B}u$$

$$y = \mathbf{C}\mathbf{x}$$

with the **A** matrices given in Exercise 3.6-4 and $\mathbf{B} = [0 \quad 1]^T$, $\mathbf{C} = [1 \quad 0]$, determine the residual response, $y_r(t)$, using state transition matrix methods, when $u = \sin \omega t$ and $\mathbf{x}(0) = [0 \quad 0]^T$. Compare this result with that obtained using (5.2-3)–(5.2-5).

5.5-5 Plot $\angle G(i\omega) = \varphi$ in a Bode plot for the system used in Figure 5.2-2.

5.5-6 For an IO system with the transfer function

$$G(s) = \frac{16}{s^2 + 4s + 16},$$

develop formulas for the Bode plot low- and high-frequency amplitude asymptotes as functions of the sinusoidal forcing frequency ω. From these formulas, determine the intersection point of the asymptotes.

5.5-7 The shaft of the armature-controlled dc motor of Exercise 1.5-16 has an unknown moment of inertia J. The motor is modeled by (1.3-48) and all other constants are known. Design an experiment that will determine J. Provide a list of the equipment required, along with a sketch of the experimental setup. Also provide a brief description of the data to be taken and how these data are to be used to calculate J.

5.5-8 The **convolution theorem** of Laplace transformation theory states that

$$\mathcal{L}\left[\int_0^t f(t - \tau)g(\tau)\, d\tau \right] = \mathcal{L}[f(t)]\mathcal{L}[g(t)],$$

where $\mathcal{L}(\cdot)$ represents the Laplace transform of the argument. Given that

$$\mathcal{L}(\sin \alpha t) = \frac{\alpha}{s^2 + \alpha^2},$$

apply the convolution theorem to (5.3-8) to obtain (5.3-9). *Hint:* Use partial fractions to express the product of two Laplace transforms as a sum of two Laplace transforms whose inverse gives (5.3-9).

5.5-9 Reproduce the results obtained in Figure 5.3-3 and then make a state-space plot $(y, \dot{y})$ of the trajectory over a sufficiently long time period to see the repeating pattern. Examine plots for other choices of ω and ω_n.

5.5-10 Substitute (5.4-7) into (5.4-6) and obtain the following limit of this expression as $T \to \infty$:

$$y(t) = \frac{1}{2\pi} \int_0^\infty \exp(i\omega t) \left[\int_{-\infty}^\infty y(\tau) \exp(-i\omega\tau) \, d\tau \right] d\omega, \quad \textbf{(5.5-1)}$$

where

$$\omega \overset{\Delta}{=} \lim_{T \to \infty} \frac{2\pi n}{T}.$$

This integral is known as the **Fourier integral**. Show that the Fourier integral can also be written as the **Fourier transform pair**

$$y(t) = \frac{1}{2\pi} \int_0^\infty F(i\omega) \exp(i\omega t) \, d\omega \qquad \textbf{(5.5-2)}$$

$$F(i\omega) = \int_0^\infty y(\tau) \exp(-i\omega\tau) \, d\tau. \qquad \textbf{(5.5-3)}$$

The term $F(i\omega)$ is known as the **Fourier transform** of $y(t)$ and (5.5-2) is the **inverse Fourier transform**.

Chapter 6

Output Feedback Control

6.1 ERROR FEEDBACK AND CLASSICAL CONTROL STRUCTURES

We have examined a number of diverse dynamical systems in terms of a set of N_x first-order state-space differential equations with control inputs. In most cases, we have discovered that these systems could be reduced to an IO system in terms of a single N_x-order differential equation. Indeed, the number of practical systems that can be reduced to IO form is large. It is for this reason that we devote this chapter to IO systems and to the classical methods for dealing with them. In Chapter 8 the IO assumption will be dropped in favor of the more general state-space representation and methods. Since IO systems represent a subset of linear systems, the design for them is simpler than that for the general state-space case.

In previous chapters we have examined IO systems subjected to initial conditions and step, sine, and uncertain inputs. For some systems, the output response to these inputs may already be acceptable. For example, a spring-mass-damper system, which is asymptotically stable, may already yield a satisfactory response for inputs of interest. For other systems, the output response may not be acceptable for any of these inputs. For example, the inverted pendulum system and the magnetic suspension system in Section 1.3 are both unstable and therefore require a controller to achieve a stable equilibrium.

The objective of this chapter is to demonstrate how system response can be modified for IO systems, by a feedback controller, so that acceptable system response can be obtained. For a point of reference, we will assume that our general **control objectives** are to have the system output follow a low-frequency input (in the limit, a step input), filter out high-frequency inputs, and be relatively insensitive to uncertain inputs.

These requirements may not be necessary or even desired for all control applications, but they do give us a way of characterizing a large class of applications as well as some concrete way of comparing systems.

Error Feedback

A very basic concept in control design is that of **error feedback**. This idea has an intuitive basis, which can be illustrated by considering the problem of controlling the speed $\omega(t)$ of a rotating object so that $\omega(t) \rightarrow r(t)$, where $r(t)$ is the desired rotational speed. One of the most famous applications to a problem of this type was made by James Watt, in his pioneering design of a governor for his steam engine.

For illustration purposes, we consider a rotating object, having a given moment of inertia J, with an applied torque Γ that we are free to choose. The system is modeled as

$$J\dot{\omega} = \Gamma,$$

or, equivalently, as

$$\dot{\omega} = u,$$

where $u = \Gamma/J$ is the control input and $\omega(t)$ is the output. The objective is to have the output $\omega(t)$ track some command input $r(t)$. If $r(t) = \bar{\omega}$ where $\bar{\omega}$ is the magnitude of a step input, then the control input function $u(\cdot)$ must be able to stabilize the system about a new equilibrium point $\bar{\omega}$. We denote the **error** by

$$e(t) = r(t) - \omega(t)$$

and its Laplace transform by $E(s) = \mathcal{L}\{e(t)\}$.

For this simple system, two qualitative observations lead to a design of an effective controller. First, since the output $\omega(t)$ depends on the control input $u(\cdot)$, through the transfer function, it seems reasonable (perhaps even axiomatic) that if a difference exists between the desired output (i.e., the command input) and the current output, then the control $u(\cdot)$ should depend on this difference. Second, we note that if the error difference is positive, because ω is too small, then a positive control should be used to increase $\omega(t)$ and thus decrease the error difference. Similarly, if the error is negative, then a negative control should be used to decrease $\omega(t)$ and move the error toward zero. The simplest possible control would be a **proportional controller** with the control determined by a positive constant K times the error e,

$$u = Ke = K[r - \omega],$$

which is illustrated in Figure 6.1-1 in terms of transfer functions.

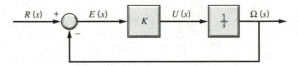

Figure 6.1-1 Angular velocity error feedback controller.

With a proportional controller, the differential equation for the controlled system is

$$\dot{\omega} = K[r - \omega].$$

Since the controlled system is stable, under a constant command input $r(t) = \bar{\omega}$ we obtain the steady-state solution $\omega(t) \rightarrow \bar{\omega}$. For a time-varying input, the residual response would have to be obtained from the general solution

$$\omega(t) = \omega(0)e^{-Kt} + \int_0^t e^{-K[t-\tau]} r(\tau)\, d\tau$$

in the limit as $t \rightarrow \infty$. The nature of the residual response clearly depends on $r(t)$. Another way to examine the effect of $r(t)$ is with the error equation

$$\dot{e} = \dot{r} - Ke,$$

which has the solution

$$e(t) = e(0)\exp(-Kt) + \int_0^t \exp(-K[t-\tau])\dot{r}(\tau)\, d\tau.$$

It follows that

$$|e(t)| \leq |e(0)\exp(-Kt)| + \left| \int_0^t \exp(-K[t-\tau])\dot{r}(\tau)\, d\tau \right|.$$

If $\dot{r}(t)$ is bounded, that is, $|\dot{r}(t)| \leq \dot{r}_m$ for all $t > 0$, it follows that

$$|e(t)| \leq |e(0)\exp(-Kt)| + \frac{\dot{r}_m}{K}|1 - \exp(-Kt)|.$$

Thus, in the limit as $t \rightarrow \infty$,

$$|e| \leq \frac{\dot{r}_m}{K}.$$

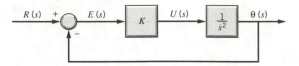

Figure 6.1-2 Angular position error feedback controller.

The error will remain small, provided that $\dot{r}_m/K$ is small. In particular, if $r(t) = \overline{\omega}$, then $\dot{r}(t) \equiv 0$ and $e(t) \to 0$ as $t \to \infty$. In summary, if $\dot{r}(t)/K$ is small, then the proportional controller is a stabilizing controller that will track the command input. The accuracy of tracking is determined by the magnitude of $\dot{r}_m/K$. Generally, we are able to vary K so that we will have control of the return time (or eigenvalue) of the controlled system.

Of course, the design process is not always this simple. As an example, consider the related problem of controlling not the speed but the angular position of the rotating object so that the object automatically points in a specified direction. Examples include pointing a telescope, "homing" a radio tuner dial to a particular frequency, and many others.

We consider the same system as before, but in terms of angular position $\theta(t)$. Thus, we have

$$\ddot{\theta} = u,$$

where $u = \Gamma/J$. The objective is to have the output $\theta(t)$ follow a command input $r(t)$. If, as before, we use the concept of error feedback as illustrated in Figure 6.1-2, then the control is given by

$$u = Ke = K[r(t) - \theta],$$

so that the overall system is governed by

$$\ddot{\theta} = K[r(t) - \theta].$$

Under a constant command input $r(t) = \overline{\theta}$, this control has the right tendency, namely, to drive $\theta(t)$ toward $\overline{\theta}$. However, without any damping the system will oscillate about $\overline{\theta}$. This difficulty could be overcome, of course, by adding some damping to the control algorithm. In the course of pursuing this concept, we will present several classical feedback control structures that involve feeding back the output and adding damping, along with some other ideas that have proved effective in various applications.

Classical Feedback Control Structures

Figure 6.1-3 illustrates a general output feedback control structure applied to an IO system, with any disturbance inputs ignored. The transfer function for the original or **primary** system (often referred to as the **plant**) is designated by $G_P(s)$. Note in Figure 6.1-3 that the input to $G_P(s)$ is still $U(s)$. The idea is to construct a controller $G_C(s)$ whose output $U(s)$ will drive the system $G_P(s)$ so that its output $Y(s)$ is acceptable. From the diagram, we see that $E(s)$ is the difference between a **scaled command input** $K_S R(s)$ and some **output feedback** function $H(s)Y(s)$. When $K_S = H(s) = 1$, then $E(s)$ is just the difference between the input and the output which we previously designated as the error. The gain K_S on the input is a **scaling factor** that is useful in obtaining a desired steady-state relationship between the input and the output. The output feedback $H(s)$ is used to modify the output signal to compensate for inadequacies in using just the output signal itself.

We say that the controller is using **error feedback** when the output feedback transfer function is given by $H(s) = 1$. Since in this case $E(s) = K_S R(s) - Y(s)$, we will now use the term "error" to designate the difference between the scaled input and the output. For proportional control the **controller** transfer function is $G_C(s) = K$, where K is a **gain** multiplier to amplify (or diminish) the effect of the error signal. When error feedback with proportional control, as in Figure 6.1-2, does not satisfy the design requirements, alternate control structures must be employed for $G_C(s)$ and $H(s)$. This process is called "compensation," and $G_C(s)$ is often referred to as a **cascade compensator**, with $H(s)$ being termed a **feedback compensator**.

We can now calculate an overall transfer function for the **controlled system** shown in Figure 6.1-3. In other words, we will determine the transfer function between the input $R(s)$ and the output $Y(s)$. It follows from Figure 6.1-3 that

$$Y(s) = G_C(s)G_P(s)[K_S R(s) - H(s)Y(s)]. \qquad (6.1\text{-}1)$$

Thus,

$$G(s) \triangleq \frac{Y(s)}{R(s)} = \frac{K_S G_C(s)G_P(s)}{1 + G_C(s)G_P(s)H(s)} = \frac{Q(s)}{P(s)}. \qquad (6.1\text{-}2)$$

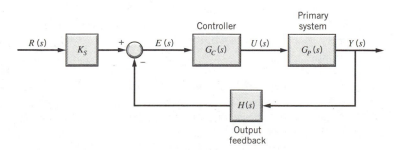

Figure 6.1-3 Block diagram for output feedback control.

The controlled system is equivalent to an IO system of the form

$$P(s)Y(s) = Q(s)R(s). \tag{6.1-3}$$

In particular, requiring that the controller and the feedback be representable in terms of Laplace transform transfer functions ensures that, like the original open-loop system $G_P(s)$, the closed-loop system $G(s)$ remains linear.

The design problem is to pick K_S, $G_C(s)$, and $H(s)$ so that a suitable response is achieved when the system is subjected to various inputs. If the system is to be a regulator, a typical command input is a constant input (i.e., a step input) that will transfer a stable system to a new equilibrium point. For this case, we want the steady-state output to be some specified multiple of the input (e.g., to accommodate different units between the input and output signals). We will also want the response to satisfy certain transient performance specifications (such as percent overshoot, return time, settling time, and so on). If the system is to be a tracking servo-system, rather than a regulator with a fixed set-point, then a typical command input is a sinusoidal input that, for stable systems, will produce a sinusoidal residual output at the forcing frequency, with an amplitude and phase shift that depend on the input. Since we want the system to "track" low-frequency inputs and to ignore high-frequency inputs associated with noise, we want the output/input amplitude ratio to be near one at low frequencies and want the amplitude to be attenuated at high frequencies. Finally, since any control system will be subject to uncertain inputs, we will want to be assured that any bounded uncertain input will produce only a displacement from equilibrium that lies within specified output bounds. The magnitude of these bounds will depend, in part, on the closed-loop eigenvalues.

We summarize typical **design objectives** as follows:

1. Asymptotic stability (with suitable closed-loop eigenvalues to satisfy transient performance specifications and requirements on uncertain inputs).
2. Steady-state output equal to a given ratio of the input.
3. Minimum sensitivity to uncertain inputs.
4. Bode plot amplitude ratio of 1 (0 db) at low frequencies.
5. Bode plot amplitude attenuation (at least, -20 db/decade) at high frequencies.

To accomplish these objectives, classical control theory focuses on the choice of the parameters in various possible structures for the controller and feedback transfer functions $G_C(s)$ and $H(s)$. We will first consider the error feedback case where $H(s) = 1$, letting $e(t)$ denote the input to the controller, with output $u(t)$. The following types of controllers are typical.

Proportional Error Control (P)

$$u(t) = Ke(t), \qquad G_C(s) = K. \tag{6.1-4}$$

This type of control action produces a control that is proportional to the error signal. As we have seen, this simplest possible controller may or may not be effective, depending on the application.

The gain K is typically provided by an electrical or mechanical device. For example, for some applications the gain may be provided by a power amplifier, whereas in other applications the gain may be provided by a field-controlled dc motor with small inductance. In any real application, there will always be a finite inductance, mass, and so on. This, in turn, will produce a finite rather than instantaneous controller response time, so that only approximate proportional control is obtained, as represented through the transfer function

$$G_C(s) = \frac{K}{1 + \tau_s s}, \qquad 0 < \tau_s << 1. \tag{6.1-5}$$

Generally, however, for such devices, τ_s is sufficiently small so that (6.1-4) may be used as the transfer function for design purposes.

Integral Error Control (I)

$$u(t) = \frac{1}{\tau_i} \int_0^t e(\tau) \, d\tau, \qquad G_C(s) = \frac{1}{\tau_i s}. \tag{6.1-6}$$

Integral control is similar to proportional control, but it builds in magnitude as long as the error persists. This property is frequently needed to achieve a zero steady output error due to an uncertain system **bias** (constant uncertain input). Note, however, that if the error $e(t)$ changes sign, say, from positive to negative, the value of the integral control will not change sign until some time later, when the accumulated positive value of the integral at the time of the error sign change has been reduced by the integral of the negative error. In practice, control designers may provide logic to reset the integrator to zero whenever the error changes sign or when some other condition (such as resetting a regulator set point) would cause the integrator to generate an incorrect response (or too large a response) over an unacceptable time period. Such a **reset control** is no longer a linear control but is, instead, a switching or relay control, which is nonlinear.

Pure integral control cannot be achieved in a physical system (because some damping, resistance, and the like always exists), but rather it is approximated in physical systems with the transfer function

$$G_C(s) = \frac{1}{\epsilon + \tau_i s}, \qquad 0 < \epsilon << 1. \tag{6.1-7}$$

A good integrator will, of course, have ϵ near zero so that for design purposes (6.1-6) may be used.

Derivative Error Control (D)

$$u(t) = \tau_d \frac{de(t)}{dt}, \qquad G_C(s) = \tau_d s. \tag{6.1-8}$$

Derivative control produces a control that is proportional to the rate of change of the error. If the error is increasing, the control is positive and vice versa. If the error is constant, the control is zero. Derivative control is used to introduce "damping" into a system and is typically applied in conjunction with other control elements, since derivative control alone will not force the error to zero. However, a pure differentiator could never be used as part of a feedback device. Any noise on top of the input signal would be differentiated and, as a consequence, the output would be chaotic even if the input is "constant." Differentiation is an inherently noisy process. However, as we will learn later, approximate derivative control can be implemented by devices that have a transfer function of the form

$$G_C(s) = \frac{\tau_d s}{1 + \tau_s s}, \qquad 0 < \tau_s << 1. \tag{6.1-9}$$

The differentiator introduces an interesting paradox. In practice, one would not use a differentiator that is too near perfection ($\tau_s \rightarrow 0$), since this would introduce the unpredictable effects mentioned previously. However, a less than perfect differentiator results in a different (but predictable) effect as given by (6.1-9). Because of this, (6.1-9) is usually employed to represent derivative control in actual design. However, for conceptual or preliminary design purposes (6.1-8) is often used.

Proportional + Integral + Derivative Control (PID)

A PID controller is a combination of proportional, integral, and derivative control types that, for the ideal cases, is given by

$$G_C(s) = K + \frac{1}{\tau_i s} + \tau_d s. \tag{6.1-10}$$

Often, the gain K (which may be the gain of a power amplifier) is factored out of expressions like (6.1-10). In doing so, the PID transfer function can be written alternately as

$$G_C(s) = K\left[1 + \frac{a}{s} + bs \right],$$

where $a = 1/(\tau_i K)$ and $b = \tau_d/K$. Proportional (P), proportional + integral (PI), and proportional + derivative (PD) controls are included as special cases. Commercial PID controllers are widely available and are used in many control applications.

Lead or Lag Control

One other control structure that we will consider in this chapter is **lead** or **lag** controllers, with the transfer function

$$G_C(s) = \frac{1 + \tau_e s}{1 + \tau_a s}. \qquad \text{(6.1-11)}$$

The names come from the phase angle frequency domain properties of this transfer function, which we will investigate later.

EXAMPLE 6.1-1   **Undamped Linearized Inverted Pendulum**

To illustrate the effect of various controllers, we will again consider the problem of pointing a rotating object. However, to make the system more interesting, we will consider an object that has a natural preferred direction of its own. In particular, we will consider an undamped inverted pendulum (consisting of a uniform rod of mass m and length 2ℓ) acted on by a torque due to gravity and by an applied torque Γ that we are free to choose. The equation of motion, as given by (1.3-49), is

$$J\ddot{\theta} = \Gamma + mg\ell \sin \theta,$$

where J is the rotational moment of inertia and θ is the angle measured positive from the upward vertical. We wish to design a controller for $\Gamma(\cdot)$ so that the pendulum points in a commanded direction $\bar{\theta}$, where $\bar{\theta}$ is a small angle, say, within $\pm 15°$ of the upward vertical. Defining the control variable as $u \triangleq \Gamma/J$ and defining $\omega_n^2 \triangleq mg\ell/J$, we can write the linearized equation of motion for small θ as

$$\ddot{\theta} - \omega_n^2\theta = u.$$

Since we are interested in how different control methods handle uncertain inputs, we will add some uncertainty to the system through an uncertain input v. Our system is then of the form

$$\ddot{\theta} - \omega_n^2\theta = u + v.$$

An error feedback control structure for this system is illustrated in Figure 6.1-4. In general, we seek a controller so that the output tracks the input, that is, $\theta(t) \rightarrow r(t)$ as $t \rightarrow \infty$. A good test function for this is a step input. Thus, we will focus on the case where $r(t) \equiv \bar{\theta} > 0$ and on the problem of stabilizing the system about a new equilibrium near the vertical. We will consider various combinations of the P, I, and D control actions.

For proportional error feedback control $[G_C(s) = K, H(s) = 1 \Rightarrow u = (K_s r - \theta)K]$, the equation of motion becomes

$$\ddot{\theta} + (K - \omega_n^2)\theta = K_s K r + v,$$

with the characteristic equation

$$\lambda^2 + K - \omega_n^2 = 0$$

yielding eigenvalues

$$\lambda_1 = \sqrt{\omega_n^2 - K}, \qquad \lambda_2 = -\sqrt{\omega_n^2 - K}.$$

The controlled equilibrium (which will exist unless $K = \omega_n^2$) under the step input $r = \bar{\theta}$ and a constant uncertainty (that is, bias) $v = \bar{v}$ is at

$$\hat{\theta} = \frac{K_s K}{K - \omega_n^2}\bar{\theta} + \frac{\bar{v}}{K - \omega_n^2}.$$

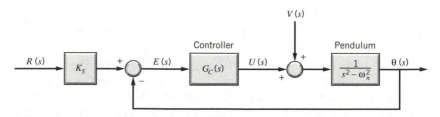

Figure 6.1-4 Inverted pendulum feedback controller.

We see that for large K, the effect of the bias can be made small. If $\bar{v} = 0$, then by choosing $K_S = (K - \omega_n^2)f/K$, we obtain $\hat{\theta} = f\bar{\theta}$. That is, the equilibrium solution can be made equal to some factor f times the input. Unfortunately, the controlled equilibrium is not asymptotically stable. For $K < \omega_n^2$, the eigenvalues are real, nonzero, and of opposite sign. Thus, the response will diverge from the controlled equilibrium point. For the other case, where $K > \omega_n^2$, the eigenvalues are imaginary, $\lambda = \pm i\omega$ with $\omega^2 = K - \omega_n^2$, and the solution will oscillate about $\hat{\theta}$, with no damping. We conclude that proportional control by itself cannot asymptotically stabilize the system.

Derivative control, as previously noted, should not be used by itself. However, we can still investigate its effects. For derivative error feedback control $[G_C(s) = \tau_d s, H(s) = 1 \Rightarrow u = (K_S \dot{r} - \dot{\theta})\tau_d]$, the equation of motion is

$$\ddot{\theta} + \tau_d \dot{\theta} - \omega_n^2 \theta = K_S \tau_d \dot{r} + v.$$

Under the step input $r(t) \equiv \bar{\theta}$ and constant uncertainty $v(t) \equiv \bar{v}$, the equilibrium occurs at $\hat{\theta} = -\bar{v}/\omega_n^2$ (for $\omega_n \neq 0$), regardless of $\bar{\theta}$. However, since the eigenvalues

$$\lambda = \tfrac{1}{2}\{-\tau_d \pm \sqrt{\tau_d^2 + 4\omega_n^2}\}$$

are real, with one positive and the other negative, the solution diverges and we conclude that derivative control, even if it could be implemented, would not asymptotically stabilize the system.

For integral control $[G_C(s) = 1/(\tau_i s), H(s) = 1 \Rightarrow u = (1/\tau_i)\int(K_S r - \theta)\,dt]$, the equation of motion, after differentiating once to remove the integral and to produce an ordinary differential equation, is given by

$$\dddot{\theta} - \omega_n^2 \dot{\theta} + \frac{1}{\tau_i}\theta = \frac{K_S}{\tau_i}r + \dot{v}.$$

Note that integral control increases the order of the system, from second- to third-order. Under the step input $r = \bar{\theta}$ and constant uncertainty $v = \bar{v}$, the controlled equilibrium occurs at $\hat{\theta} = K_S\bar{\theta}$. By choosing $K_S = f$, the output is made a specified factor f times the input. Thus, with integral control, not only do we obtain the specified ratio between input and output but also the effect of any bias on the system will be eliminated from the output. Note, however, that for this system there is always, at least, one eigenvalue λ with $\text{Re}(\lambda) > 0$. Thus, we conclude that integral control by itself cannot asymptotically stabilize this system.

This system can be stabilized, for example, by using proportional + derivative error feedback (PD) control $[G_C(s) = K + \tau_d s, H(s) = 1 \Rightarrow u =$

$K(K_s r - \theta) + (K_s \dot{r} - \dot{\theta})\tau_d]$. With PD control the equation of motion for our system becomes

$$\ddot{\theta} + \tau_d \dot{\theta} + (K - \omega_n^2)\theta = K_s K r + K_s \tau_d \dot{r} + v.$$

With $r(t) = \bar{\theta}$ and $v(t) = \bar{v}$, the new equilibrium occurs at

$$\hat{\theta} = \frac{K_s K}{K - \omega_n^2}\bar{\theta} + \frac{\bar{v}}{K - \omega_n^2}.$$

Again, if $\bar{v} = 0$, choosing $K_s = (K - \omega_n^2)f/K$, we obtain the desired relationship between input and output. Moreover, the effect of a nonzero bias can be made small by using sufficiently large K. To achieve a zero steady output error, with a nonzero bias and a finite gain K, we could add integral control. But we leave this task for later.

The eigenvalues of the system are

$$\lambda = \tfrac{1}{2}\{-\tau_d \pm \sqrt{\tau_d^2 - 4(K - \omega_n^2)}\},$$

thus the system will be asymptotically stable when

$$\tau_d > 0$$
$$K > \omega_n^2.$$

The ideas presented here are simple and conceptual. Idealized controller elements were used in the above example. As we previously noted, the pure (D) controller is not physically realizable, whereas pure (P) and (I) controllers for all practical purposes are. One will often design controllers in terms of idealized (P), (I), and (D) elements, with transfer functions given by (6.1-4), (6.1-6), and (6.1-8), with knowledge of the fact that the behavior of such systems is generally close (depending on the class of inputs) to the behavior of the real systems with transfer functions given by (6.1-5), (6.1-7), and (6.1-9). The derivative controller represents a dilemma. If "real" (P) and (I) systems are made to approach the "ideal," they remain useful (indeed, desirable from a design point of view). However, a real (D) system as it approaches the ideal becomes unusable. Generally, real (P) and (I) systems can be made close enough to the ideal to be represented by (6.1-4) and (6.1-6). However, a real derivative system must, at best, be represented by (6.1-9). Depending on the application, some redesign may be necessary when ideal components have been used in a preliminary design.

To simplify matters at this stage, we have not hesitated to use a differentiator as part of $G_C(s)$. The reader may have noted that placing the differentiator in the forward path, that is, just after the error comparitor, results in having the command signal differentiated. We will see later that a better design approach is to put derivative control in the output feedback loop [i.e., $H(s)$; see Figure 6.1-3]. This will accomplish the same desired effect (of adding damping) as having it in $G_C(s)$, but without causing differentiation of the input signal.

6.2 ANALOG COMPUTER CIRCUITS

In a proportional error feedback controller, the actual physical process for calculating the error and multiplying by a gain K may be done mechanically or electronically. Common electronic methods involve the use of analog computer **operational dc amplifiers** and/or digital control devices. Because of the close association of operational amplifier wiring diagrams and block diagrams, and since the systems we wish to control are modeled as continuous-time systems rather than digital discrete-time systems, we will examine the process in terms of operational amplifiers, which are also continuous-time devices. Operational amplifier circuits allow us to implement various control strategies in continuous time, without the added complications inherent in discrete-time digital devices.

An **operational amplifier** (called an **op amp** for short), depicted in Figure 6.2-1, may be thought of as a very high-gain high-impedance dc amplifier (Doeblin, 1966, pp. 617–622). Its transfer function, from input voltage $e_A(t)$ to output voltage $e_O(t)$, is given approximately by

$$\frac{E_O(s)}{E_A(s)} = G(s) = \frac{-A}{1 + \tau s}, \tag{6.2-1}$$

where $A \approx 10^8$ and $\tau \approx 10^{-4}$ sec. Thus, an operational amplifier behaves like a first-order dynamical system

$$\tau \dot{e}_O + e_O = -A e_A, \tag{6.2-2}$$

with a very fast return time $T = \tau$ (since τ is small). The dynamical characteristics are illustrated in the step response and Bode plots shown in Figure 6.2-2.

For our purposes, as long as the input frequencies are less than

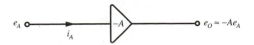

Figure 6.2-1 Operational dc amplifier.

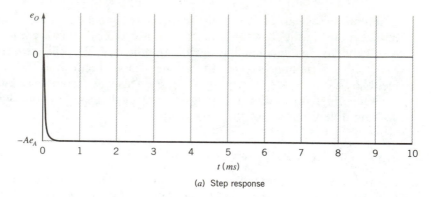

(a) Step response

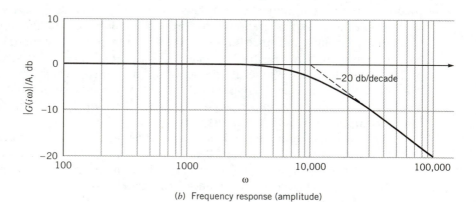

(b) Frequency response (amplitude)

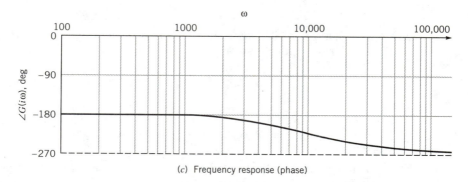

(c) Frequency response (phase)

Figure 6.2-2 Operational amplifier dynamic response.

approximately 10^4 cycles/sec, we can assume $\tau \approx 0$ and model an operational amplifier, as indicated in Figure 6.2-1, by the algebraic equation

$$e_O = -Ae_A. \tag{6.2-3}$$

In operational amplifier circuits, output voltages are usually on the order of ± 10 V. Thus, a high gain ($A \approx 10^8$) implies

$$e_A = -\frac{1}{A}e_O \approx 0, \tag{6.2-4}$$

and a high input impedance ($R_A \approx 10^6$ Ω) implies

$$i_A = \frac{e_A - e_O}{R_A} \approx 0. \tag{6.2-5}$$

Amplifier, Summer, and Comparator

If an input resistance R_1 and a feedback resistance R_f (both in megohms or mΩ) are installed as shown in Figure 6.2-3, then the relationship between an input voltage e_I and output voltage e_O is given by

$$e_O = -\frac{R_f}{R_1}e_I. \tag{6.2-6}$$

Notice that there is a sign change on any signal that passes through the amplifier. This IO relationship can be verified by writing Kirchhoff's current law at the input junction of the operational amplifier:

$$i = i_f + i_A. \tag{6.2-7}$$

Then Ohm's law yields

$$\frac{e_I - e_A}{R_1} = \frac{e_A - e_O}{R_f} + i_A, \tag{6.2-8}$$

which reduces to (6.2-6) for $e_A \approx 0$ and $i_A \approx 0$. Two amplifiers in series as shown in Figure 6.2-4 will produce a positive **gain** K by choosing $R_f/R_1 = K$.

The transfer function for the single-amplifier circuit in Figure 6.2-3 is

$$\frac{E_O(s)}{E_I(s)} = -\frac{R_f}{R_1}.$$

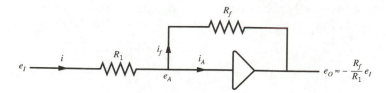

Figure 6.2-3 Amplifier circuit.

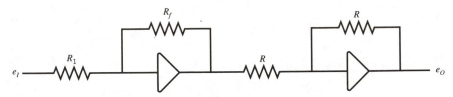

Figure 6.2-4 Gain circuit (gain $K = R_f/R_1$).

This result neglects the dynamics of the operational amplifier. If we were to account for these dynamics, solving (6.2-2) for e_A and then substituting into (6.2-8) along with $i_A \approx 0$ and $\epsilon = (R_1 + R_f)/(AR_1)$ would yield the transfer function

$$\frac{E_O(s)}{E_I(s)} = \frac{-R_f/R_1}{1 + \epsilon(1 + \tau s)} = \frac{-R_f/[(1 + \epsilon)R_1]}{1 + \epsilon\tau s/(1 + \epsilon)},$$

which is of the form of (6.1-5) where

$$K = \frac{-R_f}{(1 + \epsilon)R_1}$$

and

$$\tau_s = \frac{\epsilon\tau}{1 + \epsilon}.$$

In most instances, neglecting the dynamics of the operational amplifier, by setting $\epsilon = 0$ to obtain the simplified transfer function, will not create any difficulties.

A **summing amplifier** is achieved by adding additional inputs to the amplifier, as shown in Figure 6.2-5. It can easily be shown that the relationship between the inputs and the output is given by

$$e_O = -\left(\frac{R_f}{R_1}e_1 + \cdots + \frac{R_f}{R_n}e_n\right). \tag{6.2-9}$$

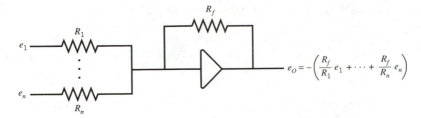

Figure 6.2-5 Summing amplifier.

$$e_O = -\left(\frac{R_f}{R_1} e_1 + \cdots + \frac{R_f}{R_n} e_n\right)$$

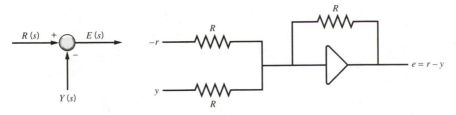

Figure 6.2-6 Comparator.

A **comparator**, for taking the difference between the command and feedback signals, can be implemented as shown in Figure 6.2-6.

Integrator

If we replace the feedback resistor in the basic operational amplifier circuit with a capacitor (in microfarads or μF), as shown in Figure 6.2-7, the circuit acts as an **integrating amplifier**.

Applying Kirchhoff's current law at the input junction of the amplifier yields

$$i = i_f + i_A,$$

or

$$\frac{e_I - e_A}{R} = C\frac{d(e_A - e_O)}{dt} + i_A. \qquad (6.2\text{-}10)$$

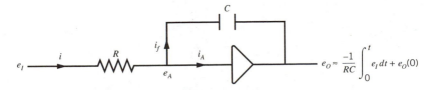

$$e_O \approx \frac{-1}{RC}\int_0^t e_I\, dt + e_O(0)$$

Figure 6.2-7 Integrator.

For $e_A \approx 0$ and $i_A \approx 0$, integrating this result yields

$$e_O = \frac{-1}{RC} \int_0^t e_I \, dt + e_O(0), \qquad \text{(6.2-11)}$$

where $e_O(0)$ is the initial voltage across the capacitor. The corresponding transfer function is

$$\frac{E_O(s)}{E_I(s)} = -\frac{1}{RCs}. \qquad \text{(6.2-12)}$$

This result neglects the dynamics of the operational amplifier. If the dynamics of e_A were included, using (6.2-2) in (6.2-10) along with $i_A \approx 0$ and the definitions $\epsilon_1 = 1/A$, $\epsilon_2 = \tau/A$, $\epsilon_3 = RC\tau/A$, $\epsilon_4 = RC/A$, we obtain the transfer function

$$\frac{E_O(s)}{E_I(s)} = -\frac{1}{(\epsilon_1 + \epsilon_3 s^2) + (RC + \epsilon_4 + \epsilon_2)s}, \qquad \text{(6.2-13)}$$

which is of the form of (6.1-7) where $\epsilon = \epsilon_1 + \epsilon_3 s^2$ and $\tau_i = RC + \epsilon_2 + \epsilon_4$. The ϵ_i are all very small quantities. Because the dynamic response of an op amp is very fast, very little accuracy is lost by using the simplified transfer function representation for the integrator obtained by setting $\epsilon_1 = \epsilon_2 = \epsilon_3 = \epsilon_4 = 0$.

A **summing integrator** can be obtained by adding additional inputs to the integrator as shown in Figure 6.2-8. It can be easily shown that the relationship between inputs and output is given by

$$e_O = -\int_0^t \left(\frac{e_1}{CR_1} + \cdots + \frac{e_n}{CR_n} \right) dt + e_O(0). \qquad \text{(6.2-14)}$$

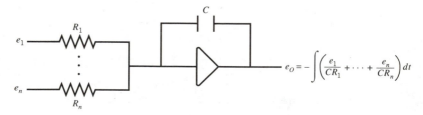

Figure 6.2-8 Summing integrator.

6.3   FEEDBACK CONTROL OF FIRST-ORDER SYSTEMS

Error Feedback Control

The block diagram for error feedback control is illustrated in Figure 6.3-1. The overall transfer function for this system is given by

$$G(s) = \frac{K_S G_C G_P}{1 + G_C G_P}.$$
(6.3-1)

We will now examine the effect of different error feedback controllers as applied to the first-order IO system

$$\dot{y} + p_0 y = q_0 u.$$
(6.3-2)

The transfer function is given by

$$G_P(s) = \frac{q_0}{s + p_0}.$$
(6.3-3)

Proportional Error Feedback

We will begin by considering **proportional** error feedback control of the form

$$u = Ke,$$
(6.3-4)

where

$$e(t) = K_S r(t) - y(t)$$

is the error between the scaled command input $K_S r(t)$ and output $y(t)$. The transfer function for the controller is given by

$$G_C(s) = K.$$
(6.3-5)

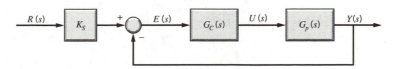

Figure 6.3-1 Error feedback control.

Using (6.3-1), we see that this results in the overall transfer function

$$G(s) = \frac{K_S K q_0}{s + p_0 + K q_0},$$ (6.3-6)

so that the controlled system is equivalent to the IO system

$$\dot{y} + (p_0 + K q_0)y = K_S K q_0 r.$$ (6.3-7)

The characteristic equation for this system is simply

$$\lambda + (p_0 + K q_0) = 0.$$ (6.3-8)

Thus by choosing an appropriate **gain** K (possibly negative), the root λ of the characteristic equation can be made to take on any real value. In particular, even if the primary system is unstable, the controlled system can be made asymptotically stable with any desired **return time** $T = 1/|\lambda|$, by choosing the gain such that $p_0 + K q_0 > 0$.

For the asymptotically stabilized system, the steady-state output response to a step input $r(t) \equiv \bar{r}$ is given by

$$y_\infty \overset{\Delta}{=} \lim_{t \to \infty} y(t) = \frac{K_S K q_0 \bar{r}}{p_0 + K q_0}.$$ (6.3-9)

By choosing $K_S = (p_0 + K q_0)f/(K q_0)$, the steady-state output is made a specified factor times the constant input. Note that to achieve a given scale factor f, K_S must be made a function of K. This is undesirable from the point of view that if K is an adjustable (variable) gain, then any adjustment of K will require an adjustment of K_S. This can be avoided with the modified error feedback control scheme shown in Figure 6.3-2.

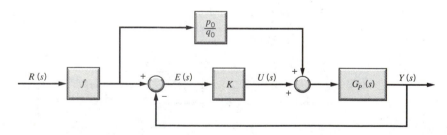

Figure 6.3-2 Modified error feedback control.

The overall transfer function for the controller of Figure 6.3-2 is given by

$$G(s) = \frac{(p_0/q_0 + K)fG_P(s)}{1 + KG_P(s)}.$$

Substituting for $G_P(s)$ from (6.3-3), we obtain

$$G(s) = \frac{(p_0 + Kq_0)f}{s + p_0 + Kq_0},$$

which is equivalent to the IO system

$$\dot{y} + (p_0 + Kq_0)y = (p_0 + Kq_0)fr.$$

With a constant input $r = \bar{r}$, the steady-state response for an asymptotically stable system is

$$y_\infty = f\bar{r}, \tag{6.3-10}$$

which is the same steady-state relationship between input and output we had before using $K_S = (p_0 + Kq_0)f/(Kq_0)$. Only now, no adjustment need be made if K is changed.

Note that a precise relationship between steady-state input and output as given by (6.3-10) could rarely be achieved in practice. This is simply because the exact values for p_0 and q_0 for a real system would not be known. For example, if p_0 and q_0 represent measured values of constants for the system (6.3-1), the actual system would likely be of the form

$$\dot{y} + (p_0 + \Delta p_0)y = (q_0 + \Delta q_0)u,$$

where Δp_0 and Δq_0 represent the differences between the actual and the measured values. The actual system can be written as

$$\dot{y} + p_0y = q_0u + (\Delta q_0u - \Delta p_0y),$$

or

$$\dot{y} + p_0y = q_0u + v, \tag{6.3-11}$$

where $v(t) = \Delta q_0u - \Delta p_0y$ represents an uncertain input to the system due to the inaccurate measurements. Under the feedback control (6.3-4) and the input $r(t) = \bar{r}$, this system has the steady-state output

$$y_{\infty} = \frac{K_S K q_0}{p_0 + K q_0} \bar{r} + \frac{v}{p_0 + K q_0}.$$

For the particular situation considered here, v will also be constant in the steady state. Choosing $K_S = (p_0 + K q_0) f / (K q_0)$ results in

$$y_{\infty} = f\bar{r} + \frac{v}{p_0 + K q_0},$$

which is the desired relationship only when $v = 0$ (no uncertain input). However, by making K large, we are able to diminish the effect of the uncertain input on the output. High gain also results in a fast system. As we can observe from (6.3-7), the output will tend to track the scaled input ($y \approx K_S r$).

It should be noted, however, that there are limits to the gain available in practical situations; hence, high gain does not mean infinite gain. In addition, since the bandwidth (Bode plot corner frequency) of the first-order system is increased with an increase in gain, the filtering effect on high-frequency noise is correspondingly diminished. This can be seen by examining the residual response to sinusoidal inputs. We observe that, since the controlled system is a first-order system, the Bode plot will be the same as in Figure 5.1-2, with p_0 and q_0 replaced by $p_0 + K q_0$ and $K_S K q_0$, respectively. The gain K determines the **corner frequency** (or **bandwidth**) $\omega_c = p_0 + K q_0$, above which the amplitude is attenuated (by -20 db/decade).

In summary, through the choice of the gain parameter K for proportional error feedback control of a first-order system, we have complete control of the eigenvalue for the resulting closed-loop system. The effect of an uncertain input on the steady output can be made arbitrarily small by choosing a sufficiently high gain, at the cost of a high bandwidth, which in turn implies less filtering of high-frequency (noisy) inputs.

Integral Error Feedback

We have previously noted that integral control can be used to eliminate errors due to an uncertain bias on the input. Integral error feedback control of the form

$$u = \frac{1}{\tau_i} \int_0^t e \, dt,$$

which has the transfer function

$$G_C(s) = \frac{1}{\tau_i s},$$

(6.3-12)

can be achieved with the op amp circuit shown in Figure 6.3-3.

Using (6.3-1), we find that integral error feedback control (6.3-12) applied to the first-order system (6.3-3) produces a closed-loop system that is second-order, with the transfer function

$$G(s) = \frac{K_s q_0/\tau_i}{s^2 + p_0 s + q_0/\tau_i}.$$

(6.3-13)

The roots of the characteristic equation

$$\lambda^2 + p_0\lambda + \frac{q_0}{\tau_i} = 0$$

yield the eigenvalues

$$\lambda = \tfrac{1}{2}(-p_0 \pm \sqrt{p_0^2 - 4q_0/\tau_i}).$$

(6.3-14)

The location of the eigenvalues in the complex plane, as a function of the parameter τ_i, is illustrated by the **root locus** plot in Figure 6.3-4, for the case $\tau_i \geq 0$ with p_0 and q_0 positive. At $1/\tau_i = 0$, the eigenvalues are at the points marked "x" in Figure 6.3-3. These points are called the **open-loop poles**, since they correspond to the eigenvalues of the open-loop transfer function $K_s G_C(s)G_P(s)$ obtained by disconnecting the feedback path from the summing junction in Figure 6.3-1. We will discuss root locus plots in detail in Section 7.1. For now we simply note that as $1/\tau_i$ is increased, the roots of the closed-loop characteristic equation move toward each other on the real axis. They meet at a "break point" when $1/\tau_i = p_0^2/(4q_0)$, where the roots are repeated, and then become complex as $1/\tau_i$ is further increased.

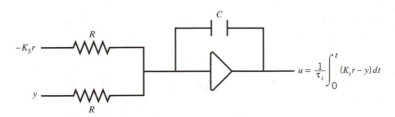

Figure 6.3-3 Integral error control $\left(\dfrac{1}{\tau_i} = \dfrac{1}{RC}\right)$.

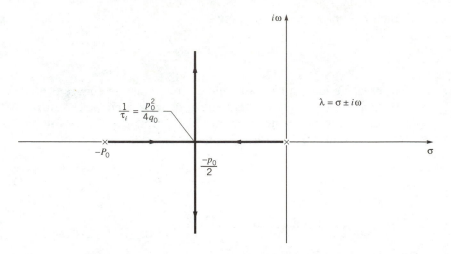

Figure 6.3-4 Root locus ($1/\tau_i \geq 0$) for integral control of a first-order system.

If p_0 is not positive (the original system is unstable), then integral error feedback control does not allow us to achieve asymptotic stability. Even when $p_0 > 0$, we do not have complete control over the eigenvalues of the closed-loop system but, instead, the roots must be chosen from somewhere on the root locus. Specifically, for real roots, the return time is

$$T = \frac{1}{\min|\lambda_i|} = \frac{2}{p_0 - \sqrt{p_0^2 - 4q_0/\tau_i}} \geq \frac{2}{p_0} \qquad \textbf{(6.3-15)}$$

and for a pair of complex roots,

$$T = \frac{1}{|\mathrm{Re}(\lambda)|} = \frac{2}{p_0}. \qquad \textbf{(6.3-16)}$$

Thus, the smallest achievable return time is fixed by p_0. This may or may not meet the performance requirements, depending on the application.

The equivalent IO system corresponding to (6.3-13) is given by

$$\ddot{y} + p_0\dot{y} + \frac{q_0}{\tau_i}y = \frac{K_s q_0}{\tau_i}r. \qquad \textbf{(6.3-17)}$$

Provided that $p_0 > 0$ and $q_0/\tau_i > 0$, the response to a step input $r(t) \equiv \bar{r}$ yields a steady-state output $y_\infty = K_s\bar{r}$, where K_s is now directly equal to the factor between input and output ($K_S = f$). Furthermore, any uncertain bias introduced into the system will be eliminated from the steady-state output. For example, substituting $u = (1/\tau_i)\int(K_s r - y)\,dt$ and $v = \bar{v}$ (constant bias) into (6.3-11) results in the system

$$\dot{y} + p_0 y = \frac{q_0}{\tau_i} \int_0^t (K_s r - y)\,dt + \bar{v}.$$

Differentiating once yields (6.3-17), which is the same result without a bias input.

For sinusoidal inputs, we note that for $q_0/\tau_i > 0$ the controlled system is in the form of a second-order system

$$\ddot{y} + 2\zeta\omega_n \dot{y} + \omega_n^2 y = K_s \omega_n^2 r(t), \tag{6.3-18}$$

with

$$\omega_n = \sqrt{q_0/\tau_i}, \qquad \zeta = \frac{p_0}{2\sqrt{q_0/\tau_i}}.$$

When $K_S = 1$, the closed-loop transfer function (6.3-13) has the same Bode plots as in Figure 5.2-1. At frequencies above the corner frequency $\omega_c = \omega_n$, the amplitude ratio attenuates at the rate of -40 db/decade.

To summarize, we have seen that integral error feedback control applied to a first-order system yields a controlled system of second order. If $q_0/\tau_i > 0$ and the original system is asymptotically stable, the controlled system will also be, but integral control cannot stabilize an unstable system. For the asymptotically stable case, integral control will eliminate an uncertain bias input, but we do not have complete control over the roots of the closed-loop characteristic equation. In particular, the minimum return time $T = 2/p_0$ is fixed by p_0 and occurs for any value of τ_i for which the system is underdamped ($4/\tau_i > p_0^2/q_0 > 0$). Increasing $1/\tau_i$ beyond critical damping increases the natural frequency of the controlled system and decreases the system's filtering ability against high-frequency inputs by increasing the Bode plot corner frequency. In short, integral error feedback control is not a bad idea when applied to stable first-order systems, but we can do better.

Proportional + Integral (PI) Error Feedback

Proportional + integral (PI) error feedback control

$$u = Ke + \frac{1}{\tau_i} \int_0^t e\,dt \tag{6.3-19}$$

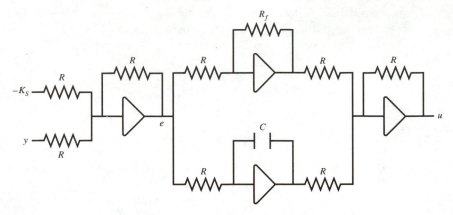

Figure 6.3-5 PI error feedback control.

can be realized with the op amp circuit shown in Figure 6.3-5, where $K = R_f$ and $\tau_i = RC$. The PI error feedback controller has the transfer function

$$G_C(s) = K + \frac{1}{\tau_i s} = \frac{K\tau_i s + 1}{\tau_i s}, \qquad \text{(6.3-20)}$$

with a corresponding Bode plot shown in Figure 6.3-6.

Using (6.3-1), we see that PI error feedback control (6.3-20) applied to

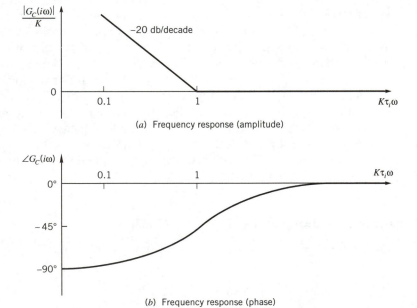

(a) Frequency response (amplitude)

(b) Frequency response (phase)

Figure 6.3-6 Bode plot for PI controller, $G_C(s) = (K\tau_i s + 1)/(\tau_i s)$.

the first-order system (6.3-3) produces a second-order overall transfer function

$$G(s) = \frac{(K\tau_i s + 1)K_S q_0/\tau_i}{s^2 + (p_0 + Kq_0)s + q_0/\tau_i} \qquad (6.3\text{-}21)$$

corresponding to the IO system

$$\ddot{y} + (p_0 + Kq_0)\dot{y} + \frac{q_0}{\tau_i}y = \frac{K_S q_0}{\tau_i}r(t) + K_S Kq_0\dot{r}(t). \qquad (6.3\text{-}22)$$

The choice of the parameters K and τ_i provides us with complete control of the coefficients in the characteristic equation

$$\lambda^2 + (p_0 + Kq_0)\lambda + \frac{q_0}{\tau_i} = 0. \qquad (6.3\text{-}23)$$

For example, if the original first-order system is unstable, not only can we make the closed-loop system asymptotically stable, by suitable choices for K and τ_i, but we can also satisfy various performance specifications, such as settling time and percent overshoot in response to a step input. Furthermore, for a step input $r(t) \equiv \bar{r}$ and for any value of $q_0/\tau_i > 0$, (6.3-22) implies that $y_\infty = K_S \bar{r}$. The steady-state output is a specified factor ($K_S = f$) times the input.

For sinusoidal inputs, the amplitude ratio of the residual response is given by

$$|G(i\omega)| = \frac{|K_S|\sqrt{1 + (K\omega_n\tau_i)^2 (\omega/\omega_n)^2}}{\sqrt{[1 - (\omega/\omega_n)^2]^2 + [(p_0 + Kq_0)/\omega_n]^2 (\omega/\omega_n)^2}}, \qquad (6.3\text{-}24)$$

where $\omega_n^2 = q_0/\tau_i > 0$. The Bode plot for the closed-loop system is somewhat similar to that for a first-order system. At low frequencies ($\omega/\omega_n \ll 1$),

$$|G(i\omega)| \to |K_S|,$$

as desired for tracking low-frequency inputs. At high frequencies ($\omega/\omega_n \to \infty$),

$$|G(i\omega)| \to \left|\frac{K_S K\omega_n\tau_i}{\omega/\omega_n}\right|.$$

Thus, the closed-loop system has a frequency ω_c, above which the amplitude ratio attenuates at a -20 db/decade rate. The PI control, while increasing the order of the system from first- to second-order, does not change the slope of the Bode plot high-frequency asymptote, because of the offsetting $+20$ db/decade term $(K\tau_i s + 1)$ in the numerator of the transfer function (6.3-21), for $K\tau_i > 0$.

To summarize, for a first-order system, PI error feedback control provides complete control of the closed-loop eigenvalues, eliminates the effect of an uncertain bias on the steady-state output, and produces a Bode plot that tracks low-frequency inputs and attenuates the output amplitude at high frequencies. The PI error feedback control satisfies all of our general performance requirements, and it represents a good choice for controlling first-order systems.

6.4   CONTROL STRUCTURES FOR SECOND-ORDER SYSTEMS

Error Feedback Control

Let us consider a second-order system of the form

$$\ddot{y} + p_1\dot{y} + p_0 y = q_0 u \qquad (6.4\text{-}1)$$

with the transfer function

$$G_P(s) = \frac{q_0}{s^2 + p_1 s + p_0}. \qquad (6.4\text{-}2)$$

The block diagram for error feedback control is again given by Figure 6.3-1 where $G_P(s)$ is now given by (6.4-2).

Proportional Error Feedback

The effect of proportional error feedback control on a second-order system is similar to its effects on first-order systems. Using proportional error feedback control

$$u = Ke = K[K_s r - y], \qquad (6.4\text{-}3)$$

we again have

$$G_C(s) = K, \qquad (6.4\text{-}4)$$

which yields the overall transfer function for the system

$$G(s) = \frac{K_s K q_0}{s^2 + p_1 s + (p_0 + K q_0)}. \qquad (6.4\text{-}5)$$

The corresponding IO representation

$$\ddot{y} + p_1\dot{y} + (p_0 + K q_0)y = K_s K q_0 r \qquad (6.4\text{-}6)$$

has the characteristic equation

$$\lambda^2 + p_1\lambda + (p_0 + Kq_0) = 0. \qquad (6.4\text{-}7)$$

The roots of the closed-loop characteristic equation, as functions of the gain parameter K, lie on the **root locus** illustrated in Figure 6.4-1 (for p_0 and p_1 positive). The eigenvalues of the primary system are at the points marked "x" in this figure. The root locus corresponds to solutions of (6.4-7) given by

$$\lambda = \tfrac{1}{2}\{-p_1 \pm \sqrt{p_1^2 - 4(p_0 + Kq_0)}\} \qquad (6.4\text{-}8)$$

for various values of K. For $q_0 > 0$, increasing K from zero moves the closed-loop eigenvalues toward the repeated root at $\lambda = -p_1/2$. An additional increase in K will then move the roots into the complex plane as shown. Thus, although a variety of roots to the characteristic equation are possible, they are limited to points on the root locus. In particular, the quickest return time is given by $2/p_1$, regardless of the gain K. Note that an unstable second-order system with $p_1 < 0$ cannot be stabilized by using proportional error feedback.

The steady-state output response to a step input $r(t) \equiv \bar{r}$ is exactly the same as for the first-order system, as given by (6.3-9), and thus by choosing $K_S = (p_0 + Kq_0)f/(Kq_0)$, the steady-state output is made to be a specified factor f times the input. Proportional error feedback does not change the character (particularly the order) of the IO system, so the Bode plot for the controlled system is similar to the second-order system

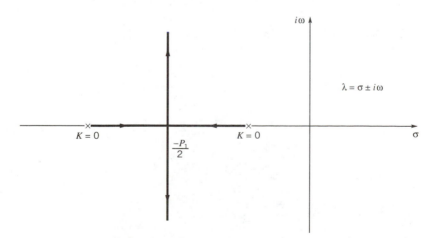

Figure 6.4-1 Root locus ($K \geq 0$) for proportional control of a second-order system.

in Figure 5.2-1, with

$$\omega_n = \sqrt{p_0 + Kq_0} \qquad\qquad (6.4\text{-}9)$$

$$\zeta = \frac{p_1}{2\sqrt{p_0 + Kq_0}}. \qquad\qquad (6.4\text{-}10)$$

The only difference is an amplitude factor that shifts the amplitude plot up or down. Specifically, at $\omega = 0$, we have

$$G(i0) = \frac{K_s Kq_0}{p_0 + Kq_0}, \qquad\qquad (6.4\text{-}11)$$

instead of $G(i0) = 1$ in Figure 5.2-1.

In summary, we note that proportional error feedback control of a second-order system cannot make a system asymptotically stable if $p_1 \leq 0$. For $p_1 > 0$, the system can be made asymptotically stable, but with limited damping. The steady output obtained in response to a step input can be made to be a specified factor f times a constant input. As with first-order systems, we can diminish the effect of an uncertain bias input on the steady-state output by using a high gain K. However, this will increase the bandwidth in the Bode plot, reducing the filtering ability against high-frequency noisy inputs.

Integral Error Feedback

Applying the integral error feedback transfer function $G_C(s) = 1/(\tau_i s)$ to the second-order system (6.4-2) yields a closed-loop system with a transfer function of the form

$$G(s) = \frac{K_s q_0/\tau_i}{s^3 + p_1 s^2 + p_0 s + q_0/\tau_i}, \qquad\qquad (6.4\text{-}12)$$

which corresponds to the third-order IO system

$$\dddot{y} + p_1 \ddot{y} + p_0 \dot{y} + \frac{q_0}{\tau_i} y = \frac{q_0}{\tau_i} K_s r(t) \qquad\qquad (6.4\text{-}13)$$

with characteristic equation

$$\lambda^3 + p_1 \lambda^2 + p_0 \lambda + \frac{q_0}{\tau_i} = 0.$$

We see that integral error feedback control provides us with control of only the last coefficient in the characteristic equation. We cannot use integral error alone to stabilize an unstable system. The root locus for $1/\tau_i \geq 0$, for a primary system whose eigenvalues are complex with negative real parts, is shown in Figure 6.4-2. Note that, in spite of the fact that the primary system is stable, the controlled system will be unstable if the controller gain $1/\tau_i$ is made too large. This type of behavior can be induced in many control applications where one can adjust the gain. For example, in some analog x-y plotters, if the "gain" knob is turned too high, the arm of the plotter begins to chatter.

In response to a step input $r(t) \equiv \bar{r}$, the controlled system (assumed stable) yields a steady output response $y_\infty = K_S\bar{r}$, regardless of the gain $1/\tau_i$. For sinusoidal inputs, the Bode plot has a low-frequency asymptote of $20 \log_{10}|K_S|$ db (0 db if $K_S = 1$), a corner frequency given by $\omega_c^3 = q_0/\tau_i$, and a high-frequency attenuation of -60 db/decade.

Thus, we observe that the integral error feedback control of a second-order system produces a third-order system that is stable only if the original system is stable and the gain $1/\tau_i$ is not set too high. As with the first-order system, integral error feedback will eliminate the effect of an uncertain bias on the steady-state output. The frequency response characteristics provide good low-frequency tracking and good high-frequency noise rejection. The deficiencies are that we do not have control of the closed-loop eigenvalues and the system becomes unstable

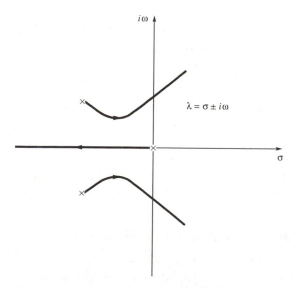

Figure 6.4-2 Root locus $(1/\tau_i \geq 0)$ for integral control of a second-order system.

as the gain increases. The gain limitation also limits the bandwidth of the system for low-frequency tracking.

Proportional + Integral (PI) Error Feedback

Since PI control proved very effective with first-order systems, we will investigate its effects when applied to a general second-order system. The PI controller and its transfer function are given by (6.3-19) and (6.3-20), respectively, and can be implemented by using the op amp circuit in Figure 6.3-5. The overall closed-loop transfer function for the controlled system is

$$G(s) = \frac{(K_S q_0 / \tau_i) + K_S K q_0 s}{s^3 + p_1 s^2 + (p_0 + K q_0)s + q_0 / \tau_i}, \qquad (6.4\text{-}14)$$

which corresponds to the third-order IO system

$$\dddot{y} + p_1 \ddot{y} + (p_0 + K q_0)\dot{y} + \frac{q_0}{\tau_i} y = \frac{K_S q_0}{\tau_i} r(t) + K_S K q_0 \dot{r}(t) \qquad (6.4\text{-}15)$$

with characteristic equation

$$\lambda^3 + p_1 \lambda^2 + (p_0 + K q_0)\lambda + \frac{q_0}{\tau_i} = 0.$$

Through K and τ_i we now have control of two of the three coefficients in the characteristic equation, but if $p_1 < 0$ the original system is unstable and we cannot stabilize it. Therefore, unlike the first-order case, PI control cannot be used in all instances. Of course, in many applications the single coefficient that we cannot adjust in the characteristic equation may have a value such that the system response is adequate under PI control. However, PI control is not sufficient for controlling a general second-order system. Note that, under PI error feedback in the asymptotically stable case, we obtain a $20 \log_{10}|K_S|$ db (0 db if $K_S = 1$) low-frequency Bode plot asymptote, a corner frequency given by $\omega_c^2 = q_0 / \tau_i$, and a high-frequency attenuation slope of -40 db/decade.

Proportional + Derivative Error (PD) Feedback

For the second-order system (6.4-1), PI feedback does not provide any additional damping. To introduce damping, some form of derivative feedback is needed. Here we will investigate the effect of a proportional + derivative (PD) type of control.

Consider a controller that uses proportional + derivative error feedback of the form

$$u = K[K_s r - y] + \tau_d[K_s \dot{r} - \dot{y}], \qquad (6.4\text{-}16)$$

with $K > 0$ and $\tau_d > 0$. The first term in the controller tends to drive $y(t)$ toward $K_s r(t)$, for $q_0 > 0$ and $p_0 > 0$ in (6.4-1). The second term augments this "restoring force" if the error $e(t) = K_s r - y$ is diverging ($de^2/dt > 0 \rightarrow e\dot{e} > 0$) and opposes the force if the error is converging ($e\dot{e} < 0$). As an example of the usefulness of this controller, recall the undamped inverted pendulum discussed at the beginning of this chapter. With only proportional error feedback, the pendulum can be stabilized about the vertical, but it will continue to oscillate because the system contains no damping. In effect, the proportional controller only knows that an error condition exists, which it tries to null, but it does not know whether the error is converging or diverging. The controller in (6.4-16) has more information and can apply a modified restoring force, to eliminate the overshoot and oscillation.

The transfer function for this PD controller is given by

$$G_c(s) = K + \tau_d s$$

and, if it is implemented as in Figure 6.3-1 with the primary system given by the second-order transfer function (6.4-2), we obtain the following overall transfer function:

$$G(s) = \frac{K K_s q_0 + K_s \tau_d q_0 s}{s^2 + (p_1 + \tau_d q_0)s + (p_0 + K q_0)}, \qquad (6.4\text{-}17)$$

which is the transfer function for the second-order system

$$\ddot{y} + (p_1 + \tau_d q_0)\dot{y} + (p_0 + K q_0)y = K_s K q_0 r(t) + K_s \tau_d q_0 \dot{r}(t) \qquad (6.4\text{-}18)$$

with characteristic equation

$$\lambda^2 + (p_1 + \tau_d q_0)\lambda + (p_0 + K q_0) = 0, \qquad (6.4\text{-}19)$$

in which we can arbitrarily place the closed-loop eigenvalues with appropriate choices for K and τ_d. Except for not being able to eliminate a nonzero uncertain bias input (which can be eliminated by adding integral error feedback), it would appear that the PD or perhaps a PID controller "solves" the second-order system control problem. Unfortunately, this is not the case, since both the PD and PID controllers employ derivative control that is not physically realizable. However, as we shall see in the next section, controllers can be built that come close to a PD or PID system. Thus, the main advantage of the derivative feedback concept can still be used.

6.5 USING THE DERIVATIVE IN FEEDBACK CONTROL DESIGN

Pseudo-Derivative Analog Circuit

To implement any derivative control, such as the PD error feedback compensator,

$$u = Ke + \tau_d \dot{e}, \tag{6.5-1}$$

we need a device to **differentiate** the signal $e(t)$ to obtain $\dot{e}(t)$. As a candidate, consider the op amp circuit shown in Figure 6.5-1. The governing equations are

$$e_I - e_A = \int_0^t \frac{i}{C} dt$$

$$i = i_f + i_A = \frac{e_A - e_1}{R} + i_A = \frac{e_A + e_O}{R} + i_A.$$

Substituting for i in the first equation, using $e_A \approx 0$ and $i_A \approx 0$, and then differentiating the result yields

$$e_O = \tau_d \dot{e}_I, \tag{6.5-2}$$

where $\tau_d = RC$.

However, this circuit (or any other representation of it, including numerical differentiation of the signal) is so sensitive to noise that it cannot be used in practice. To understand this, examine the Bode amplitude plot in Figure 6.5-2 for the transfer function

$$G(s) = \tau_d s. \tag{6.5-3}$$

With a constant slope of 20 db/decade, all high-frequency inputs will be amplified. Thus, even for a "constant" input, which will always contain a

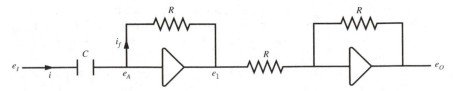

Figure 6-5-1 Derivative circuit ($\tau_d = RC$).

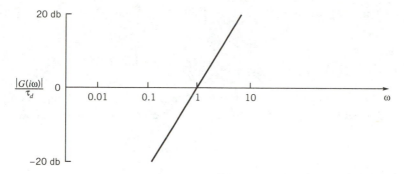

Figure 6.5-2 Bode plot for $G(s) = \tau_d s$.

small amount of high-frequency noise, the output will be very noisy and unusable as a derivative estimate.

These difficulties can be overcome by adding a small resistance R_s in series with the input capacitor, producing the **pseudo-derivative** circuit shown in Figure 6.5-3. The governing equations are

$$e_I - e_1 = \int_0^t \frac{i}{C}\, dt$$

$$\frac{e_1 - e_A}{R_s} = i$$

$$\frac{e_A - e_2}{R} = i_f = i - i_A.$$

If we use $e_A \approx 0$ and $i_A \approx 0$, the first two equations yield

$$e_I - e_1 = \int_0^t \frac{e_1}{R_s C}\, dt$$

and the second two yield

$$e_1 = -\frac{R_s}{R} e_2 = \frac{R_s}{R} e_O.$$

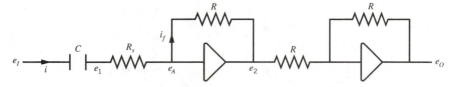

Figure 6.5-3 Pseudo-derivative circuit with $R_s \ll 1$.

Thus, we have

$$e_I - \frac{R_s}{R} e_O = \frac{1}{RC} \int_0^t e_O \, dt.$$

Differentiating this result, multiplying through by RC, and then replacing RC with τ_d and R_sC with $\tau_s \ll 1$, we get the **pseudo-derivative** system

$$\tau_s \dot{e}_O + e_O = \tau_d \dot{e}_I, \tag{6.5-4}$$

which has the transfer function

$$\frac{E_O(s)}{E_I(s)} = G(s) = \frac{\tau_d s}{1 + \tau_s s}. \tag{6.5-5}$$

An alternate way of generating a pseudo-derivative was provided by Example 2.1-1. In that case, the pseudo-derivative was given by

$$\epsilon \dot{y} + y = \dot{u},$$

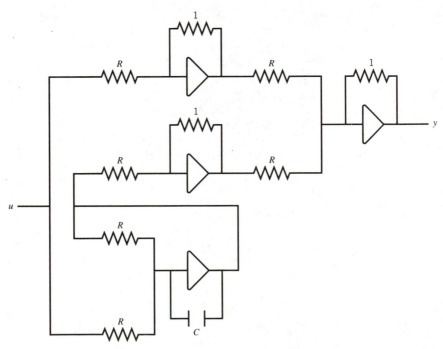

Figure 6.5-4 An alternate pseudo-derivative circuit with $R = C = \sqrt{\epsilon}$, $\epsilon \ll 1$ (e.g., $\epsilon = 0.01$).

which is equivalent to (6.5-4) with $\tau_s = \epsilon$, $\tau_d = 1$, $e_0 = y$, and $e_i = u$. This first-order system has an equivalent state-space representation given by

$$\dot{x} = \frac{-x + u}{\epsilon}$$

$$y = \frac{-x + u}{\epsilon}.$$

An op amp circuit that solves these equations is illustrated in Figure 6.5-4, where the pseudo-derivative is generated by using only integrators and summers. Both pseudo-derivative circuits will produce the same results. The advantage of the pseudo-derivative circuit of Figure 6.5-4 is that it conforms with standard analog computer wiring practice.

The Bode plot for the first pseudo-derivative device (6.5-4) is shown in Figure 6.5-5. The device is a phase **lead** circuit that acts like a differentiator over the frequency range $0 < \omega\tau_s < 1$. A small value of τ_s can be implemented by using a small input resistance R_s, leaving us free to choose the gain, $\tau_d = RC$. Note that, unlike a true differentiator, high-frequency inputs to the pseudo-derivative device will not be amplified without bound. However, they will be present in the output, with an amplitude of approximately τ_d/τ_s.

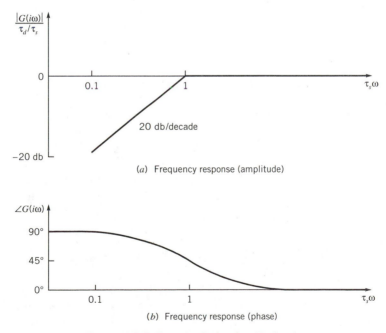

(a) Frequency response (amplitude)

(b) Frequency response (phase)

Figure 6.5-5 Pseudo-derivative Bode plot.

Pseudo-Derivative Feedback Control

Being able to construct a pseudo-derivative device, one would be tempted to simply use this device anywhere derivative control seems appropriate. However, this could lead to unexpected performance results. For example, if the pseudo-derivative is used in the PD error feedback controller, then under high-frequency inputs the input signal would not be differentiated. As suggested by Figure 6.5-5, it is better to use derivative control at points in the system where signals are not rapidly changing. Since the output signal is always filtered from the input by the (stable) system itself, derivative control is usually used only in conjunction with the output signal. For example, consider again the second-order system whose transfer function is given by (6.4-2). One can secure the same characteristic equation obtained previously with PD error feedback [that is, $G_C(s) = K + \tau_d s$ in Figure 6.3-1] by using instead the design shown in Figure 6.5-6.

The overall transfer function in this case is

$$G(s) = \frac{K_S K q_0}{s^2 + (p_1 + \tau_d K q_0)s + (p_0 + K q_0)}, \qquad (6.5\text{-}6)$$

which is the transfer function for the IO system

$$\ddot{y} + (p_1 + \tau_d K q_0)\dot{y} + (p_0 + K q_0) = K_S K q_0 r(t). \qquad (6.5\text{-}7)$$

This IO system indeed has the same characteristic equation as (6.4-18), but without the derivative of the command input.

Replacing the differentiator $\tau_d s$ in Figure 6.5-6 with the pseudo-derivative transfer function (6.5-5) produces a lead compensator (6.1-11) with $\tau_e = \tau_s + \tau_d$ and $\tau_a = \tau_s$. That is, a **proportional + pseudo-derivative** feedback control block is a lead compensator, as shown in Figure 6.5-7. This compensator will be discussed in more detail in the next section. The overall transfer function of the control system depicted in Figure 6.5-7,

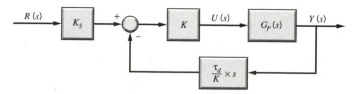

Figure 6.5-6 Using derivative control in the feedback loop.

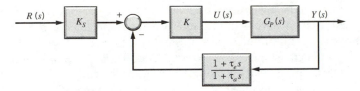

Figure 6.5-7 Proportional error plus pseudo-derivative.

with $G_P(s)$ again given by (6.4-2), is

$$G(s) = \frac{(1 + \tau_a s)K_S Kq_0}{\tau_a s^3 + (1 + \tau_a p_1)s^2 + (p_1 + \tau_a p_0 + \tau_e Kq_0)s + (p_0 + Kq_0)},$$
(6.5-8)

which corresponds to the third-order IO system

$$\tau_a \dddot{y} + (1 + \tau_a p_1)\ddot{y} + (p_1 + \tau_a p_0 + \tau_e Kq_0)\dot{y} + (p_0 + Kq_0)y$$
$$= K_S Kq_0[r(t) + \tau_a \dot{r}(t)].$$
(6.5-9)

Note that with pseudo-derivative control in the feedback loop, we once again will have the input signal differentiated. However, for small τ_a (set $\tau_a = 0$) we have

$$\ddot{y} + (p_1 + \tau_e Kq_0)\dot{y} + (p_0 + Kq_0)y = K_S Kq_0 r(t),$$
(6.5-10)

which, if we set $\tau_e = \tau_d$, is identical to (6.5-7). Thus, for small τ_a the effect of the pseudo-derivative in the feedback loop will be similar to PD control in the feedback loop. Provided that τ_a is small, we can still design the feedback gains as though we were using PD feedback compensation. However, the actual eigenvalues will be slightly different depending on τ_a. In Section 7.1 we will learn how to draw a root locus for this system with τ_a as a parameter, so that the eigenvalues for the actual system can be determined.

Lead or Lag Feedback Compensation

The proportional + pseudo-derivative feedback controller shown in Figure 6.5-7 is a lead compensator, which is a special case of a more general **lead** ($\tau_e > \tau_a$) or **lag feedback** ($\tau_e < \tau_a$) controller. An operational amplifier circuit for generating lead or lag compensation is shown in Figure 6.5-8. It has essentially the structure of a proportional (lower-branch) plus pseudo-derivative circuit (upper-branch), except that the pseudo-derivative is multiplied by a constant β, using a potentiometer. On an analog computer a potentiometer allows us to multiply a signal by a fraction. The same result, of course, could be accomplished by selecting a different input resistor from the pseudo-derivative circuit to the summer. On an analog computer, however, there are generally only a few values of input resistors available, so that a potentiometer is used instead.

From our previous analysis of the pseudo-derivative circuit [see Figure 6.5-3 and Equation (6.5-4)], if we note that $e_O = -e_2$, we then have

$$\tau_s \dot{e}_2 + e_2 = -\tau_d \dot{e}_I,$$

where $\tau_s = R_s C$ and $\tau_d = RC$. The corresponding transfer function is

$$\frac{E_2(s)}{E_I(s)} = \frac{-\tau_d s}{1 + \tau_s s}. \tag{6.5-11}$$

It follows from Figure 6.5-8 that

$$e_O = -\beta e_2 - e_1,$$

and $e_1 = -e_I$. Thus we obtain

$$e_O = e_I - \beta e_2.$$

Taking Laplace transforms of both sides yields

$$E_O(s) = E_I(s) - \beta E_2(s).$$

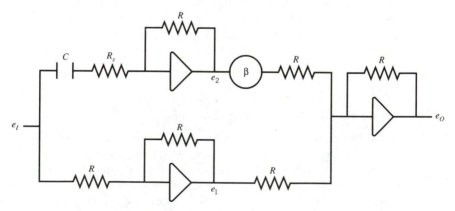

Figure 6.5-8 Lead or lag circuit.

Thus,

$$\frac{E_O(s)}{E_I(s)} = 1 - \beta \frac{E_2(s)}{E_I(s)}.$$

Substituting from (6.5-11), we obtain

$$\frac{E_O(s)}{E_I(s)} = \frac{1 + (\tau_s + \beta\tau_d)s}{1 + \tau_s s},$$

which is a transfer function of the same form as (6.1-11) with $\tau_e = (\tau_s + \beta\tau_d)$ and $\tau_a = \tau_s$, yielding

$$\frac{E_O}{E_I} = H(s) = \frac{1 + \tau_e s}{1 + \tau_a s}.$$

Note that we have replaced τ_s with τ_a, since R_s need not be small for lag compensation, and we can identify the lead or lag properties of this device according to the relative size of the two time constants.

For the case where $\tau_e > \tau_a$ (as in proportional + pseudo-derivative feedback), we have a phase **lead** device. This can be seen from the Bode plot of Figure 6.5-9. As $\omega \to \infty$, the amplitude approaches $H_\infty =$

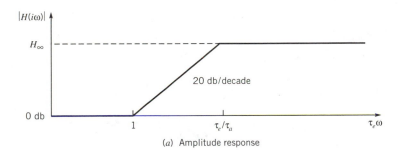

(a) Amplitude response

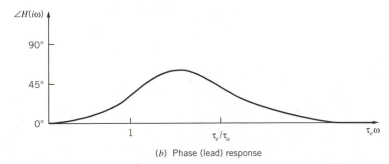

(b) Phase (lead) response

Figure 6.5-9 Bode plot for a lead device ($\tau_e > \tau_a$): $H(s) = (1 + \tau_e s)/(1 + \tau_a s)$.

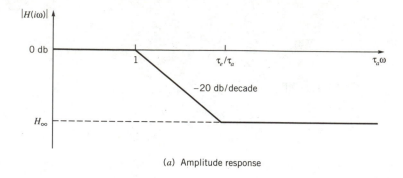

(a) Amplitude response

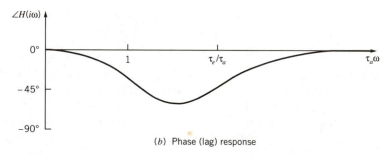

(b) Phase (lag) response

Figure 6.5-10 Bode plot for a lag device ($\tau_e < \tau_a$): $H(s) = (1 + \tau_e s)/(1 + \tau_a s)$.

$20 \log(\tau_e/\tau_a)$ db. For $\tau_e < \tau_a$, we have a phase **lag** device. This can be seen from the Bode plot of Figure 6.5-10.

Low-Pass Filter

The op amp circuit in Figure 6.5-11 is called a **low-pass filter**. Summing the currents at the input junction yields

$$\frac{e_I - e_A}{R} = \frac{e_A - e_1}{R} + C\frac{d(e_A - e_1)}{dt}.$$

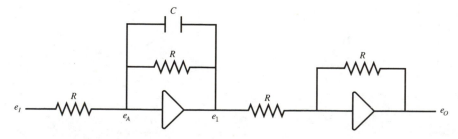

Figure 6.5-11 Low-pass filter circuit ($\tau = RC$).

Setting $e_A = 0$ and multiplying through by R and noting that $e_O = -e_1$, we obtain

$$e_I = -e_1 - RC\frac{de_1}{dt} = e_O + RC\frac{de_O}{dt}.$$

This system has the transfer function

$$\frac{E_O}{E_I} = H(s) = \frac{1}{1 + \tau s}, \qquad (6.5\text{-}12)$$

where $\tau = RC$. The corresponding Bode plot is shown in Figure 6.5-12. Thus, the system passes sinusoidal inputs undiminished over the bandwidth $0 < \omega\tau < 1$. This low-pass filter is a useful device in many control applications when one wishes to eliminate a high-frequency component from a signal. Often, the high-frequency component will be noise.

One way to *think* of lead or lag feedback control is in terms of "filtering" the output before using proportional + derivative feedback, as indicated in Figure 6.5-13. From this point of view, pseudo-derivative feedback and phase lead feedback are equivalent to **filtered PD feedback** compensation. However, one cannot construct a lead or lag compensator in this way. In practice, only the combination shown in Figure 6.5-7 will work. If one were to try to implement the controller of Figure 6.5-13 in a

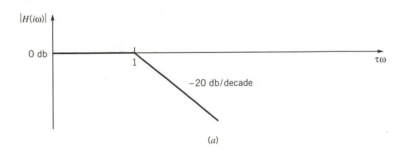

(a)

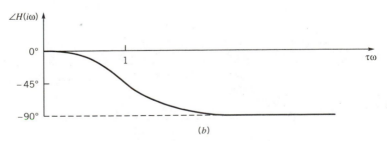

(b)

Figure 6.5-12 Bode plot for a low-pass filter: $H(s) = 1/(1 + \tau s)$.

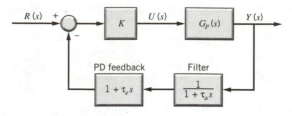

Figure 6.5-13 PD feedback with filtering will not work.

system, it would still fail. This is simply because any even slightly noisy input into a pure differentiator will produce a random output.

PID Compensation

In view of the previous discussion, PID control for practical applications means proportional, integral, and pseudo-derivative. Instead of using a differentiator in the feedback loop, we will employ lead-lag compensation there with $\tau_a = \tau_s$ and small τ_s. The remaining PID elements are shown in Figure 6.5-14.

The overall transfer function for the system is given by

$$G(s) = \frac{(1 + K\tau_i s)(1 + \tau_a s)G_P(s)}{(1 + \tau_a s)\tau_i s + (1 + K\tau_i s)(1 + \tau_e s)G_P(s)}.$$

For the second-order system (6.4-2), this reduces to

$$G(s) = \frac{(1 + \tau_a s)(1 + K\tau_i s)q_0/\tau_i}{(1 + \tau_a s)\{s^3 + p_1 s^2 + p_0 s\} + (1 + K\tau_i s)(1 + \tau_e s)q_0/\tau_i} \qquad \textbf{(6.5-13)}$$

with the corresponding fourth-order IO system

$$\tau_a y^{(4)} + (1 + \tau_a p_1)\dddot{y} + (p_1 + \tau_a p_0 + Kq_0 \tau_e)\ddot{y}$$

$$+ (p_0 + Kq_0 + q_0\tau_e/\tau_i)\dot{y} + \frac{q_0}{\tau_i}y$$

$$= \frac{q_0}{\tau_i}[r(t) + (\tau_a + K\tau_i)\dot{r}(t) + K\tau_a\tau_i\ddot{r}(t)]. \qquad \textbf{(6.5-14)}$$

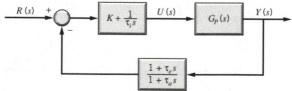

Figure 6.5-14 PID control compensation.

The parameters K, τ_a, τ_i, and τ_e provide us with complete control over the location of the closed-loop eigenvalues, so that not only can we stabilize the system but we can also control its transient response characteristics. For a step input we get zero steady output error, regardless of the parameter choices for an asymptotically stable system. For sinusoidal inputs the transfer function has the low-frequency asymptote

$$|G(i\omega)|_{\omega \to 0} = 1$$

and the high-frequency asymptote

$$|G(i\omega)|_{\omega \to \infty} = \frac{|Kq_0|}{\omega^2}.$$

Therefore, high-frequency inputs will be attenuated at -40 db/decade for frequencies above the corner frequency

$$\omega_c = \sqrt{|Kq_0|}. \tag{6.5-15}$$

Notice that the corner frequency does not depend on τ_a. Under the assumption that the frequency range of interest satisfies $\omega\tau_a \ll 1$, we set $\tau_a = 0$, and the overall transfer function (6.5-13) is approximated by

$$G(s) = \frac{(1 + K\tau_i s)(q_0/\tau_i)}{s^3 + (p_1 + Kq_0\tau_e)s^2 + (p_0 + Kq_0 + q_0\tau_e/\tau_i)s + q_0/\tau_i}, \tag{6.5-16}$$

which has the corresponding third-order IO system

$$\dddot{y} + (p_1 + Kq_0\tau_e)\ddot{y} + (p_0 + Kq_0 + q_0\tau_e/\tau_i)\,\dot{y} + \frac{q_0}{\tau_i} y$$

$$= \frac{q_0}{\tau_i}[r(t) + K\tau_i \dot{r}(t)]. \tag{6.5-17}$$

We have complete control over eigenvalue placement, zero steady output error for a step input, and the same frequency response asymptotes and corner frequency as when $\tau_a \neq 0$. The PID compensation as illustrated in Figure 6.5-14 satisfies all of our basic design requirements for controlling a general second-order system.

Higher-Order Systems

All of the controller design examples presented thus far have been for first- or second-order IO systems with right-hand sides containing no derivatives. Fortunately, the basic design philosophy can be carried over to the more general system given by (2.1-10). We will first discuss higher-order systems with the assumption $q_0 \neq 0$, $q_1 = q_2 = \cdots = q_{Nx} = 0$.

Then we will discuss how to deal with systems with more complicated right-hand sides.

The transfer function for a higher-order system with no input derivatives will be of the form

$$G(s) = \frac{q_0}{P(s)}. \qquad (6.5\text{-}18)$$

Pole Cancellation

In general, the higher-order polynomial in (6.5-18) could be factored into lower-order polynomials of the form

$$G(s) = P_1(s)P_2(s). \qquad (6.5\text{-}19)$$

Suppose that $P_1(s)$ is a first- or second-order polynomial and $P_2(s)$ is what remains of $P(s)$ after the first- or second-order factor has been removed. Since we know how to deal with first- or second-order systems, why not place a compensator after $G_C(s)$ with the transfer function $P_2(s)$ as shown in Figure 6.5-15? The effect would be to replace the primary system with one having the transfer function

$$G_P(s) = \frac{q_0 P_2(s)}{P_1(s)P_2(s)} \to \frac{q_0}{P_1(s)}. \qquad (6.5\text{-}20)$$

We could then design $G_C(s)$ and $H(s)$ as though the primary system had the transfer function (6.5-20). In this way, we could reduce all problems to one of designing controllers for first- or second-order systems.

As appealing as this idea seems, called **pole cancellation**, it will not work! The reasons are quite simple. As discussed in Section 2.2, exact pole cancellation, in reducing the order of the system, would hide the fact that the pole is still there in the actual system. In particular, pole cancellation for a pole in the right half of the complex plane would hide an unstable mode in the free response of the system. A second reason is apparent from Figure 6.5-15 where we see that $P_2(s)$ is a transfer function

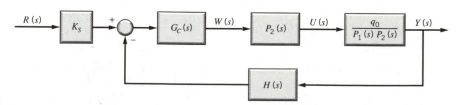

Figure 6.5-15 A pole cancellation method that is not realizable.

with $W(s)$ as an input and $U(s)$ as an output. That is,

$$U(s) = P_2(s)W(s). \qquad \text{(6.5-21)}$$

Thus, $P_2(s)$ must be a device that would take one or more derivatives of its input signal $w(t)$. We have already demonstrated that this is not possible to do by any real device. However, the basic idea of (approximate) pole cancellation can still be used if we (1) avoid canceling poles with positive real parts, and (2) either replace $P_2(s)$ with appropriate pseudo-derivative compensators or place appropriate phase lead compensators in series with $H(s)$.

Zero Cancellation

Whereas pole cancellation cannot be used to simplify higher-order systems, a **zero cancellation** method can be used to simplify the design of minimum phase IO systems with derivatives on the right-hand side of (2.1-10). The transfer function for the primary system is

$$G_P(s) = \frac{Q(s)}{P(s)}. \qquad \text{(6.5-22)}$$

Consider placing a compensator after $G_C(s)$ with the transfer function $q_0/Q(s)$ as shown in Figure 6.5-16. In this case, we would be effectively replacing the primary system with one with the transfer function

$$G_P(s) = \frac{q_0 Q(s)}{Q(s)P(s)} \rightarrow \frac{q_0}{P(s)} \qquad \text{(6.5-23)}$$

Then $G_C(s)$ and $H(s)$ could be designed on this basis, with a caution that the "hidden" poles and zeros of $G_P(s)$ are still actually present. The extra compensator must be designed to produce the transfer function $q_0/Q(s)$. Thus, the physical device must satisfy a differential equation whose transfer function is given by

$$Q(s)W(s) = q_0 U(s). \qquad \text{(6.5-24)}$$

This can be done by using integrating and summing amplifiers.

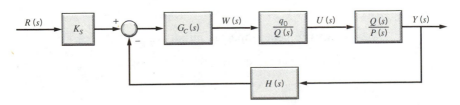

Figure 6.5-16 A zero cancellation method that is realizable.

Multiple Time Scales

Finally, it should be noted that higher-order systems may often have two or more time scales. What this means is that the system eigenvalues may be grouped in such a way that each group has a return time significantly different from the other groups. Control design may then focus on each group separately, thereby reducing the dimension of the problem.

For example, consider a variation of the ball and beam apparatus in Exercise 1.5-10. A voice-coil type actuator mounted on the beam is used to position a sliding mass on a beam in order to balance the beam at a specified orientation. The beam is centered on a knife edge with the actuator and mass near one end and an equal fixed weight at the other end. Separately, both the actuator plus mass and the weighted beam are second-order systems. Together, they constitute a fourth-order system. However, under control, this system may be made to have two time scales by making the return time of the actuator plus mass fast and by making the return time of the weighted beam slow. In so doing, the overall control design can be divided into two parts. The first part consists of designing a controller for the beam as if the sliding mass could be positioned instantaneously (see Exercise 6.6-12). The second part consists of designing a controller for the actuator so that the actuator plus sliding mass eigenvalues result in a return time that is, indeed, much faster than the return time of the beam. With sufficient separation of the eigenvalues for the two subsystems, the overall controlled system will have long-term behavior similar to that of the second-order beam.

6.6 EXERCISES

6.6-1 An IO system is described by the transfer function

$$G_P(s) = \frac{1}{s(s + 5)}.$$

(a) Calculate the open-loop response of this system to a unit step input with all initial conditions equal to zero.

(b) Calculate the closed-loop response of this system under an error feedback controller of the form

$$U(s) = 10[R(s) - Y(s)]$$

with the same input as in part (a).

6.6-2 Consider a water tank modeled by

$$A\frac{dh}{dt} = q_i - q_o,$$

where A = surface area, h = height of water in tank, q_i = inflow, q_o = outflow. The outflow can be approximated by

$$q_o = k_1 h,$$

where k_1 is an outlet valve constant.

(a) Obtain the transfer function for the system.

(b) A float valve is installed so that the inflow is given by

$$q_i = k(h_n - h),$$

where h_n is a reference height. Is this an error feedback system? Draw a block diagram of the controlled system and obtain the overall transfer function.

(c) Determine the response of this system with $h(0) = h_n = 1$, $A = 1$, $k_1 = 0.5$, and $k = 1$.

(d) Determine the steady-state height.

6.6-3 The modern toilet (water closet) is an automatic control device in which the flush may be thought of as a resetting of initial conditions (tank empty). The water level is a measurable quantity and may be thought of as the "output."

(a) Determine the response of a typical toilet (if necessary, remove the top and see how it works).

(b) Can you think of any way of improving performance? How would you do it?

6.6-4 (a) Determine the damping ratio, natural frequency, and the damped frequency for the control system shown in Figure 6.6-1, with $K = 25$.

(b) Determine the response of this system to a unit step input with all initial conditions zero.

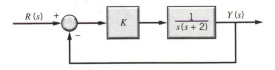

Figure 6.6-1 The control system for Exercise 6.6-4.

6.6-5 To increase the damping ratio of the system of Exercise 6.6-4, the control system in Figure 6.6-2 is proposed. With $K = 25$, determine the value of h to give a damping ratio of 0.5. Is the proposed system practical? If not, how would you change it?

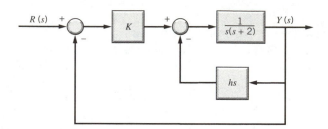

$$R(s) \quad + \quad K \quad + \quad \frac{1}{s(s+2)} \quad Y(s)$$

$$hs$$

Figure 6.6-2 Control system for Exercise 6.6-5.

6.6-6 A control force whose line of action lies in the horizontal direction is attached to the mass of the system depicted in Figure 3.2-2, where $m = 1$, $k = 0.5$, and $\beta = 0.1$. An error feedback controller was designed for this system as in Figure 6.3-1 with $G_C = K$ and $K_S = 1$. Under this controller, the undamped natural frequency for the controlled system is $\omega_n = 2$ rad/sec.
(a) Determine the value of the gain K.
(b) What is the damping ratio of the controlled system?
(c) Given $r(t) = \sin 2t$, what is the amplitude of the output as $t \to \infty$?

6.6-7 A recording device is available in which a force is applied directly to the pen to move it as shown in Figure 6.6-3. Assume negligible friction between the pen point and the paper. The pen is of mass m. Design a feedback control system so that under a constant reference input r, the output (displacement of the pen from a fixed reference) is $y = 10r$ when the system reaches steady state. The system should have a settling time of 0.1 sec (with $\epsilon = 0.01$) and have a flat Bode magnitude response over the bandwidth of the system. What is the resulting bandwidth?

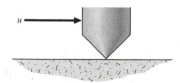

u

Figure 6.6-3 Recording device for Exercise 6.6-7.

6.6-8 Consider the feedback control system in Figure 6.1-3, with

$$G_P(s) = \frac{1}{s-1}$$

and PI cascade compensation

$$G_C(s) = K + \frac{1}{\tau_i s}, \qquad H(s) = 1.$$

(a) Determine parameter values K_S, K, and τ_i so that the controlled system has eigenvalues $\lambda = -2 \pm i$ and $y_\infty = 0.1\bar{r}$ in response to a step input $r(t) = \bar{r}$.

(b) Construct the (approximate, asymptotic) Bode amplitude plot for sinusoidal inputs.

(c) Determine the system bandwidth and return time.

6.6-9 Consider the same system as in Exercise 6.6-8, but with the (I) control placed in a feedback compensator. That is,

$$G_P(s) = \frac{1}{s-1}$$

with

$$G_C(s) = K, \qquad H(s) = 1 + \frac{1}{\tau_i s}.$$

(a) Determine parameter values K_S, K, and τ_i so that the controlled system has eigenvalues $\lambda = -2 \pm i$ and $y_\infty = \alpha$ in response to a **ramp** input $r(t) = \alpha t$, where $R(s) = \alpha/s^2$.

(b) Determine the steady output response y_∞ to a step input $r(t) = \bar{r}$.

(c) Construct the (approximate, asymptotic) Bode amplitude plot.

6.6-10 Consider the control system in Figure 6.1-3, with

$$G_P(s) = \frac{1}{s^2 - 4}, \qquad G_C(s) = K(1 + as), \qquad H(s) = 1.$$

Determine K and a so that the system has repeated eigenvalues with real part as negative as possible, subject to the constraints $0 \le K \le 20$ and $a > 0$.

6.6-11 In the absence of friction the motion of a voice coil type of actuator is given by

$$m\ddot{z} = k_f k_p u,$$

where $m = 0.75$ slugs is the mass of the actuator and the attached load, $k_f = 1.67$ lb/amp is a force constant, $k_p = 0.036$ amp/V is a power amplifier constant, z is displacement (ft), and u is the applied voltage (V). A PID controller is proposed for this system, of the form

$$u = h_1[r - z] - h_2\dot{z} + h_3 \int_0^t (r - z)dt,$$

where $r(t)$ is a command input voltage used to position the load mass.

(a) Draw a block diagram illustrating the controlled system.
(b) Determine the overall transfer function for the controlled system.
(c) Determine values for the feedback gains h_1, h_2, and h_3 to yield controlled system eigenvalues $\lambda_1 = -0.01$, $\lambda_{2,3} = -2 \pm 6i$.

6.6-12 A uniform beam with moment of inertia J is supported by a knife-edge at its midpoint. A fixed mass m is attached to the beam at a distance ℓ from the support, with an identical movable mass on the opposite side of the support. Let u be the displacement of the movable mass from the balance point and assume that the movable mass can be positioned instantaneously. Let θ be the angular rotation of the beam from the horizontal. For small θ the equation of motion is given by

$$J\ddot{\theta} = mgu,$$

where $J = 3.82$ slug/ft^2 and $mg = 0.75$ lb.

(a) Determine the transfer function $G_P(s)$ for this system.
(b) Is the system stable?
(c) For the feedback controller in Figure 6.5-7, determine the overall transfer function for the controlled system. Reduce your answer to a ratio of polynomials.
(d) Determine K, τ_e, and τ_a so that the controlled system has a characteristic equation with roots at $\lambda = -1, -2 \pm i$.
(e) Determine K_S so that for a constant input $r(t) = \bar{r}$ the steady state output is $\theta_\infty = 0.1\bar{r}$.

6.6-13 A given system is described by

$$\ddot{y} + \dot{y} - 2y = u.$$

(a) Determine the transfer function $G_P(s)$ for this sytem.
(b) Determine if the primary system is stable.
(c) For the feedback controller in Figure 6.5-7, determine the overall transfer function for the controlled system. Reduce your answer to a ratio of polynomials.
(d) Determine the equivalent IO differential equation describing the overall controlled system, with output $y(t)$ and input $r(t)$.

(e) Determine values for K, τ_e, τ_a so that the overall controlled system has a characteristic equation with roots at $\lambda = -5$, $-2 \pm i$.

6.6-14 The linearized inverted pendulum system controlled by a low-inductance dc motor was approximated in Example 2.3-3 by an IO system of the form

$$\ddot{\theta} + p_1\dot{\theta} + p_0\theta = q_0u.$$

A particular realization of this system resulted in the following set of parameters:

$$p_0 = -17.627/\text{sec}^2, \quad p_1 = 0.187/\text{sec}, \quad q_0 = 0.6455 \text{ rad/sec}^2\text{-V}.$$

(a) For what values of K will error feedback of the form

$$U(s) = K[R(s) - \theta(s)]$$

make this system stable?

(b) Determine the value of K so that the closed-loop system will have an undamped natural frequency of 2π rad/sec.

6.6-15 The inverted pendulum IO system defined in Exercise 6.6-14 is to be controlled as in Figure 6.5-7. Determine

(a) The transfer function for the overall system.

(b) The characteristic equation for the overall system.

(c) The values for K, τ_e, and τ_a so that the overall characteristic equation is given by

$$(\lambda + 10)(\lambda + \pi - i5.44)(\lambda + \pi + i5.44) = 0.$$

Is $H(s) = (1 + \tau_e s)/(1 + \tau_a)$ a phase lead or a phase lag device?

Chapter 7

Stability Analysis in Output Feedback Systems

7.1 ROOT LOCUS

Given the characteristic equation for a system, the Routh–Hurwitz analysis procedure in Section 3.5 only provides information on the number of characteristic roots (eigenvalues) with positive real parts. It does not tell us the location of the eigenvalues, although some additional information can be learned by shifting the origin of the complex plane while using the Routh procedure.

In this section we will be concerned with finding and plotting the actual location of the eigenvalues as a function of the parameters in the dynamical system. As we have learned in previous examples, as we vary a parameter, the eigenvalues will move, in the complex plane, along some curve that we wish to find. Such a plot is termed a **root locus** (Evans, 1948), and it can be an effective tool for the design of feedback control systems. We will focus on the problem of varying one parameter at a time, although the basic results could also be used to investigate the root locations for variations in more than one parameter at a time. For example, if we vary two parameters, we obtain a two-dimensional surface, whereas three parameters yield a three-dimensional surface, and so on.

The Characteristic Equation in Root Locus Form

As a motivation for the approach that we will take and as a means for introducing some terminology, consider the feedback control system illustrated in Figure 7.1-1. The **closed-loop transfer function** $G(s)$ is

$$G(s) \triangleq \frac{Y(s)}{R(s)} = \frac{G_C(s)G_P(s)}{1 + G_C(s)G_P(s)H(s)} = \frac{Q(s)}{P(s)} \qquad \textbf{(7.1-1)}$$

and the polynomial characteristic equation $P(s) = 0$ corresponds to setting the denominator to zero

$$0 = 1 + F(s), \qquad \textbf{(7.1-2)}$$

where

$$F(s) \triangleq G_C(s)G_P(s)H(s). \qquad \textbf{(7.1-3)}$$

The function $F(s)$ is often called the **open-loop transfer function** which, for the control system in Figure 7.1-1, is equivalent to the system obtained by disconnecting the feedback from the summing junction and considering the feedback signal as the output of the resulting open-loop system.

Consider now the general problem of plotting the location, in the complex plane, of the roots of an nth-order polynomial equation

$$0 = P(s) = s^n + p_{n-1}s^{n-1} + p_{n-2}s^{n-2} + \cdots + p_1s + p_0, \qquad \textbf{(7.1-4)}$$

where the coefficients $p_0, p_1, \ldots, p_{n-1}$ (assumed to be real) depend on some parameter. It should be emphasized that the results we will present are general and are not restricted to characteristic equations that result from feedback control systems.

We will be concerned with the effects of a **parameter** K (usually positive) on the location of the roots of the polynomial equation (7.1-4). Note that K can be any parameter of interest and is not confined to the gain K used in many block diagrams. By convention the phrase **root locus** refers not only to the general problem of plotting the roots as a function of K but also to the particular case where $K \geq 0$. The case where $K \leq 0$ is called the **complimentary root locus**, and the **complete root locus** deals with both cases, namely, $-\infty < K < \infty$. Unless otherwise noted, we will be concerned with the $K \geq 0$ case.

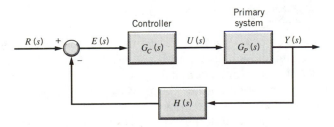

Figure 7.1-1 Block diagram for output feedback control.

For the purpose of constructing a root locus for a parameter K, we transform the polynomial equation (7.1-4) to the form (7.1-2), yielding the **root locus equation**

$$0 = 1 + F(s) = 1 + K\frac{\mathcal{Q}(s)}{\mathcal{P}(s)}, \qquad (7.1\text{-}5)$$

where the open-loop transfer function

$$F(s) \triangleq K\frac{\mathcal{Q}(s)}{\mathcal{P}(s)} = K\frac{\displaystyle\prod_{i=1}^{N_z}(s - z_i)}{\displaystyle\prod_{j=1}^{N_p}(s - p_j)} \qquad (7.1\text{-}6)$$

is written as the product of the parameter K, called the **open-loop gain**, times the ratio of two polynomials $\mathcal{Q}(s)$ and $\mathcal{P}(s)$ of order N_z and N_p, respectively. We do not necessarily assume that $N_z \leq N_p$, although this is usually true for feedback control systems. The complex numbers p_j [the roots of $\mathcal{P}(s) = 0$] are referred to as the **poles** of the open-loop transfer function, since the magnitude of $F(s)$ becomes infinite at $s = p_j$. Similarly, the complex numbers z_i [the roots of $\mathcal{Q}(s) = 0$] are called the **zeros** of the open-loop transfer function.

The poles and zeros of the open-loop transfer function $F(s)$ are fixed numbers, independent of the parameter K. In a feedback control system they can often be determined easily from the block diagram of the control system, since they frequently arise from a network of first- or second-order elements. In any case, even if the open-loop poles and zeros have to be determined by numerically solving higher-order polynomial equations, such a procedure will only need to be done twice [we will also need the roots of $F'(s) = 0$] in the process of sketching an accurate plot of the root locus for the parameter K.

EXAMPLE 7.1-1   **Parametric Root Locus Equation**

The parameter K, used for generating the root locus, is often a gain parameter in a feedback control system, but it need not be. For example, consider the control system shown in Figure 7.1-2, which could be the controller for regulating the speed of a rotating object, using a dc field-controlled motor as the actuator.

If we are interested in the gain k, with the motor time constant τ fixed, then we can write the characteristic equation in root locus form directly,

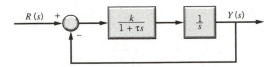

Figure 7.1-2 A speed control system.

using (7.1-3), as

$$0 = 1 + K \frac{1}{s(\tau s + 1)}, \qquad (7.1\text{-}7)$$

with $K = k$. If we want to study the effect of varying the parameter τ with k held constant, we can transform the polynomial characteristic equation

$$0 = \tau s^2 + s + k \qquad (7.1\text{-}8)$$

into the form (7.1-5), either as

$$0 = 1 + K \frac{s^2}{s + k}, \qquad (7.1\text{-}9)$$

in terms of the parameter $K = \tau$, or as

$$0 = 1 + K \frac{s + k}{s^2}, \qquad (7.1\text{-}10)$$

in terms of the parameter $K = 1/\tau$.

   This example highlights the fact that a root locus representation of the characteristic equation is not unique, even though the polynomial characteristic equation itself is unique. The root locus representation depends on the choice of the root locus parameter K, and there may be more than one possible representation for investigating the effects of varying a particular system parameter in the characteristic equation. For the case of the parameter τ above, the root locus representation in (7.1-10) has the desirable property that the numerator of $F(s)$ is a lower-order polynomial than the denominator, so that $F(s)$ is a physically realizable transfer function. However, such an interpretation is not required for root locus analysis.

Root Locus Conditions

In the root locus equation (7.1-5), $F(s)$ is a function of a complex variable s. As indicated in Figure 7.1-3, if $s = \sigma + i\omega$ is viewed as a point in a

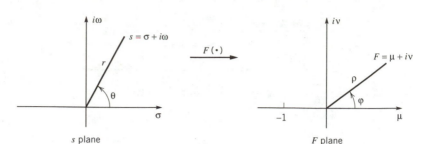

Figure 7.1-3 Functions of a complex variable.

complex s plane, then $F(s) = \mu + iv$ is a point in a corresponding complex F plane.

In polar form a point in the s plane has the representation

$$
\begin{aligned}
s &= \sigma + i\omega \\
&= r\,(\cos\theta + i\sin\theta) \\
&= re^{i\theta},
\end{aligned}
\tag{7.1-11}
$$

where r is the **modulus** (or **magnitude**) of s, given by

$$
r = |s| \triangleq \sqrt{s^*s} = \sqrt{\sigma^2 + \omega^2},
\tag{7.1-12}
$$

with $()^*$ denoting the complex conjugate (i replaced by $-i$), and θ is the **angle** of s measured from the positive real axis, which is given by the two-argument arctangent function

$$
\theta = \angle s \triangleq \tan^{-1}\left(\frac{\mathrm{Im}\{s\}}{\mathrm{Re}\{s\}}\right) = \tan^{-1}\left(\frac{\omega}{\sigma}\right),
\tag{7.1-13}
$$

that is, $\cos\theta = \sigma/r$ and $\sin\theta = \omega/r$. The corresponding point in the F plane is given by

$$
\begin{aligned}
F(s) &= \mu + iv \\
&= \rho\,(\cos\varphi + i\sin\varphi) \\
&= \rho e^{i\varphi},
\end{aligned}
\tag{7.1-14}
$$

where ρ is the modulus of $F(s)$, given by

$$
\rho = |F(s)| \triangleq \sqrt{F^*(s)F(s)} = \sqrt{\mu^2 + v^2},
\tag{7.1-15}
$$

and φ is the angle of $F(s)$, given by

$$\varphi = \angle F(s) \triangleq \tan^{-1} \left(\frac{\text{Im}\{F(s)\}}{\text{Re}\{F(s)\}} \right) = \tan^{-1} \left(\frac{\nu}{\mu} \right). \qquad \textbf{(7.1-16)}$$

If we view $F(s)$ as a function of a complex variable s, the root locus form of the characteristic equation can be written as

$$0 + i0 = 1 + F(s) = 1 + \frac{K\mathcal{Q}(s)}{\mathcal{P}(s)}. \qquad \textbf{(7.1-17)}$$

Thus, a point s is on the root locus if and only if the following conditions are satisfied:

$$\angle F(s) = 180° + N \times 360°, \qquad N = 0, \pm 1, \dots \qquad \textbf{(7.1-18)}$$

$$|F(s)| = 1. \qquad \textbf{(7.1-19)}$$

For $F(s)$ of the form (7.1-6), the root locus can be constructed by finding the points s that satisfy (7.1-18), where

$$\angle F(s) = \angle K + \sum_{i=1}^{N_z} \angle (s - z_i) - \sum_{j=1}^{N_p} \angle (s - p_j), \qquad \textbf{(7.1-20)}$$

and then determining the root locus parameter K, from (7.1-19), by

$$|K| = \frac{\displaystyle\prod_{j=1}^{N_p} |s - p_j|}{\displaystyle\prod_{i=1}^{N_z} |s - z_i|}. \qquad \textbf{(7.1-21)}$$

To illustrate these results, consider the s plane shown in Figure 7.1-4, with the poles and zeros of $F(s)$ denoted by "x" and "o", respectively. If we assume that the real and imaginary axes have the same scale, $|s - p_j|$ is the distance from the pole at p_j to the point s and $\angle (s - p_j)$ is the angle from the positive real-axis direction to the line from the point p_j to the point s. For the situation illustrated in Figure 7.1-4, we have from (7.1-20)

$$\angle F(s) = \angle K + \alpha_1 - (\beta_1 + \beta_2 + \beta_3),$$

where $\angle K = 0°$ if $K \geq 0$ and $\angle K = 180°$ if $K < 0$.

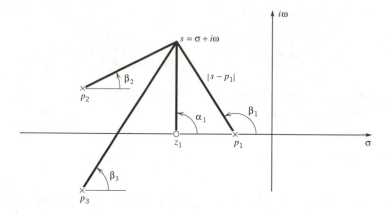

Figure 7.1-4 Root locus angles and magnitudes.

Properties of the Root Locus

The angle criterion in (7.1-18) can be used to numerically plot a root locus, for example, by using Newton's method to find the real and imaginary parts of s to satisfy (7.1-18), while changing K by some step size to move along the root locus. In the next section we will consider this numerical algorithm. For now we will develop several properties that allow us to sketch the approximate root locus. In addition, these properties will be useful in connection with the numerical scheme (Ash and Ash, 1968) that we will employ.

To illustrate various properties of a root locus, consider the system shown in Figure 7.1-5. For reference, the root locus for $K \geq 0$ is shown in Figure 7.1-6, but for the purposes of discussion we will assume that the root locus is being constructed as we proceed in the discussion.

In root locus form the characteristic equation for the closed-loop system can be written as

$$0 = 1 + F(s), \tag{7.1-22}$$

where the open-loop transfer function

$$F(s) = K \frac{1}{s(s + 4)(s + 5)} \tag{7.1-23}$$

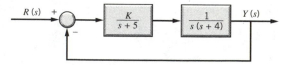

Figure 7.1-5 Example system for root locus properties.

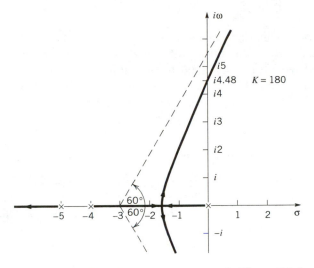

Figure 7.1-6 Root locus for the system in Figure 7.1-5, $K \geq 0$.

has no zeros ($N_z = 0$) and $N_p = 3$ poles, located at $s = 0, -4, -5$ as indicated by the points marked "×" in Figure 7.1-6. From $F(s)$ we construct the root locus for $K \geq 0$ using the properties presented in the following discussion.

1. Symmetry

The root locus is symmetric about the real axis, since complex roots occur in conjugate pairs. For this reason we will often only show the upper half of the root locus, along with the real-axis portion. More generally, the root locus is symmetric about any line of symmetry associated with the poles and zeros of $F(s)$ (Kuo, 1982, p. 391).

2. Real-Axis Segments

For $K \geq 0$ ($K \leq 0$), sections of the real axis are part of the root locus if the number of real poles and zeros of $F(s)$ to the right (left) of a point in question is odd (even).

This result follows directly from the angle criterion (7.1-18), since real poles or zeros to the left of s in (7.1-20) contribute $0°$, a real pole to the right contributes $-180°$, and a real zero to the right contributes $180°$. Note that complex pairs of poles or zeros do not affect the result, since their angular contributions to $F(s)$ cancel on the real axis.

For our example problem, by looking for the real numbers that satisfy (7.1-18), we see that the real-axis segments $-\infty < \sigma < -5$ and $-4 < \sigma < 0$ lie on the root locus for $K \geq 0$.

3. Branches

The root locus contains $N = \max\{N_z, N_p\}$ branches. For the usual case where $N_p \geq N_z$, the number of branches is equal to the number of poles N_p of $F(s)$, which is equivalent to the number of eigenvalues of the closed-loop characteristic equation and the order of the characteristic polynomial. For our example there are three branches, as illustrated in Figure 7.1-6.

4. Starting and Ending Points

Each branch starts ($K = 0$) at a pole of $F(s)$ and ends ($K = \pm\infty$) at a zero of $F(s)$. If $N_p > N_z$ (the usual case), there are $N_p - N_z$ branches that end at infinity. If $N_p < N_z$, there are $N_z - N_p$ branches that start at infinity.

The result follows from (7.1-22), rewritten as

$$K = \frac{-\mathcal{P}(s)}{\mathcal{Q}(s)}, \tag{7.1-24}$$

where $F(s) = K\mathcal{Q}(s)/\mathcal{P}(s)$. At $K = 0$, the characteristic roots (i.e., points on the root locus) must also be roots of $\mathcal{P}(s)$ and poles of $F(s)$. For $K = \pm\infty$ points on the root locus must correspond to the roots of $\mathcal{Q}(s)$, which are the zeros of $F(s)$.

For our illustrative example, with three poles and no zeros, there are three branches (all of the branches, in this case) that go to infinity.

5. Asymptotes

At large values of s, the $|N_p - N_z|$ branches that go to or come from infinity are asymptotic to straight lines that intersect at a common point $\bar{s}$ on the real axis, called the **centroid of the asymptotes**, given by

$$\bar{s} = \frac{\displaystyle\sum_{j=1}^{N_p} p_j - \sum_{i=1}^{N_z} z_i}{N_p - N_z}. \tag{7.1-25}$$

Using Newton's identities (3.5-2), this result follows from the observation that (7.1-24), with (7.1-6), can be written as

$$-K = \frac{(s - p_1)(s - p_2)\cdots}{(s - z_1)(s - z_2)\cdots} = \frac{s^{N_p} - (p_1 + p_2 + \ldots)s^{N_p-1} + \ldots}{s^{N_z} - (z_1 + z_2 + \ldots)s^{N_z-1} + \ldots}.$$

From the binomial theorem

$$(1 + x)^{-1} = 1 - x + x^2 - x^3 + \ldots \quad (|x| < 1),$$

with $x = -(z_1 + z_2 + \ldots)/s + \ldots$ and large $|s|$, we have

$$-K = s^{N_p - N_z} - [(p_1 + p_2 + \ldots) - (z_1 + z_2 + \ldots)]s^{N_p - N_z - 1} + \ldots.$$

Now for large $|s|$

$$s - p_i \approx s - \bar{s} \qquad \text{and} \qquad s - z_1 \approx s - \bar{s},$$

where $\bar{s}$ is the intersection of the asymptotes with the real axis. Thus, we can also write

$$-K \approx \frac{(s - \bar{s})^{N_p}}{(s - \bar{s})^{N_z}} = (s - \bar{s})^{N_p - N_z}$$

$$= s^{N_p - N_z} - (N_p - N_z)\bar{s}s^{N_p - N_z - 1} + \ldots.$$

Comparing the last two expressions for $-K$ yields (7.1-25).

The **angles of the asymptotes** with respect to the positive real axis are given by

$$\theta_M = \begin{cases} \dfrac{2M + 1}{N_p - N_z} 180° & \text{if } K \geq 0 \\[4mm] \dfrac{2M}{N_p - N_z} 180° & \text{if } K \leq 0 \end{cases} \qquad M = 0, 1, \ldots, |N_p - N_z| - 1.$$

$$(7.1\text{-}26)$$

These results follow from expanding $F(s)$ for large s, as was done for $\bar{s}$, but we refer the reader to Kuo (1982, pp. 393–396) for the proofs.

For our illustrative problem, we have $N_p - N_z = 3$ asymptotes. Their centroid is at

$$\bar{s} = \frac{(0 - 4 - 5) - 0}{3 - 0} = -3$$

and their angles are

$$\theta_M = \frac{2M + 1}{3} 180°, \qquad M = 0, 1, 2$$

$$= 60°, 180°, 300°.$$

This last result can be verified by noting that (7.1-24), at large values of s, implies

$$-K = s(s + 4)(s + 5) \approx s^3.$$

Then for $K \geq 0$, the angle criterion (7.1-18) requires that

$$\angle s^3 = 3\angle s = -180° \pm N \times 360°, \qquad N = 0, 1, \ldots,$$

which yields only three distinct values $\angle s = -60°, 60°, 180°$.

6. Breakpoints

Breakpoints (where branches of the root locus intersect, corresponding to multiple roots) may occur at points on the root locus where

$$\frac{dK(s)}{ds} = 0, \tag{7.1-27}$$

where $K(s)$ is given by (7.1-24). If N branches approach or leave a breakpoint, their arrival and departure angles must be $180°/N$ apart (Kuo, 1982, p. 403).

Condition (7.1-27) applies not only on the real axis, where s and both sides of (7.1-27) are real-valued, but also at complex points s, by setting the real and imaginary parts of dK/ds to zero. Also, (7.1-27) is only a necessary (not a sufficient) condition for a point on the root locus to be a breakpoint.

Breakpoints correspond to local minima, maxima, or saddle points in the value of K along the root locus. This is most easily seen for the case of a breakpoint on the real axis. For our illustrative example, we solve the characteristic equation for $K(s)$, yielding

$$K(s) = -s(s + 4)(s + 5) = -(s^3 + 9s^2 + 20s). \tag{7.1-28}$$

Then

$$0 = \frac{dK}{ds} = -(3s^2 + 18s + 20)$$

yields the possible breakpoints

$$s = -1.472 + i0, \qquad -4.528 + i0.$$

Examining the real-axis segments of the root locus for $K \geq 0$, we observe that the first point is on the root locus, and is therefore a candidate breakpoint, but the second point is not on the root locus. Furthermore, from Figure 7.1-6, we see that the two branches starting (at $K = 0$) from $s = 0$ and $s = -4$ must meet at some point in between, corresponding to a local maximum of $K(s)$ along the real axis, as illustrated in Figure 7.1-7. Thus, we conclude that a breakpoint occurs at $s = -1.472$. The value of K at

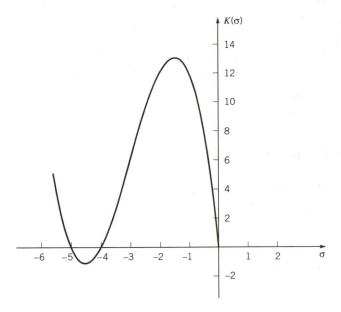

Figure 7.1-7 $K(s)$ on the real axis for the system in Figure 7.1-5, $K \geq 0$.

the breakpoint is given by (7.1-28) as

$$K = -s(s + 4)(s + 5) = 13.128.$$

Also, we note that, since two branches approach the breakpoint, each will turn through 90° as it departs the real axis. That is, in this example, the root loci leave perpendicular to the real axis. This is very frequently the case with breakpoints on the real axis, but not always (see Exercise 7.5-3).

7. Imaginary-Axis Intercepts

The intersection of the root locus with the imaginary axis can be found using the Routh–Hurwitz stability criteria of Section 3.5.

For our illustrative system, the root locus equation (7.1-22), with $F(s)$ given by (7.1-23), yields the polynomial characteristic equation

$$0 = s^3 + 9s^2 + 20s + K$$

and the Routh array

$$
\begin{array}{c|cc}
s^3: & 1 & 20 \\
s^2: & 9 & K \\
s^1: & \dfrac{180 - K}{9} & 0 \\
s^0: & K. &
\end{array}
$$

At $K = 180$, we get a row of zeros. From the preceding row we obtain the auxiliary equation

$$0 = 9s^2 + 180$$

whose roots

$$s = \pm i\sqrt{20} = \pm i4.472$$

are also roots of the original characteristic equation, at which the root locus intersects the imaginary axis.

For our illustrative example, the information gathered thus far is sufficient to make a fairly accurate sketch of the root locus. However, in general, there is one more property that is important for systems having complex poles or zeros in the open-loop transfer function $F(s)$.

8. Angle at a Complex Pole or Zero

If we consider a point s on the root locus and very close to a complex pole ($s \approx p_k$), then the **angle of departure** θ_d of the root locus from a complex pole at p_k is just $\angle(s - p_k)$. Since s is close to p_k, using (7.1-20), $\angle F(s)$ is approximately given by

$$\angle F(s) = \angle K + \sum_{i=1}^{N_z} \angle(p_k - z_i) - \sum_{\substack{j=1 \\ j \neq k}}^{N_p} \angle(p_k - p_j) - \theta_d.$$

From condition (7.1-18) and choosing $\angle F(s) = -180°$, we have

$$\theta_d = 180° + \angle K + \sum_{i=1}^{N_z} \angle(p_k - z_i) - \sum_{\substack{j=1 \\ j \neq k}}^{N_p} \angle(p_k - p_j). \quad \textbf{(7.1-29)}$$

Similarly, if we select $\angle F(s) = 180°$ in (7.1-18), the **angle of arrival** θ_a of the root locus to a complex zero at $s = z_k$ is given by

$$\theta_a = 180° - \angle K + \sum_{j=1}^{N_p} \angle(z_k - p_j) - \sum_{\substack{i=1 \\ i \neq k}}^{N_z} \angle(z_k - z_i). \quad \textbf{(7.1-30)}$$

For example, consider a system with the characteristic equation

$$0 = 1 + \frac{K}{s(s^2 + 2s + 2)} = 1 + F(s),$$

having open-loop poles at

$$s = p_1 = 0$$
$$s = p_2 = -1 + i$$
$$s = p_3 = -1 - i.$$

For $K \geq 0\,(\angle K = 0°)$, the angle of departure from the pole at $p_2 = -1 + i$ is given by

$$\theta_d = 180° + \{0°\} - \{135° + 90°\} = -45°,$$

as illustrated in Figure 7.1-8.

Numerical Computation of the Root Locus

We present here a numerical algorithm for plotting the root locus. The algorithm is a modification of the one in Ash and Ash (1968). In terms of a complex variable $s = x + iy$, the algorithm steps along branches of the root locus at a fixed step size $|\Delta s| = \delta$ and uses Newton's root-finding

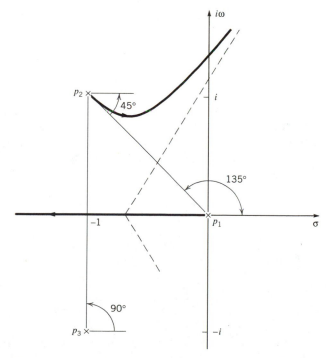

Figure 7.1-8 Angle of departure from complex pole for $0 = 1 + K/[s(s^2 + 2s + 2)]$, $K \geq 0$.

method to satisfy the root locus angle criterion given by (7.1-18) and (7.1-20).

For the open-loop transfer function

$$F(s) = K\frac{\mathcal{Q}(s)}{\mathcal{P}(s)} = K\frac{\displaystyle\prod_{m=1}^{N_z}(s - z_m)}{\displaystyle\prod_{j=1}^{N_p}(s - p_j)}, \qquad (7.1\text{-}31)$$

let $p_j = PR_j + iPI_j$ and $z_m = ZR_m + iZI_m$ denote the open-loop poles and zeros, respectively. The root locus angle criterion can be written as

$$\angle F(s) = n\pi, \qquad (7.1\text{-}32)$$

where

$$n = \begin{cases} \pm 1, \pm 3, \pm 5, \ldots & \text{for } K > 0 \\ 0, \pm 2, \pm 4, \ldots & \text{for } K < 0. \end{cases} \qquad (7.1\text{-}33)$$

We define the angle function

$$\hat{f}(s) = f(x,y) \overset{\Delta}{=} \sum_{m=1}^{N_z} \text{ATAN2}(y - ZI_m, x - ZR_m)$$

$$- \sum_{j=1}^{N_p} \text{ATAN2}(y - PI_j, x - PR_j) - n\pi, \quad (7.1\text{-}34)$$

where $\text{ATAN2}(y,x) = \tan^{-1}(y/x)$ is the Fortran two-argument arctangent function and n is chosen according to (7.1-33) so that $-\pi < \hat{f}(s) \le \pi$. A point s in the complex plane is on the root locus if and only if $\hat{f}(s) = 0$.

Along the root locus, the slope of a branch can be computed using Taylor's theorem

$$f(x + \Delta x, y + \Delta y) = f(x,y) + f_x(x,y)\Delta x + f_y(x,y)\Delta y + O(|\Delta s|),$$

where $O(|\Delta s|)/|\Delta s| \to 0$ as $\Delta s \to 0$ and subscripts denote partial derivatives. The slope is given by

$$\tan \theta = \lim_{|\Delta s| \to 0} \frac{\Delta y}{\Delta x} = -\frac{f_x(x,y)}{f_y(x,y)}$$

and the slope angle, in the direction of increasing gain magnitude $|K|$, is

$$\theta(x,y) = \text{ATAN2}[-f_y(x,y), f_x(x,y)]. \qquad (7.1\text{-}35)$$

For the angle in the direction of decreasing gain magnitude, reverse the signs of the arguments in (7.1-35). The partial derivatives can be determined by differentiating $f(x,y)$, yielding

$$
\begin{aligned}
f_x(x,y) = &- \sum_{m=1}^{N_z} \frac{y - ZI_m}{(x - ZR_m)^2 + (y - ZI_m)^2} \\
&+ \sum_{j=1}^{N_p} \frac{y - PI_j}{(x - PR_j)^2 + (y - PI_j)^2}
\end{aligned}
\tag{7.1-36}
$$

$$
\begin{aligned}
f_y(x,y) = &\sum_{m=1}^{N_z} \frac{x - ZR_m}{(x - ZR_m)^2 + (y - ZI_m)^2} \\
&- \sum_{j=1}^{N_p} \frac{x - PR_j}{(x - PR_j)^2 + (y - PI_j)^2}.
\end{aligned}
\tag{7.1-37}
$$

The plotting algorithm generally starts at an open-loop pole and moves in the direction of increasing gain magnitude, although the open-loop gain K may be either postive or negative. If there are more open-loop zeros than poles, one can start at the zeros and move in the direction of decreasing gain magnitude.

To allow the plotting algorithm to "zoom in" on any portion of the root locus, the plot boundary should be scanned for sign changes in $\hat{f}(s)$. When this occurs, say, between $s = s_1$ and $s = s_2$, a one-dimensional root-finding scheme such as *regula falsi* (Press et al., 1988, p. 263) can be used to find α such that $0 = \bar{f}(\alpha) \stackrel{\Delta}{=} \hat{f}[(1 - \alpha)s_1 + \alpha s_2]$. If the resulting point $s = (1 - \alpha)s_1 + \alpha s_2$ corresponds to a point at which the root locus enters the plotting region (that is, if the appropriate $\theta(x,y)$ points into the plotting region), then this entry point should be added to the list of starting points. If a portion of the plot boundary coincides with the real axis, the sign-change search should be performed slightly above or below the real axis, since $\hat{f}(s)$ is discontinuous across the negative real axis.

From a given point $s = x + iy$ on the root locus, a step of size $|\Delta s| = \delta$ is taken in the direction θ, yielding a point

$$
s + \Delta s = (x + \delta \cos \theta) + i(y + \delta \sin \theta),
\tag{7.1-38}
$$

which is approximately on the root locus. Then Newton's method is used to find the angle θ to satisfy the angle criterion precisely. Defining

$$
h(\theta) \stackrel{\Delta}{=} f(x + \delta \cos \theta, y + \delta \sin \theta),
$$

we have

$$
h(\theta + \Delta \theta) = h(\theta) + h_\theta(\theta)\Delta \theta + O(|\Delta \theta|).
$$

Setting the left-hand side equal to zero and neglecting higher-order terms, we obtain Newton's iteration

$$\theta \rightarrow \theta + \Delta\theta, \tag{7.1-39}$$

where

$$\Delta\theta = -\frac{h(\theta)}{h_\theta(\theta)} \tag{7.1-40}$$

and

$$h_\theta(\theta) = \frac{\partial f}{\partial x}\frac{dx}{d\theta} + \frac{\partial f}{\partial y}\frac{dy}{d\theta}$$

$$= -f_x(x + \delta\cos\theta, y + \delta\sin\theta)\delta\sin\theta \tag{7.1-41}$$

$$+ f_y(x + \delta\cos\theta, y + \delta\sin\theta)\delta\cos\theta.$$

The iteration is repeated until $|h(\theta)|$ is smaller than some specified tolerance.

To detect breakpoints, the slope angles of the root locus are computed at the current point s and at the next point $s + \Delta s$. When the slope angle changes by too large an amount, say,

$$|\theta(s + \Delta s) - \theta(s)| > \min\left\{\frac{\pi}{10}, \frac{\pi}{3|N_p - N_z|}\right\}, \tag{7.1-42}$$

the step size is quartered and the step Δs is recomputed. If the large change in slope angle persists after the step size has been quartered four times ($\delta/4$, $\delta/16$, $\delta/64$, $\delta/256$), then a nearby breakpoint has been detected. Newton's method is then used to determine the precise breakpoint location. Plotting is then continued beyond the breakpoint by restoring the original step size and taking the departing branch that corresponds to the first right-hand turn. Both of these procedures will be explained in more detail.

At a breakpoint, with the gain $K(s)$ expressed in terms of the ratio of two polynomials

$$K(s) = -\frac{\mathcal{P}(s)}{\mathcal{Q}(s)}, \tag{7.1-43}$$

we have

$$0 + i0 = \frac{dK(s)}{ds} = \frac{\mathcal{P}(s)\mathcal{Q}'(s) - \mathcal{Q}(s)\mathcal{P}'(s)}{\mathcal{Q}(s)^2},$$

where $()' = d()/ds$. To determine the precise breakpoint location, we apply a complex-valued Newton's method to find a neighboring point s

such that

$$0 + i0 = g(s) \triangleq \mathcal{P}(s)\mathcal{Q}'(s) - \mathcal{Q}(s)\mathcal{P}'(s). \qquad \textbf{(7.1-44)}$$

The iteration is given by

$$s \rightarrow s + \Delta s, \qquad \textbf{(7.1-45)}$$

where

$$\Delta s = -\frac{g(s)}{g'(s)}. \qquad \textbf{(7.1-46)}$$

Since $\mathcal{P}(s)$ and $\mathcal{Q}(s)$ are polynomials, defined in (7.1-31), the first- and second-order derivatives appearing in (7.1-46) can be computed easily, by analytically differentiating the two polynomials. This procedure differs from the approach in Ash and Ash (1968), in which a stationary point is sought for the angle function $\hat{f}(s)$ instead of for $K(s)$. We choose the gain $K(s)$ because the angle function is insensitive to changes in s far away from the open-loop poles and zeros.

Once a breakpoint has been reached, at $s = s_b$, plotting is continued by taking the departing branch that corresponds to the first right-hand turn. This notion is based on the fact that if N branches meet at a breakpoint, then they leave the breakpoint at angles that are π/N rad apart. Since the order N of the breakpoint is not known, a sequence of test points

$$s_k = s_b + \frac{\delta(\cos \varphi_k + i \sin \varphi_k)}{100}$$

on a small circle centered at s_b is checked to find the value of k yielding the closest match between the slope angle θ_k at s_k and the angle $\angle(s_k - s_b)$, where

$$\varphi_k = \theta_b - \pi + \frac{\pi}{k}, \qquad k = 1, 2, 3, \ldots$$

is the kth trial direction of departure from the breakpoint at s_b, and θ_b is the direction of arrival at the breakpoint.

7.2   ROOT LOCUS DESIGN

PID Design for a Second-Order System

As an application of root locus design, we will consider the problem of designing a controller for the rod and motor inverted pendulum system discussed in Section 1.3 and in Example 2.3-3. We will assume that the armature inductance is small, so that the device acts like a second-order system, with the transfer function

$$\frac{\theta(s)}{U(s)} = G_P(s) = \frac{q_0}{s^2 + p_1 s + p_0}, \qquad (7.2\text{-}1)$$

where $\theta(t)$ is the angle of the rod from the upward vertical and $u(t)$ is the voltage applied to the motor. For a particular implementation, the parameters in the transfer function have been determined experimentally and have the following values:

$$p_0 = -17.627/\text{sec}^2$$
$$p_1 = 0.187/\text{sec} \qquad (7.2\text{-}2)$$
$$q_0 = 0.6455 \text{ rad/sec}^2\text{-V}.$$

We know the system is unstable, and we want to design a controller that will not only stabilize the system but will also produce a zero steady output error in response to a step input. Furthermore, we would like the eigenvalues of the closed-loop system to have real parts less than some specified value, say, $\text{Re}\{\lambda\} < -2$.

We will begin with a PI error feedback controller, as illustrated in Figure 7.2-1, where a $[= 1/(K\tau_i)]$ is constant. However, this controller may not be adequate, since we have observed that it does not provide complete control of the closed-loop eigenvalues when it is applied to second-order systems.

Let $\bar{K} = Kq_0 \geq 0$. Then the root locus form of the characteristic equation is

$$0 = 1 + F(s) = 1 + \bar{K} \frac{s + a}{s(s^2 + p_1 s + p_0)}, \qquad (7.2\text{-}3)$$

where the open-loop transfer function $F(s)$ has poles at $s = -4.293, 0, 4.106$, and one zero at $s = -a$.

As indicated by the plot in Figure 7.2-2, for $s = \sigma + i\omega$, the real-axis segments $0 \leq \sigma \leq 4.106$ and $-4.293 \leq \sigma \leq -a$ are on the root locus if we assume that the positive parameter a is not greater than 4.293. The root

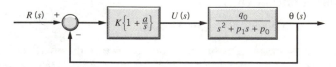

Figure 7.2-1 PI controller for inverted pendulum.

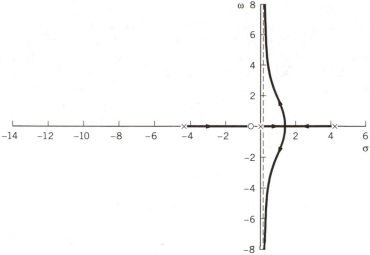

Figure 7.2-2 PI root locus for inverted pendulum.

locus asymptotes occur at angles

$$\theta_M = \frac{2M + 1}{3 - 1} \, 180°, \qquad M = 0, 1$$

$$= \pm 90°,$$

and the asymptotes intersect at

$$\bar{s} = \frac{\Sigma p - \Sigma z}{N_p - N_z} = \frac{(-4.293 + 0 + 4.106) - (-a)}{3 - 1} = \frac{-0.187 + a}{2}. \qquad \textbf{(7.2-4)}$$

By inspection of Figure 7.2-2, we observe that there must be a breakpoint on the real axis between $\sigma = 0$ and $\sigma = 4.106$, with two root loci leaving perpendicular to the real axis since, for $N = 2$, each branch will turn through $180°/2 = 90°$.

From (7.2-4) we see that for the system to be stable for some sufficiently large value of $\bar{K}$, we must have $a < 0.187$. In addition, integral control requires $a \geq 0$, for otherwise, the system will be unstable for all values of $\bar{K}$. Finally, we note that the smaller we make a, the more the asymptotes move to the left. However, we can move the asymptotes no further to the left than $s = -0.0935$, which will not produce the desired closed-loop eigenvalues with a real part less than -2.

It is clear that we need some additional compensation in the feedback controller. It should also be evident from our previous observations that derivative feedback will solve the problem. Derivative feedback is introduced by means of a phase lead feedback circuit, such as the proportional + pseudo-derivative or lead circuits discussed in Section 6.5. Combined with PI control, this will produce the PID control compensator shown in Figure 6.5-14, with $1/\tau_i = Ka$. We have already learned that such a controller, applied to a general second-order system, will allow us to choose any desired closed-loop eigenvalues.

To see the effect of PID compensation on the root locus of our inverted pendulum system, we note that the new open-loop transfer function

$$F(s) = Kq_0 \frac{s + a}{s(s^2 + p_1 s + p_0)} \left(\frac{1 + \tau_e s}{1 + \tau_a s}\right) \tag{7.2-5}$$

has an additional pole and zero, with the pole ($s = -1/\tau_a$) to the left of the zero ($s = -1/\tau_e$), since $\tau_e > \tau_a$ for a lead circuit.

For the new root locus the angles of the asymptotes will remain as before, at $\pm 90°$, since the net number of open-loop poles minus zeros has not changed. The new centroid of the asymptotes is given by

$$\bar{s} = \frac{\Sigma p - \Sigma z}{N_p - N_z} = \frac{(-4.293 - 1/\tau_a + 0 + 4.106) - (-a - 1/\tau_e)}{4 - 2}$$

$$= \frac{-0.187}{2} + \frac{1}{2}\left(a + \frac{1}{\tau_e} - \frac{1}{\tau_a}\right) \tag{7.2-6}$$

and it can be shifted as far left as we like by choosing $\tau_a > 0$ sufficiently small. Shifting the asymptotes to the left of $s = -2$ will not satisfy the design requirements completely. In order to have *all* of the eigenvalues to the left of $s = -2$, we should not introduce any real-valued poles or zeros in the interval $-2 \leq s \leq 0$; for otherwise, a corresponding portion of that region on the real axis will lie on the root locus.

One suitable design is given by the parameter values

$$a = 3$$

$$\tau_e = 0.2 \tag{7.2-7}$$

$$\tau_a = 0.08,$$

which yields the open-loop transfer function

$$F(s) = \overline{K} \frac{s + 3}{s(s^2 + p_1 s + p_0)} \left(\frac{s + 5}{s + 12.5}\right), \tag{7.2-8}$$

where now $\overline{K} = 2.5Kq_0$. The resulting root locus is shown in Figure 7.2-3.

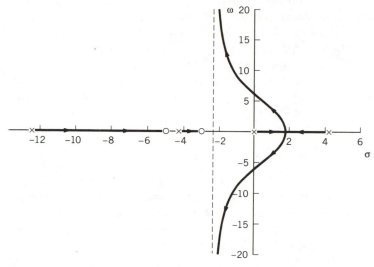

Figure 7.2-3 Root locus for inverted pendulum with PID feedback control.

7.3   ROBUST PARAMETRIC STABILITY

At various points in this text, we have considered both state-space systems and IO systems subject to uncertain inputs $v(t)$. In this section we will consider systems with uncertain parameters. This type of uncertainty differs in two ways from an uncertain input $v(t)$. First, the uncertain parameters in the system are constants, rather than unknown functions of time. We will, however, assume that the unknown parameters are bounded and that the upper and lower bounds are known. Second, the unknown parameters are assumed to be one or more elements of the matrices in a state-space representation

$$\dot{x} = Ax + Bu$$

$$y = Cx + Du$$

or the coefficients p_i or q_j in an IO representation

$$\frac{d^n y}{dt^n} + p_{n-1} \frac{d^{n-1} y}{dt^{n-1}} + \cdots + p_1 \dot{y} + p_0 y = q_0 u + q_1 \dot{u} + \cdots + q_n \frac{d^n u}{dt^n}.$$

Note that the unknown parameters are multiplicative coefficients, rather than additive uncertain inputs, which we have previously considered. Of course, uncertainties in parameters can be converted to an equivalent representation involving the types of uncertain inputs that we discussed previously in connection with (6.3-11).

Also note that, regardless of the system representation, the uncertain parameters appear linearly, although they themselves may be nonlinear functions of various system parameters, such as a subsystem gain, time constant, damping ratio, natural frequency, and so on. This linearity assumption leads to a very elegant result dealing with guaranteed asymptotic stability.

Systems with Uncertain Parameters

As we learned in Section 4.4, when a system is subjected to a bounded but otherwise unknown time-varying input $v(t)$, one cannot guarantee that the origin (or other equilibrium point) can be made asymptotically stable by a suitable choice of a control $u(\cdot)$. Usually, the best one can do is to achieve **guaranteed ultimate boundedness**, that is, find a control $u(\cdot)$ that will keep the state near (perhaps arbitrarily near) the equilibrium. For results in this area, including systems with uncertain parameters, see Gutman (1979); Leitmann (1981); Corless and Leitmann (1981); Barmish and Leitmann (1982); Ryan et al. (1985).

When the uncertainties in a system are constants rather than unknown time-varying inputs, it is possible a controller can still be designed that achieves not just guaranteed ultimate boundedness but also **guaranteed asymptotic stability**, that is, asymptotic stability regardless of the uncertain parameters. For example, in an unforced second-order IO system in standard form,

$$\ddot{y} + 2\zeta\omega_n\dot{y} + \omega_n^2 y = 0,$$

the origin is asymptotically stable for all damping ratios $\zeta > 0$.

In this section we will consider the stability of feedback control systems with bounded uncertain parameters. The uncertain parameters may occur in the primary system, the measurement system, or in the controller. Regardless of where the uncertain parameters occur and whether the system is considered in state-space or IO form, stability is determined by the roots of the characteristic equation, an nth-order polynomial with real coefficients of the general form

$$0 = f(\lambda) \triangleq c_0 + c_1\lambda + c_2\lambda^2 + \cdots + c_n\lambda^n, \tag{7.3-1}$$

with the coefficients unknown except for the bounds

$$\alpha_i \leq c_i \leq \beta_i, \qquad i = 0, \ldots, n. \tag{7.3-2}$$

Kharitonov's Four Stability Polynomials

For a particular set of coefficient values, a polynomial of the form (7.3-1) will be termed (asymptotically) **stable** if all of the roots of (7.3-1) have negative real parts. The **guaranteed asymptotic stability** problem is to determine conditions under which all of the polynomials (7.3-1)–(7.3-2) are stable, within the bounds given for the uncertain coefficients.

If even one of the coefficients in (7.3-1) is uncertain, then there are an infinite number of polynomials formed by (7.3-1)–(7.3-2). Thus, it is not possible to check directly the stability of all of the uncertain polynomials. Fortunately, this is not required. Even though the coefficients may be highly nonlinear functions of the various system parameters, it turns out that a necessary and sufficient condition for all of the polynomials (7.3-1)–(7.3-2) to be stable is that the set of all polynomials with coefficients at their upper or lower bounds be stable. The necessity of this condition is obvious. For a proof of sufficiency (based on linearly appearing coefficients and the resulting convexity of the family of characteristic polynomials), see Kharitonov (1978).

For an nth-order system with all coefficients uncertain, a brute-force check of all of the polynomials having coefficients at their upper or lower bounds would involve 2^{n+1} polynomials. For example, a second-order system would involve 8 polynomials, and a third-order system 16, and so on. The really elegant result in Kharitonov (1978) is that *one only needs to check four polynomials* involving coefficients at their upper or lower bounds. These four polynomials are of the form (7.3-1), and their coefficients are defined in a pairwise fashion, $c_{2k}, c_{2k+1}, k = 0, 1, \ldots, \hat{k}$, $\hat{k} = n/2$ if $n =$ even, $\hat{k} = (n-1)/2$ if $n =$ odd. Based on the pairwise assignment of coefficients at upper or lower bounds, we assign each of the four **Kharitonov polynomials** with a mnemonic $\{k =$ even; $k =$ odd$\}$ name:

$f_1(\lambda)$ {max, max; min, min}

$$c_{2k} = \begin{cases} \beta_{2k} & k \text{ even} \\ \alpha_{2k} & k \text{ odd} \end{cases}$$

$$c_{2k+1} = \begin{cases} \beta_{2k+1} & k \text{ even} \\ \alpha_{2k+1} & k \text{ odd} \end{cases}$$

$f_2(\lambda)$ {min, min; max, max}

$$c_{2k} = \begin{cases} \alpha_{2k} & k \text{ even} \\ \beta_{2k} & k \text{ odd} \end{cases}$$

$$c_{2k+1} = \begin{cases} \alpha_{2k+1} & k \text{ even} \\ \beta_{2k+1} & k \text{ odd} \end{cases}$$

$f_3(\lambda)$ {min, max; max, min}

$$c_{2k} = \begin{cases} \alpha_{2k} & k \text{ even} \\ \beta_{2k} & k \text{ odd} \end{cases}$$

$$c_{2k+1} = \begin{cases} \beta_{2k+1} & k \text{ even} \\ \alpha_{2k+1} & k \text{ odd} \end{cases}$$

$f_4(\lambda)$ {max, min; min, max}

$$c_{2k} = \begin{cases} \beta_{2k} & k \text{ even} \\ \alpha_{2k} & k \text{ odd} \end{cases}$$

$$c_{2k+1} = \begin{cases} \alpha_{2k+1} & k \text{ even} \\ \beta_{2k+1} & k \text{ odd.} \end{cases}$$

The set of polynomials $f(\lambda)$ in (7.3-1) with bounded coefficients (7.3-2) have no roots λ with $\text{Re}(\lambda) \geq 0$ if and only if the four polynomials $f_i(\lambda)$, $i = 1, \ldots, 4$, have no roots λ with $\text{Re}(\lambda) \geq 0$. Proofs of this remarkable theorem can be found in Kharitonov (1978) and Bose (1985).

As an example, which illustrates the naming convention, consider the third-order characteristic equation

$$0 = f(\lambda) = c_0 + c_1\lambda + c_2\lambda^2 + c_3\lambda^3$$

with coefficient bounds $\alpha_i \leq c_i \leq \beta_i$, $i = 0, \ldots, 3$. We have $n = 3$ so that $\hat{k} = 1$, and the Kharitonov polynomials are

$$
\begin{array}{cc}
k = 0 & k = 1 \\
\overbrace{\qquad\qquad} & \overbrace{\qquad\qquad}
\end{array}
$$

$$
\begin{aligned}
f_1(\lambda) &= \beta_0 + \beta_1\lambda + \alpha_2\lambda^2 + \alpha_3\lambda^3 \\
f_2(\lambda) &= \alpha_0 + \alpha_1\lambda + \beta_2\lambda^2 + \beta_3\lambda^3 \\
f_3(\lambda) &= \alpha_0 + \beta_1\lambda + \beta_2\lambda^2 + \alpha_3\lambda^3 \\
f_4(\lambda) &= \beta_0 + \alpha_1\lambda + \alpha_2\lambda^2 + \beta_3\lambda^3
\end{aligned}
$$

For $f_1(\lambda)$ the coefficients appear in a {max, max; min, min} sequence. For higher-order systems, requiring additional coefficients, this sequence or partial sequence would simply be repeated until all coefficients have been determined. Similar coefficient upper- and lower-bound sequences occur for the other Kharitonov polynomials, leading to the mnemonic names chosen for the polynomials.

Routh–Hurwitz Criteria

The stability of the Kharitonov polynomials can be tested by using the Routh–Hurwitz criteria.

EXAMPLE 7.3-1   **Control of a Gun Turret**

A field-controlled dc motor implements error feedback control of the angular position $\theta(t)$ of a damped rotating gun turret, as shown in Figure 7.3-1. The nominal system parameters and their uncertainty ranges are $a = 4 \pm 1$, $\tau = 0.15 \pm 0.1$, and $K = 2.5 \pm 0.5$. The characteristic equation for this system is

$$0 = f(\lambda) = c_0 + c_1\lambda + c_2\lambda^2 + c_3\lambda^3,$$

where, from the parameter ranges, we have

$$c_0 = \frac{K}{\tau} \in [8, \quad 60]$$

$$c_1 = \frac{a}{\tau} \in [12, \quad 100]$$

$$c_2 = \frac{1 + \tau a}{\tau} \in [7, \quad 25]$$

$$c_3 = 1.$$

The four Kharitonov polynomials are

$$f_1(\lambda) = 60 + 100\lambda + \ 7\lambda^2 + \lambda^3$$

$$f_2(\lambda) = \ 8 + \ 12\lambda + 25\lambda^2 + \lambda^3$$

$$f_3(\lambda) = \ 8 + 100\lambda + 25\lambda^2 + \lambda^3$$

$$f_4(\lambda) = 60 + \ 12\lambda + \ 7\lambda^2 + \lambda^3.$$

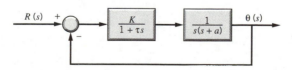

Figure 7.3-1 Example system for Kharitonov polynomials.

For each $f_i(\lambda)$, reading alternating coefficients from right to left, we determine the following Routh arrays:

$f_1(\lambda)$		$f_2(\lambda)$		$f_3(\lambda)$		$f_4(\lambda)$	
1	100	1	12	1	100	1	12
7	60	25	8	25	8	7	60
$\dfrac{640}{7}$	0	$\dfrac{292}{25}$	0	$\dfrac{2492}{25}$	0	$\dfrac{24}{7}$	0
60		8		8		60	

Since column 1 for each Kharitonov polynomial contains no sign changes, we conclude that all of the roots for each $f_i(\lambda) = 0$ have negative real parts. Thus, the control system is stable for all parameter values in the specified ranges. That is, the control system is guaranteed asymptotically stable.

Root Locus Criteria

The root locus is a plot of the location of the roots of the characteristic equation in terms of a single parameter $K \geq 0$. By considering K as an unknown parameter, with $K_{min} \leq K \leq K_{max}$, Kharitonov's technique can be used to determine if the system is asymptotically stable for a range of K values.

For an nth-order system the characteristic equation in root locus form can be written as

$$0 = 1 + K\frac{q_0 + q_1 s + \cdots + q_n s^n}{p_0 + p_1 s + \cdots + p_n s^n},$$

which yields the equivalent polynomial representation

$$0 = f(\lambda) = c_0 + c_1\lambda + \cdots + c_n\lambda^n,$$

where

$$c_i = p_i + Kq_i, \quad i = 0, 1, \ldots, n.$$

Since the c_i are linear in K, it follows that the upper and lower bounds of each c_i will occur at the corresponding K bounds if $q_i > 0$, or at the opposite bounds if $q_i < 0$. Note that c_i is fixed if $q_i = 0$ (or if K is fixed). In any event, the four Kharitonov polynomials are functions of only one uncertain parameter K and, hence, they reduce to only two polynomials, which correspond to $f(\lambda)$ itself, evaluated at $K = K_{min}$ and $K = K_{max}$. Thus, in the case of a single parameter the usual root locus procedure for imaginary-axis crossings, using the Routh array for $f(\lambda)$ with parameter K, completely determines the range of K values for guaranteed asymptotic stability.

For the more general case, where there is more than one uncertain parameter, Kharitonov's four polynomials coupled with the Routh–Hurwitz criteria can determine stability bounds on the parameters more easily than plotting several root loci. Of course, the Routh–Hurwitz stability criteria alone could be used, but at the cost of having possibly some complicated, generally nonlinear, functions of the parameters appear in column one of the Routh array.

EXAMPLE 7.3-2   **Turret System with Uncertain Gain K**

Consider the same system as in Example 7.3-1, but with $a = 4$, $\tau = 0.2$, and $0 < K_{min} \le K \le K_{max}$. The characteristic equation is

$$0 = f(\lambda) = 5K + 20\lambda + 9\lambda^2 + \lambda^3$$

and Kharitonov's polynomials are

$$f_1(\lambda) = f_4(\lambda) = 5K_{max} + 20\lambda + 9\lambda^2 + \lambda^3$$
$$f_2(\lambda) = f_3(\lambda) = 5K_{min} + 20\lambda + 9\lambda^2 + \lambda^3.$$

Thus, for this example we only need to check the stability of $f(\lambda)$ itself at the upper and lower bounds on K, instead of plotting the entire root locus. That is, we are assured in this case that if $f(\lambda)$ is stable at both $K = K_{min}$ and $K = K_{max}$, then the root locus will not cross (or wander back and forth across) the $Re(\lambda) = 0$ imaginary axis.

The Routh array for $f(\lambda)$ is given by

$$\begin{array}{cc} 1 & 20 \\ 9 & 5K \\ \dfrac{180 - 5K}{9} & 0 \\ 5K & \end{array}$$

and we see that the roots of $f(\lambda) = 0$ have negative real parts for $0 < K < 36$. This system has the same root locus as in Figure 7.1-6, with K in Figure 7.1-6 being $5K$ here.

7.4 FREQUENCY RESPONSE METHODS

Until now all of our stability results have dealt with the time domain and the eigenvalues corresponding to the free response of a system. However, we have learned in Section 5.2 that the transfer function $G(s)$ for an IO system, evaluated, as a function of frequency, at $s = i\omega$, can also serve to determine the (residual) response to sinusoidal inputs. In this section we will see that similar frequency response methods can also yield information about the stability of a system. Historically, these frequency response methods were developed for the analysis of ac electrical systems, since they typically deal with sinusoidal inputs. However, frequency response methods are also directly applicable to vibrating mechanical systems. In fact, the analysis techniques are also applicable to general stability analysis for constant coefficient linear systems, even if the actual planned inputs are not sinusoidal.

As in root locus analysis, it is convenient to write the characteristic equation, say, for the system in Figure 7.1-1, in the form

$$0 = \Psi(s) \triangleq 1 + F(s), \qquad (7.4\text{-}1)$$

where $\Psi(s)$ corresponds to the closed-loop characteristic polynomial, and $F(s)$ is the open-loop transfer function, of the form

$$F(s) \triangleq K \frac{\mathcal{Q}(s)}{\mathcal{P}(s)} = K \frac{\displaystyle\prod_{i=1}^{N_z} (s - z_i)}{\displaystyle\prod_{j=1}^{N_p} (s - p_j)}, \qquad (7.4\text{-}2)$$

where $s = p_j$ are the open-loop poles and $s = z_i$ are the open-loop zeros.

Note that the poles of $F(s)$, $s = p_j, j = 1, \ldots , N_p$, are the poles of the function $\Psi(s) = [\mathcal{P}(s) + K\mathcal{Q}(s)]/\mathcal{P}(s)$ and that the zeros of $\Psi(s)$, $s = \lambda_k$, $k = 1, \ldots , n \triangleq \max\{N_p, N_z\}$, are the n eigenvalues of the closed-loop system. The function $\Psi(s)$ can also be written in terms of its poles and zeros in a form similar to (7.4-2). However, all of the pertinent information is contained in the open-loop transfer function $F(s)$ and we will work with the characteristic equation in the form

$$F(s) = -1 + i0. \qquad (7.4\text{-}3)$$

Polar Plots

Frequency response methods are associated with sinusoidal inputs. Thus we will be concerned with $F(s)$ evaluated at $s = i\omega$. In particular, it will be informative to construct a **polar plot** of $F(i\omega)$ in the complex plane, by plotting (ρ, φ) for $0 \leq \omega < \infty$, with

$$
\begin{aligned}
F(i\omega) &= \mu + iv \\
&= \rho\,(\cos \varphi + i \sin \varphi) \qquad (7.4\text{-}4) \\
&= \rho e^{i\varphi},
\end{aligned}
$$

where ρ is the modulus of $F(i\omega)$, given by

$$\rho = |F(i\omega)| \triangleq \sqrt{F(i\omega)F(-i\omega)} = \sqrt{\mu^2 + v^2}, \qquad (7.4\text{-}5)$$

and φ is the angle of $F(i\omega)$, given by

$$\varphi = \angle F(i\omega) \triangleq \tan^{-1}\left(\frac{\mathrm{Im}\{F(i\omega)\}}{\mathrm{Re}\{F(i\omega)\}}\right) = \tan^{-1}\left(\frac{v}{\mu}\right). \qquad (7.4\text{-}6)$$

EXAMPLE 7.4-1 **First-Order System**

Consider a first-order system with the open-loop transfer function

$$F(s) = \frac{1}{1 + Ts} \Rightarrow F(i\omega) = \frac{1}{1 + i\omega T}. \qquad (7.4\text{-}7)$$

Then

$$|F(i\omega)| = \frac{1}{\sqrt{1 + (\omega T)^2}} \qquad (7.4\text{-}8)$$

$$\angle F(i\omega) = \tan^{-1}(-\omega T). \qquad (7.4\text{-}9)$$

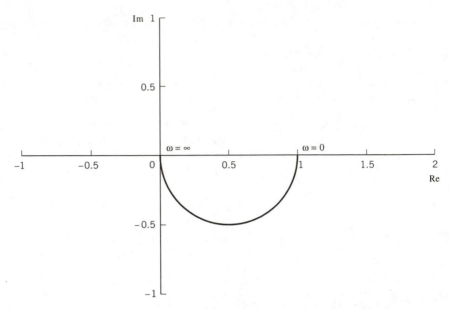

Figure 7.4-1 Polar plot of $F(i\omega)$ for $F(s) = 1/(1 + Ts)$.

The polar plot of $F(i\omega)$ is shown in Figure 7.4-1 for $0 \le \omega < \infty$ and $T > 0$. The plot for negative ω is symmetric about the real axis and is not shown in the figure.

EXAMPLE 7.4-2 **Second-Order System**

Consider a second-order system with the open-loop transfer function

$$F(s) = \frac{\omega_n^2}{s^2 + 2\zeta\omega_n s + \omega_n^2}. \tag{7.4-10}$$

Then

$$|F(i\omega)| = \frac{1}{\sqrt{[1 - (\omega/\omega_n)^2]^2 + (2\zeta\omega/\omega_n)^2}}, \tag{7.4-11}$$

and

$$\angle F(i\omega) = \tan^{-1}\left[\frac{-2\zeta\omega/\omega_n}{1 - (\omega/\omega_n)^2}\right]. \tag{7.4-12}$$

The polar plot of $F(i\omega)$, $0 \le \omega < \infty$, is shown in Figure 7.4-2 for various values of the damping ratio ζ. It can be shown that (1) the resonant peak in

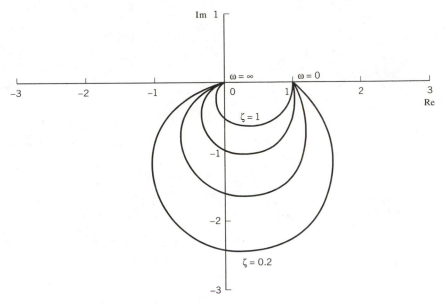

Figure 7.4-2 Polar plots of $F(i\omega)$ for $F(s) = \omega_n^2/(s^2 + 2\zeta\omega_n s + \omega_n^2)$.

$F(i\omega)$, given by (5.2-10)–(5.2-11), corresponds to $\max|F(i\omega)|$, and (2) the negative imaginary-axis intercept in Figure 7.4-2 corresponds to $\omega = \omega_n$. We leave the proof as an exercise for the reader.

EXAMPLE 7.4-3   **Phase Lead or Lag System**

The lead and lag systems considered previously in Section 6.5 have a transfer function of the form

$$F(s) = \frac{1 + \tau_e s}{1 + \tau_a s}, \tag{7.4-13}$$

where $\tau_e > \tau_a$ for a phase lead system and $\tau_e < \tau_a$ for a phase lag system. For a polar plot of

$$F(i\omega) = \frac{1 + i\omega\tau_e}{1 + i\omega\tau_a},$$

we have

$$|F(i\omega)| = \sqrt{\frac{1 + (\omega\tau_e)^2}{1 + (\omega\tau_a)^2}}$$

$$\varphi \triangleq \angle F(i\omega) = \tan^{-1}(\omega\tau_e) - \tan^{-1}(\omega\tau_a).$$

Since

$$\tan(\alpha - \beta) = \frac{\tan \alpha - \tan \beta}{1 + \tan \alpha \tan \beta},$$

we have

$$\tan \varphi = \frac{\omega \tau_e - \omega \tau_a}{1 + \omega^2 \tau_e \tau_a}.$$

The maximum or minimum value φ_m of φ can be determined from

$$0 = \frac{d\varphi}{d\omega} = \frac{\tau_e}{1 + (\omega \tau_e)^2} - \frac{\tau_a}{1 + (\omega \tau_a)^2},$$

which yields

$$\omega_m = \frac{1}{\sqrt{\tau_e \tau_a}} \tag{7.4-14}$$

and

$$\varphi_m = \tan^{-1} \sqrt{\frac{\tau_e}{\tau_a}} - \tan^{-1} \sqrt{\frac{\tau_a}{\tau_e}}.$$

This result can be put in the more useful form

$$\sin \varphi_m = \frac{\alpha - 1}{\alpha + 1}, \qquad \alpha = \frac{\tau_e}{\tau_a}, \tag{7.4-15}$$

where $\alpha > 1$ and $\varphi_m > 0$ for a phase lead device and $\alpha < 1$ and $\varphi_m < 0$ for a phase lag device. The result in (7.4-15) follows from the expression for $\tan \varphi$. The geometry, shown in Figures 7.4-3 and 7.4-4 for a lead and a lag

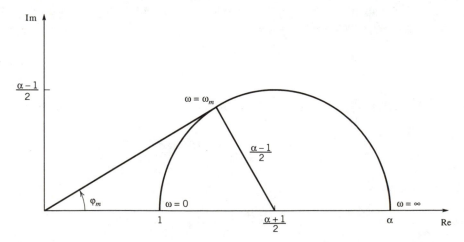

Figure 7.4-3 Polar plot for a phase lead device: $F(s) = (1 + \tau_e s)/(1 + \tau_a s)$, $\alpha = \tau_e/\tau_a > 1$.

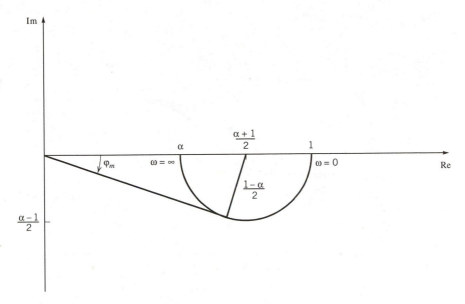

Figure 7.4-4 Polar plot for a phase lag device: $F(s) = (1 + \tau_e s)/(1 + \tau_a s)$, $\alpha = \tau_e/\tau_a < 1$.

device, respectively, can be verified by writing $F(i\omega) = \mu + i\nu$ and showing that

$$\left(\mu - \frac{\alpha + 1}{2}\right)^2 + \nu^2 = \frac{(\alpha - 1)^2}{4}. \tag{7.4-16}$$

Note that, from (7.4-15), the maximum value possible for φ_m for a lead device is $\varphi_m \to \pi/2$ as $\alpha \to \infty$. Similarly, for a lag device the most negative value possible for φ_m is $\varphi_m \to -\pi/2$ as $\alpha \to 0$. Also note that, on a $\log_{10}(\omega)$ scale, the frequency ω_m from (7.4-14) at which φ_m occurs is given by

$$\log_{10}(\omega_m) = \frac{1}{2}\left[\log_{10}\left(\frac{1}{\tau_e}\right) + \log_{10}\left(\frac{1}{\tau_a}\right)\right]. \tag{7.4-17}$$

Thus, on a Bode plot, the maximum (minimum) phase of a lead (lag) device occurs at a frequency corresponding to the geometric midpoint between the two corner frequencies, as illustrated in Figures 7.4-5 and 7.4-6.

Contour Mapping

In order to develop stability criteria in the frequency domain, we will be interested in what happens to a closed contour Γ_s in the complex s plane when points on Γ_s are mapped to corresponding points Γ_ψ in the complex

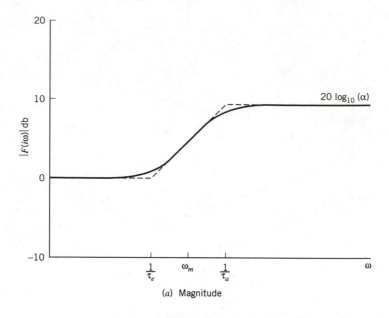

(a) Magnitude

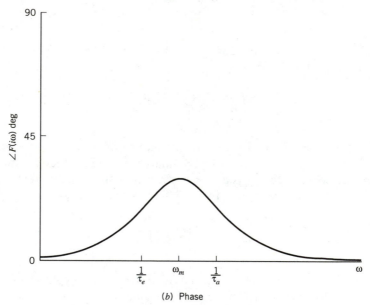

(b) Phase

Figure 7.4-5 Bode plot for a phase lead device: $F(s) = (1 + \tau_e s)/(1 + \tau_a s)$, $\alpha = \tau_e/\tau_a > 1$.

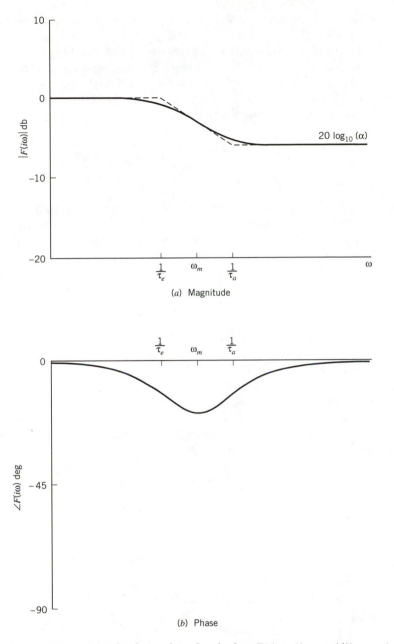

(a) Magnitude

(b) Phase

Figure 7.4-6 Bode plot for a phase lag device: $F(s) = (1 + \tau_e s)/(1 + \tau_a s)$, $\alpha = \tau_e/\tau_a < 1$.

Ψ plane by the function $\Psi(s) = \mu + i\nu$ defined by (7.4-1). The situation is illustrated in Figure 7.4-7.

To proceed, we will first present some definitions from the theory of complex variables (e.g., Kreyszig, 1983). The presentation will be in terms of the closed-loop function $\Psi(s)$, although the final result will be applied using the open-loop transfer function $F(s)$.

The function $\Psi(s)$ is **analytic** at a point s_0 in the complex plane if and only if

$$\left. \frac{d\Psi(s)}{ds} \right|_{s=s_0} \overset{\Delta}{=} \lim_{\Delta s \to 0} \frac{\Psi(s_0 + \Delta s) - \Psi(s_0)}{\Delta s}$$

exists uniquely as $\Delta s \to 0$ from any direction in the complex plane.

To test for an analytic function, let $s = x + iy$ and $\Psi(s) = \mu(x,y) + i\nu(x,y)$. Then $\Psi(s)$ is analytic at s if and only if $\Psi(s)$ satisfies the **Cauchy–Reimann equations**

$$\frac{\partial \mu}{\partial x} = \frac{\partial \nu}{\partial y} \quad \text{and} \quad \frac{\partial \mu}{\partial y} = -\frac{\partial \nu}{\partial x}. \tag{7.4-18}$$

If $\Psi(s)$ is analytic at s, then all of the usual rules for differentiation apply and all derivatives of $\Psi(s)$ of all orders exist and are continuous in a neighborhood of s. For example, $\Psi(s) = 1/s$ is analytic everywhere except at $s = 0 + i0$. Thus, except at the origin, $\Psi'(s) = -1/s^2$. **Rational functions,** defined by the ratio of two polynomials $\Psi(s) = N(s)/D(s)$, are analytic except where $D(s) = 0$. At such a point, $\Psi(s)$ has a pole. More precisely, a function $\Psi(s)$ has a **pole of order** k at $s = p$, if and only if $\overline{\Psi}(s) \overset{\Delta}{=} (s - p)^k\Psi(s)$ is analytic at $s = p$ and k is the smallest positive integer for which this is true.

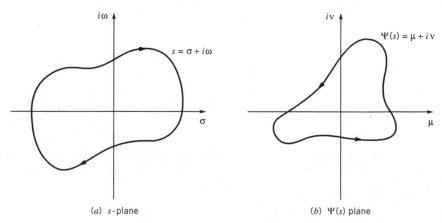

(a) s-plane   (b) $\Psi(s)$ plane

Figure 7.4-7 Contour mapping.

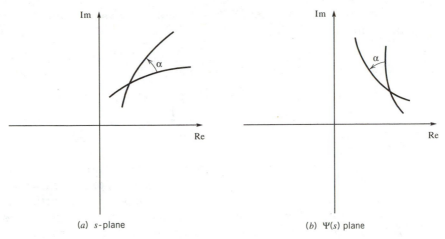

(a) s-plane   (b) Ψ(s) plane

Figure 7.4-8 Conformal mapping preserves angles, in both magnitude and direction.

A function $\Psi(s)$ is a **conformal mapping** on a region $\mathcal{S}$ in the s plane if and only if it preserves angles in both magnitude and orientation, as illustrated in Figure 7.4-8. It can be shown (Kreyszig, 1983, p. 603) that $\Psi(s)$ is conformal on a region $\mathcal{S}$ if (1) $\Psi(s)$ is analytic on $\mathcal{S}$ and (2) $\Psi'(s) \neq 0$ on $\mathcal{S}$.

Consider a contour Γ_s in the s plane, as illustrated in Figure 7.4-9. We take a clockwise traverse around Γ_s as the positive direction. All points to the right of Γ_s as it is traversed in the positive direction are said to be **enclosed** by Γ_s, indicated by the shaded regions in Figure 7.4-9.

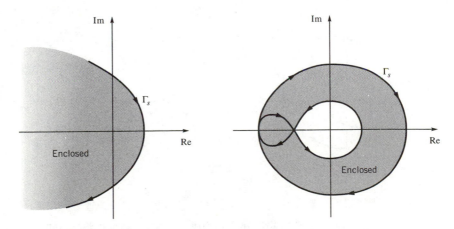

Figure 7.4-9 Points "enclosed" by a contour in the s-plane.

The closed-loop function $\Psi(s)$ will be a single-valued rational function of the form

$$\Psi(s) = \overline{K} \frac{\displaystyle\prod_{k=1}^{n} (s - \lambda_k)}{\displaystyle\prod_{j=1}^{N_p} (s - p_j)}.$$

Let Γ_s be an arbitrary closed path in the s plane that does not pass through any poles or zeros of $\Psi(s)$. Let Γ_Ψ be the corresponding closed path of point $\Psi(s)$ as s traverses Γ_s in the positive (clockwise) direction.

The path Γ_Ψ in the Ψs plane makes N **positive encirclements** of a point α if the phasor (the line from α to a moving point Ψs on Γ_Ψ) rotates through $N \times 360°$ clockwise as s traverses Γ_s in the positive (clockwise) direction. As illustrated in Figure 7.4-10 with α as the origin, the net number N of encirclements of α can be computed by drawing an arbitrary ray from α. Then N is the net number of times that the closed curve Γ_Ψ crosses the ray, with clockwise crossings being positive and counterclockwise crossings being negative.

Cauchy's **principle of the argument** states that if Γ_s encircles Z zeros and P poles of $\Psi(s)$, then Γ_Ψ encircles the origin in the $\Psi(s)$ plane N times in the positive direction, where

$$N = Z - P. \tag{7.4-19}$$

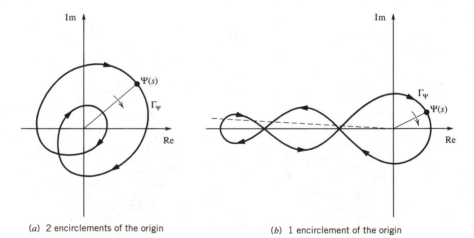

(a) 2 encirclements of the origin   (b) 1 encirclement of the origin

Figure 7.4-10 Encirclements of the origin in the $\Psi(s)$ plane.

This result follows from the fact that

$$\angle \Psi(s) = \angle \overline{K} + \sum_{k=1}^{n} \angle(s - \lambda_k) - \sum_{j=1}^{N_p} \angle(s - p_j).$$

For a zero, $s - \lambda_k$, there will be $+1$ encirclements of the origin by Γ_Ψ if λ_k is inside Γ_s and 0 encirclements of the origin if λ_k is outside Γ_s. For a pole, $1/(s - p_j)$, there will be -1 encirclements of the origin by Γ_Ψ if p_j is inside Γ_s and 0 encirclements of the origin if p_j is outside Γ_s. Figure 7.4-11 illustrates several applications of (7.4-19).

Nyquist Stability Criterion

To translate the result in (7.4-19) into a stability criterion, we use a path that encircles the right half of the complex s plane. The **Nyquist path** Γ_s, illustrated in Figure 7.4-12a (used in Example 7.4-4 that we will discuss shortly), consists of the imaginary axis $s = \pm i\omega$ and a semicircle of infinite radius covering the right-half s plane. If there are any poles or zeros of $\Psi(s)$ on the imaginary axis, the path Γ_s is perturbed infinitesimally to go around these points to the right of the points, that is, so that they are not "enclosed" by Γ_s.

We can determine the number Z of zeros of $\Psi(s)$ in the right-half s plane, that is, the number of closed-loop eigenvalues with a positive real part, by traversing the Nyquist path in a clockwise direction and determining the number N of times that the resulting $\Psi(s)$ contour Γ_Ψ encircles the origin clockwise in the $\Psi(s)$ plane. Then Z can be computed from (7.4-19).

Since $\Psi(s) = 1 + F(s)$, the origin in the $\Psi(s)$ plane corresponds to the point $F(s) = -1 + i0$ in the $F(s)$ plane. Let Γ_F be the contour generated in the $F(s)$ plane as a point s moves clockwise around the Nyquist path Γ_s in the s plane. The **Nyquist stability criterion** is that a system will be asymptotically stable provided

$$N + P = 0, \tag{7.4-20}$$

where P = number of poles p of $F(s)$ with $\text{Re}(p) > 0$, and N = net number of encirclements (clockwise = positive, counterclockwise = negative) of the $F(s) = -1 + i0$ point by the contour Γ_F in the $F(s)$ plane. The number N can be determined, using a ray from the $F(s) = -1 + i0$ point, as the net number of times that the Γ_F contour crosses the ray, with clockwise being positive and counterclockwise negative.

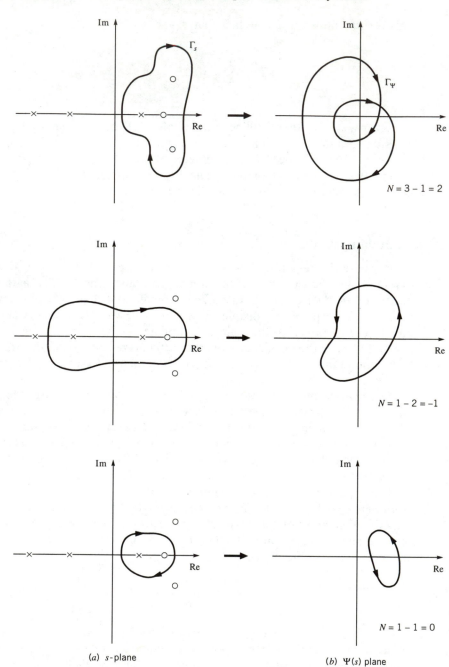

(a) s-plane

(b) $\Psi(s)$ plane

Figure 7.4-11 Examples of the application of (7.4-19).

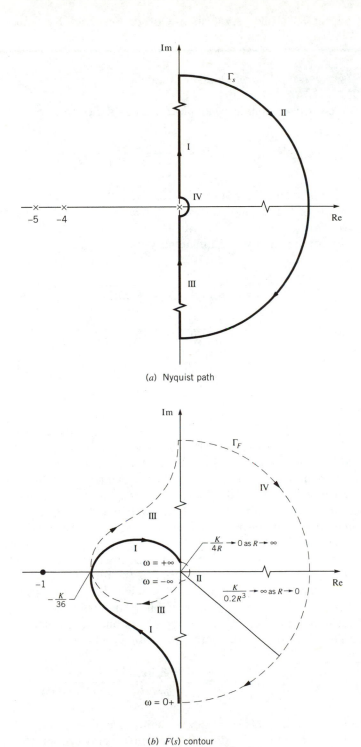

(a) Nyquist path

(b) F(s) contour

Figure 7.4-12 Nyquist stability criterion for Example 7.4-4.

EXAMPLE 7.4-4   **Turret Control System**

The control system in Example 7.3-2 has the open-loop transfer function

$$F(s) = \frac{K}{s(s + 4)(1 + 0.2s)}.$$

As illustrated in Figure 7.4-12a, the Nyquist path for this system consists of four segments: two on the $s = \pm i\omega$ axis, with $0 < \omega < \infty$, and two semicircles $s = Re^{i\varphi}$, with $R \to 0$ or $R \to \infty$ and $-\pi/2 \le \varphi \le \pi/2$. On the positive $s = i\omega$ imaginary axis, we have

$$F(i\omega) = \frac{K}{i\omega(4 + i\omega)(1 + i0.2\omega)}$$

$$|F(i\omega)| = \frac{K}{\omega\sqrt{(16 + \omega^2)(1 + 0.04\omega^2)}}$$

$$\angle F(i\omega) = -\left[\frac{\pi}{2} + \tan^{-1}\left(\frac{\omega}{4}\right) + \tan^{-1}(0.2\omega)\right].$$

The corresponding $F(s)$ contour for $s = i\omega$ is plotted as segment I in Figure 7.4-12b. The contour for $s = -i\omega$ is symmetric about the real axis and is shown as segment III in Figure 7.4-12b.

In lieu of actually plotting the $F(i\omega)$ contour I, a sketch of it could have been constructed as follows. The beginning of the $F(i\omega)$ contour, corresponding to $\omega = 0+$, occurs at $|F(i\omega)| \to \infty$ and $\angle F(i\omega) \to -\pi/2$, that is, moving upward from $-\infty$ along the negative imaginary axis in the $F(s)$ plane. Similarly, the end of the $F(i\omega)$ contour, corresponding to $\omega \to +\infty$, occurs at $|F(i\omega)| \to 0$ and $\angle F(i\omega) \to -3\pi/2$, that is, downward at the origin along the positive imaginary axis in the $F(s)$ plane. One additional point, where the $F(i\omega)$ contour crosses the real axis, is important not only for sketching $F(i\omega)$ but also, as we will learn, because it determines whether or not the system is stable. The real-axis intercept(s) can be determined by multiplying the numerator and denominator of $F(i\omega)$ by the conjugate of the denominator, to write $F(i\omega)$ in the form

$$F(i\omega) = \mu + i\nu = \frac{-1.8K\omega^2 + iK\omega(0.2\omega^2 - 4)}{\omega^2(16 + \omega^2)(1 + 0.04\omega^2)}.$$

Setting $\nu = 0$ yields $\omega^2 = 20$, which in turn gives $\mu = -K/36$. Since this intercept point is unique, the $F(i\omega)$ contour only crosses the real axis once in the $F(s)$ plane.

For the two semicircular segments of the Nyquist path, corresponding to $s = Re^{i\varphi}$, we have

$$F(s) = \frac{K}{Re^{i\varphi}(4 + Re^{i\varphi})(1 + 0.2Re^{i\varphi})}.$$

For $R \to \infty$,

$$|F(s)| \to \frac{K}{0.2R^3} \to 0$$

and for $R \to 0$,

$$|F(s)| \to \frac{K}{4R} \to \infty.$$

Thus, the corresponding $F(s)$ contours, indicated by II and IV in Figure 7.4-12b, are semicircles of radius $|F(s)|$ = constant $\to 0$ and ∞, respectively.

The semicircles II and IV in Figure 7.4-12b are typical of many open-loop transfer functions and are often not drawn explicitly on an $F(s)$ plot. In addition, the $F(-i\omega)$ contour is often not drawn, since it is symmetric to $F(i\omega)$ about the real axis. Thus, an $F(s)$ plot along the Nyquist path may only show the $F(i\omega)$ portion. However, to apply the Nyquist stability criterion, the implied portions of the entire $F(s)$ plot are required, even if they are not explicitly plotted.

To test the stability of this system using the Nyquist criterion (7.4-20), we first note that the open-loop transfer function $F(s)$ has no poles in the right-half of the s plane. Thus, $P = 0$ and the Nyquist criterion reduces to $N = 0$. That is, the $F(s)$ contour must have zero net encirclements of the $-1 + i0$ point in the $F(s)$ plane. From Figure 7.4-12b we conclude that the system will be asymptotically stable if $K < 36$, marginally stable (an eigenvalue with zero real part) if $K = 36$, and unstable if $K > 36$. These results agree with those of Example 7.3-2.

Gain and Phase Margins

The Nyquist stability criterion can be simplified for a large class of closed-loop control systems. If the open-loop transfer function $F(s)$ has no poles in the right-half s plane, which we will assume for the remainder of this section, then $P = 0$ in (7.4-20). For this case, as illustrated in Figure 7.4-12b for Example 7.4-4, stability can be determined simply from the $F(i\omega)$ contour with $0 < \omega < \infty$. The closed-loop system will be *asymptotically stable ($N = 0$) if the $F(s) = -1 + i0$ point is to the left of* the $F(i\omega)$ contour as the contour is traversed in the direction of increasing ω. That is, the $F(i\omega)$ contour must not "enclose" the $-1 + i0$ point.

The closeness of $F(i\omega)$ to the $-1 + i0$ point is a measure of the **relative stability** of a system. The situation is illustrated in Figure 7.4-13, which shows s-plane paths and the corresponding $F(s)$ contours for $s = \sigma + i\omega$, with the s-plane paths being either lines of constant σ or ω. For a particular constant σ path, the corresponding $F(s)$ contour will pass exactly through the $-1 + i\omega$ point, at a particular frequency ω. This case corresponds to $s = \sigma \pm i\omega$ being an eigenvalue of the closed-loop system. The value of σ for this $F(s)$ contour determines the horizontal location of the closed-loop eigenvalue in the s plane. Thus, as illustrated in Figure 7.4-14, the distance between the $F(s) = -1 + i0$ point and the $F(i\omega)$ contour indicates how close the closed-loop eigenvalues are to the imaginary axis.

Figure 7.4-15 shows a typical $F(i\omega)$ plot (and corresponding Bode plots) for a stable closed-loop system with an open-loop transfer function $F(s)$ of the form (7.4-2). The frequency $\omega = \omega_\varphi$ where $F(i\omega)$ crosses the negative real axis is termed the **phase cross-over frequency**. The frequency $\omega = \omega_k$ at which $F(i\omega)$ crosses the unit circle is called the **gain cross-over frequency**.

Since the gain K in (7.4-2) is a multiplier, increasing (decreasing) K at a particular frequency ω would simply change $|F(i\omega)|$ and move $F(i\omega)$ away from (toward) the origin along a ray from the origin. In particular, there is

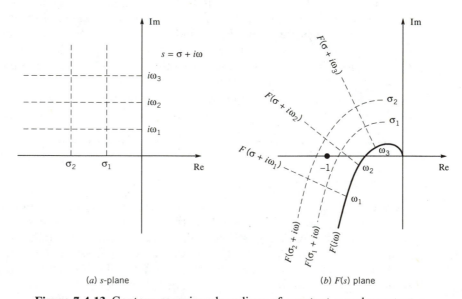

(a) s-plane (b) F(s) plane

Figure 7.4-13 Contour mapping along lines of constant σ and constant ω.

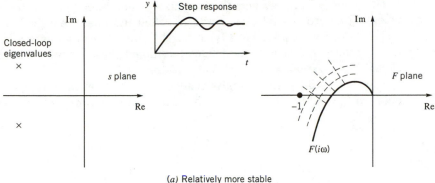

(a) Relatively more stable

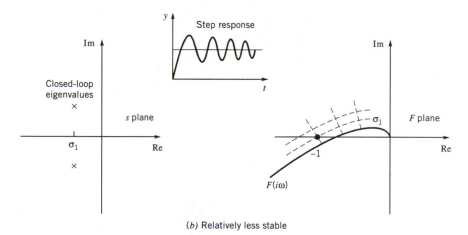

(b) Relatively less stable

Figure 7.4-14 Polar plots and relative stability.

a certain value K_g that the gain K could be multiplied by, given by

$$K_g = \frac{1}{|F(i\omega_\varphi)|},\qquad(7.4\text{-}21)$$

so that the $F(i\omega)$ contour would pass through the $-1 + i0$ point. If the gain were multiplied by more than this value, the closed-loop system would become unstable. The **gain margin** GM is defined as K_g in decibels. That is,

$$\text{GM} = 20\log_{10}(K_g) = -20\log_{10}(|F(i\omega_\varphi)|).\qquad(7.4\text{-}22)$$

The angle γ from the negative real axis counterclockwise to the point where $F(i\omega)$ crosses the unit circle is called the **phase margin** PM. If a phase lag of γ were added to $F(s)$ at fixed K, the $F(i\omega)$ contour would pass through the $-1 + i0$ point, and if any additional phase lag were added, the closed-loop system would become unstable.

The **simplified Nyquist stability criteria**, for a system with no open-loop poles in the right-half plane, are that *both the gain margin* GM *and the*

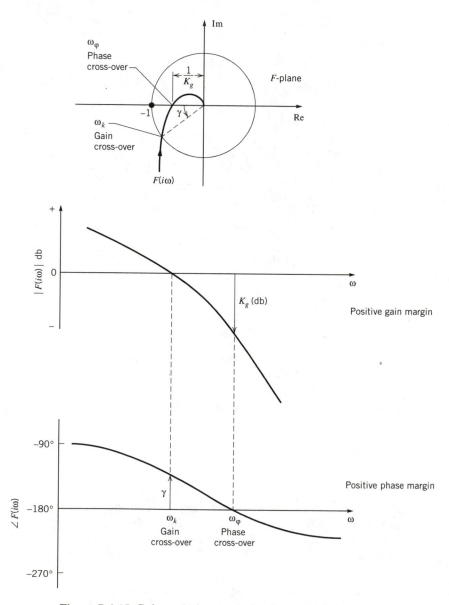

Figure 7.4-15 Gain and phase margins for a stable system.

phase margin PM *must be positive* for the closed-loop system to be asymptotically stable. Typical design requirements are GM $\geq$ 8 db and PM $\geq$ 30°.

In practice, the gain and phase margins are usually determined from a Bode plot of $F(i\omega)$, rather than from a polar plot. As illustrated in Figure 7.4-15 for a stable system and Figure 7.4-16 for an unstable system, the

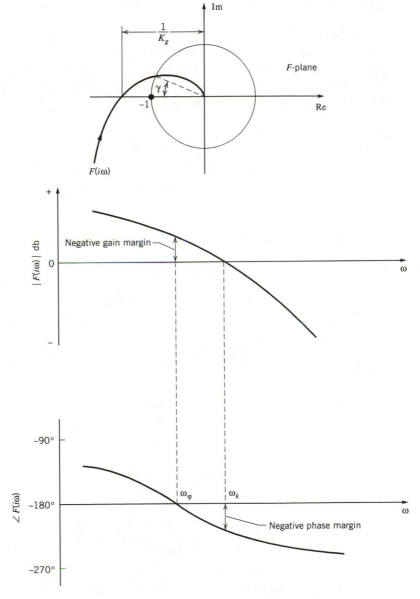

Figure 7.4-16 Gain and phase margins for an unstable system.

gain margin is measured at the phase cross-over frequency ω_φ, where $\angle F(i\omega)$ crosses the $-180°$ line. The gain margin is positive when $|F(i\omega_\varphi)|_{db} < 0$ db. The phase margin on a Bode plot is measured at the gain cross-over frequency ω_k, where $|F(i\omega_k)|_{db} = 0$ db. The phase margin is positive if $\angle F(i\omega_k) > -180°$.

Lead or Lag Compensation

When performance specifications on a feedback control system are specified in terms of frequency domain criteria, such as gain margin or phase margin, Bode plots provide a useful analysis tool; phase lead or lag compensators can often be used effectively, since they provide a means to alter the phase of an open-loop transfer function $F(s)$ and, hence, the phase and gain margins. In particular, they provide the capability to rotate an $F(i\omega)$ contour so that the $-1 + i0$ point is to the left as ω increases, thereby stabilizing an unstable system, or to simply alter the phase or gain margins to meet specified Nyquist stability requirements. We illustrate a typical design process with an example.

EXAMPLE 7.4-5 **Phase Lead Compensation**

Consider the system illustrated in Figure 7.4-17, with

$$G_P(s) = \frac{K}{s(1 + s)(1 + 0.01s)}.$$

We wish to design a lead or lag compensator

$$G_C(s) = \frac{1 + \tau_e s}{1 + \tau_a s}, \qquad (7.4\text{-}23)$$

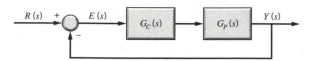

Figure 7.4-17 Error feedback control system.

for which the open-loop transfer function becomes

$$F(s) = \frac{K}{s(1 + s)(1 + 0.01s)} \left(\frac{1 + \tau_e s}{1 + \tau_a s}\right), \qquad \text{(7.4-24)}$$

to yield an asymptotically stable closed-loop system with (1) a steady-state error, in response to a unit **ramp** input $r(t) = t$, of $e_\infty \leq 0.001$ and (2) a phase margin $\geq 25°$.

1. **Steady Error and Nominal Gain.** For the closed-loop transfer function

$$\frac{Y(s)}{R(s)} = \frac{G_C(s)G_P(s)}{1 + G_C(s)G_P(s)}, \qquad \text{(7.4-25)}$$

the steady output error $e(t) \overset{\Delta}{=} r(t) - y(t)$ satisfies

$$E(s) = \frac{1}{1 + G_C(s)G_P(s)} R(s). \qquad \text{(7.4-26)}$$

For a unit ramp input, $R(s) = 1/s^2$, the Laplace transform final value theorem (2.2-5) yields

$$0.001 \geq e_\infty = \lim_{s \to 0} \frac{1}{s[1 + G_C(s)G_P(s)]} = \frac{1}{K} \Rightarrow K \geq 1000.$$

2. **Uncompensated System.** For $G_C(s) = 1$ and $K = 1000$, the Bode plot in Figure 7.4-18 yields

$$GM = -20 \text{ db}$$
$$PM = -14°,$$

which corresponds to an unstable system, with gain and phase cross-over frequencies of

$$\omega_k = 32 \text{ rad/sec}$$
$$\omega_\varphi = 10 \text{ rad/sec}.$$

Thus, some form of compensation is required. For this example we will use phase lead compensation, to increase the phase margin to a positive value.

3. **Estimate Phase Lead Required.** We need to adjust the phase of the open-loop transfer function by an amount

$$\Delta\varphi = 25 - PM = 25 - (-14) = 39°,$$

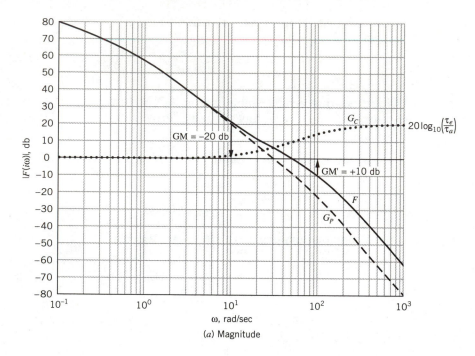

(a) Magnitude

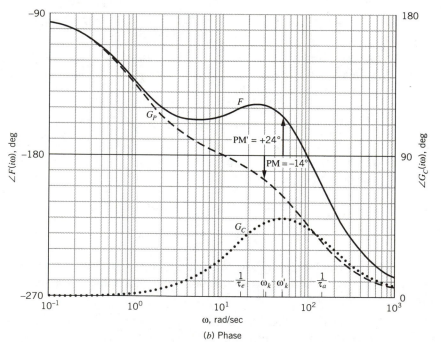

(b) Phase

Figure 7.4-18 Bode plots for Example 7.4-5, with $G_P(s) = K/[s(1 + s)(1 + 0.01s)]$, $G_C(s) = (1 + 0.055s)/(1 + 0.00073s)$, and $F(s) = G_C(s)G_P(s)$.

which implies that we need a phase lead compensator in the vicinity of the gain cross-over frequency $\omega_k = 32$ rad/sec. However, a phase lead compensator will shift the gain cross-over to a higher frequency ω_k', where the phase is less than at the original gain cross-over; in this case, considerably less. Thus, we try a phase lead device with a maximum phase value $\varphi_m = 50°$. From (7.4-15)

$$\alpha = \frac{1 + \sin \varphi_m}{1 - \sin \varphi_m} = 7.449,$$

where $\alpha = \tau_e/\tau_a$.

4. *Locate the phase lead corner frequencies.* We want to position the phase lead corner frequencies $\omega_e = 1/\tau_e$ and $\omega_a = 1/\tau_a$ so that φ_m occurs at the new (unknown) gain cross-over frequency ω_k'. Recall from (7.4-17) that φ_m occurs at a frequency φ_m that is the geometric mean of ω_e and ω_a on a logarithmic scale.
   (a) Locate ω_m at a frequency where the uncompensated magnitude is one half of the high-frequency phase lead magnitude, so that the compensated system has a gain cross-over at $\omega_k' = \omega_m$:

$$|G_P(\omega_m)| = -10 \log_{10}(\alpha) = -8.78 \text{ db.}$$

From Figure 7.4-18, this magnitude occurs at $\omega_m \approx 50$ rad/sec.

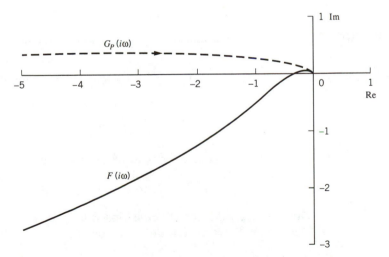

Figure 7.4-19 Polar plots for Example 7.4-5, with $G_P(s) = K/[s(1 + s)(1 + 0.01s)]$, $G_C(s) = (1 + 0.055s)/(1 + 0.00073s)$, and $F(s) = G_C(s)G_P(s)$.

(b) For corner frequencies from (7.4-14)

$$\tau_e = \frac{\sqrt{\alpha}}{\omega_m} = 0.055 \Rightarrow \omega_e = \frac{1}{\tau_e} = 18.2 \text{ rad/sec.}$$

Then

$$\tau_a = \frac{\tau_e}{\alpha} = 0.0073 \Rightarrow \omega_a = \frac{1}{\tau_a} = 137.0 \text{ rad/sec.}$$

Thus, the phase lead compensator is given by

$$G_C(s) = \frac{1 + 0.055s}{1 + 0.0073s}.$$

The resulting feedback control system has the open-loop transfer function $F(s) = G_C(s)G_P(s)$, whose Bode plot is shown in Figure 7.4-18. From this Bode plot we have a compensated gain margin of GM′ = 10 db and a compensated phase margin of PM′ = 24°. Since both of these are positive, the closed-loop system is asymptotically stable.

Figure 7.4-19 shows the polar plots of the open-loop transfer functions with and without the phase lead compensator. This figure clearly illustrates the effect of a phase lead device, as a counterclockwise rotation of the polar plot for the open-loop transfer function.

7.5 EXERCISES

7.5-1 A certain IO system is described by the transfer function

$$G_P(s) = \frac{1}{s(s - 1)}.$$

An error feedback controller is to be used of the form

$$U(s) = K[R(s) - Y(s)].$$

(a) Draw a block diagram for the closed-loop system.
(b) Determine the transfer function for the overall closed-loop system.
(c) Determine the roots of the characteristic equation in terms of K.
(d) For what values of K is the closed-loop system stable?

7.5-2 A system with a transfer function

$$G_P(s) = \frac{1}{s(s^2 + 4s + 5)}$$

is to be controlled by means of output error feedback, as in Figure 6.3-1, with $K_S = 1$ and $G_C(s) = K$. For $K > 0$, verify the root locus shown in Figure 7.5-1. In particular, determine values for the
(a) Asymptote angles
(b) Intersection point of the asymptotes
(c) Angle of departure
(d) Intersection with the imaginary axis
(e) Break-out point
(f) Break-in point

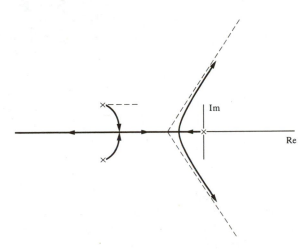

Figure 7.5-1 Root locus for Exercise 7.5-2.

7.5-3 Draw a root locus for $K > 0$, and determine the applicable quantities listed in Exercise 7.5-2, for each of the following characteristic equations:
(a) $s(s + 4)(s^2 + 4s + 5) + K = 0$
(b) $s(s + 4)(s^2 + 4s + 8) + K = 0$
(c) $s(s + 4)(s^2 + 4s + 13) + K = 0$

7.5-4 Draw a root locus for the overall characteristic equation for the system shown in Figure 6.1-3 when
(a) $G_P(s) = 1/[s(s^2 + 6s + 13)]$, $H(s) = 1$, $G_C(s) = K$, $K_S = 1$
(b) $G_P(s) = 1/[(s + 1)(s^2 + 6s + 13)]$, $H(s) = s + 4$, $G_C(s) = K$, $K_S = 1$
and determine the values of K for which the systems will be stable.

7.5-5 Sketch a root locus for the system depicted in Figure 6.1-3 where $G_P(s) = 1/[s(s^2 + 4s + 5)]$, $H(s) = 1 + s$, $G_C(s) = K$, $K_s = 1$.

7.5-6 Consider the inverted beam (Example 2.3-3) as a second-order IO system of the form

$$\ddot{\theta} + p_1\dot{\theta} + p_0\theta = q_0 u,$$

where $p_0 = -17.627/\text{sec}$, $p_1 = 0.187/\text{sec}^2$, $q_0 = 0.6455 \text{ rad/sec}^2$-V. For the control system shown in Figure 7.5-2, draw a root locus for
(a) $K > 0$ with $G_C(s) = 1$ and $H(s) = 0$
(b) $K > 0$ with $G_C(s) = 0.3/s$ and $H(s) = 1$
(c) $1/\tau_i > 0$ with $G_C(s) = 1/(\tau_i s)$, $H(s) = 1$, and $K = 62.17$
(d) $K > 0$ with $G_C(s) = (\tau_i s + 1)/(\tau_i s)$, $H(s) = 0$, and $\tau_i = 0.1$

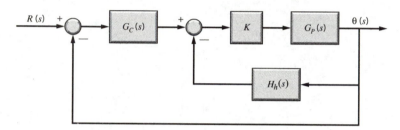

Figure 7.5-2 Control system for Exercise 7.5-6.

7.5-7 Accurately sketch the root locus ($K \geq 0$) for the system shown in Figure 7.5-3. Determine the following quantities:
(a) Asymptotes
(b) Breakpoints
(c) Departure and arrival angles
(d) Imaginary intercepts
(e) Range of K values for stability.

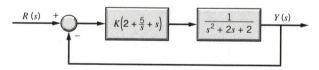

Figure 7.5-3 Control system for Exercise 7.5-7.

7.5-8 A system has the characteristic equation

$$s^2 + Ks + 2K = 0.$$

Show that the complex points on the root locus ($K \geq 0$) lie on a circle and determine the center and radius of the circle.

7.5-9 A system has the characteristic equation

$$0 = 1 + \frac{K(s^2 + 4s + a)}{s(s + T)(s^2 + 2s + 2)}.$$

Accurately sketch the root locus for $K \geq 0$ with $a = 20$ and $T = 8$. Determine the location and K value for each imaginary-axis crossing.

7.5-10 For the system in Exercise 7.5-9, use Kharitonov's technique to determine the range(s) of parameter values for K for guaranteed asymptotic stability with $10 \leq a \leq 20$ and $5 \leq T \leq 8$.

7.5-11 For the first-order system in Example 7.4-1 and the corresponding polar plot in Figure 7.4-1, show that the
(a) Plot is a portion of a circle.
(b) Bottom point on the circle corresponds to $\omega = 1/T$.

7.5-12 For the second-order system in Example 7.4-2 and the corresponding polar plot in Figure 7.4-2, show that for a given damping ratio ζ and natural frequency ω_n the
(a) Resonant frequency, given by (5.2-10), corresponds to $\max|F(i\omega)|$.
(b) Imaginary-axis intercept corresponds to $\omega = \omega_n$.

7.5-13 For the lead or lag system in Example 7.4-3, verify Equation (7.4-16).

7.5-14 From the polar plots in Figure 7.4-2, determine the damping ratio ζ for the two unlabeled plots.

7.6-15 Construct accurate polar and Bode plots of $F(i\omega)$, and on each show the location and value for the gain and phase margins, for a closed-loop system with the open-loop transfer function

$$F(s) = \frac{10}{(1 + s)(1 + 0.5s)(1 + 0.1s)}.$$

Is the closed-loop system stable?

7.5-16 Construct accurate polar and Bode plots of $F(i\omega)$, and on each show the location and value for the gain and phase margins, for a closed-loop system with the open-loop transfer function

$$F(s) = \frac{100(s + 10)}{(s + 1)^2(s^2 + 8s + 100)}.$$

Is the closed-loop system stable?

7.5-17 Consider an error feedback control system of the form shown in Figure 7.4-17 with

$$G_P(s) = \frac{K}{1+s}.$$

Design a lead or lag compensator

$$G_C(s) = \frac{1 + \tau_e s}{1 + \tau_a s},$$

so that the following performance specifications are satisfied: (1) steady output error $e_\infty \leq 0.02$ in response to a unit step input and (2) phase margin of approximately 60°. Construct a Bode plot, as in Figure 7.4-18, showing the amplitude and phase plots and the gain and phase margins before and after compensation.

7.5-18 A unity feedback system, with a minimum phase open-loop transfer function $F(s)$, has Bode plot shown in Figure 7.5-4.

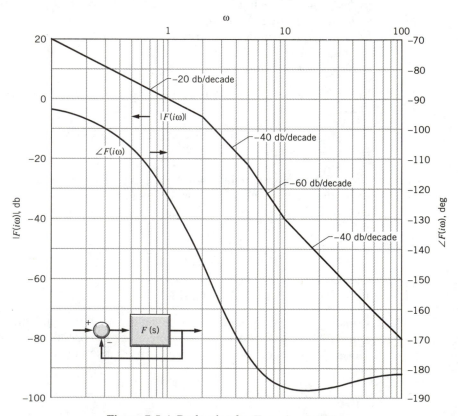

Figure 7.5-4 Bode plot for Exercise 7.5-18.

(a) Determine the gain margin GM (in db) and the phase margin PM (in degrees). Is the closed-loop system stable?

(b) Determine $F(s)$.

7.5-19 Consider a unity feedback system with open-loop transfer function

$$F(s) = \frac{K}{s(s + 1)(s + 2)}.$$

For $K = 2$ the system has the polar plot shown in Figure 7.5-5.

(a) Determine the gain margin GM (in db) and the phase margin PM (in degrees) for $K = 2$. Is the closed-loop system stable?

(b) Determine the value of K so that the closed-loop system has phase margin PM $= 20°$. Determine the corresponding gain margin GM (in db).

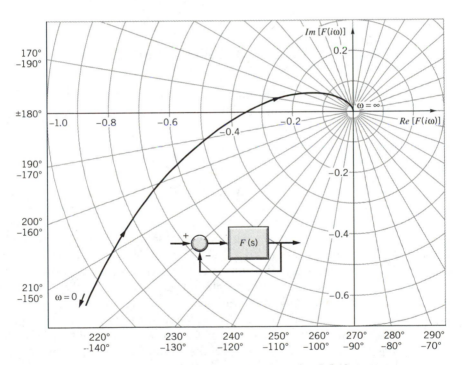

Figure 7.5-5 Polar plot for Exercise 7.5-19.

Chapter 8

State Variable Feedback Design

8.1 CONTROLLABILITY AND OBSERVABILITY OF LINEAR SYSTEMS

In this chapter we will develop state variable feedback control methods for the deterministic linear state-space systems previously defined by (2.1-3) and (2.1-4), namely,

$$\dot{\mathbf{x}} = \mathbf{A}\mathbf{x} + \mathbf{B}\mathbf{u} \qquad (8.1\text{-}1)$$

$$\mathbf{y} = \mathbf{C}\mathbf{x} + \mathbf{D}\mathbf{u}, \qquad (8.1\text{-}2)$$

where $\mathbf{x}$ is the state vector (dimension N_x), $\mathbf{u}$ is the control vector (dimension N_u), $\mathbf{y}$ is the measurement vector (dimension N_y), and $\mathbf{A}$, $\mathbf{B}$, $\mathbf{C}$, and $\mathbf{D}$ are real constant matrices of size $N_x \times N_x$, $N_x \times N_u$, $N_y \times N_x$, and $N_y \times N_u$, respectively. In contrast to Chapters 6 and 7, we do not restrict the systems to be single-input single-output. If $N_u > 1$, we have a multiple-input system, and if $N_y > 1$, we have a multiple-output system. As was true with single-input single-output systems, the basic control objective will be to transfer the system to some new operating condition, based on a command input, and to stabilize the system about the new operating condition.

Controllable and Observable Systems

Depending on the dynamical system and the available measurements, it may not always be possible to achieve a desired operating condition. For example, a swimmer cannot swim upstream against a strong current, or a weight lifter cannot lift a 1000-lb weight. These are examples of systems

286

that are not controllable because the controller does not have sufficient "force" to accomplish the objective. Control limits are factors that affect controllability, but it is not the factor of concern in this chapter. The control **u** in (8.1-1) is assumed to have no constraints or limits on it. The control can be chosen as large as necessary. A more basic type of controllability will be discussed here.

The topics of controllability and observability were introduced briefly in Section 2.1, when we discussed various representations of a control system. In this section we will consider these topics in more detail.

The state-space system (8.1-1) is **controllable** if and only if for every initial state **x**(0) in state space and any desired target state such as **x**(T) = **0**, there exists a finite time T and a control **u**(t), $t \in [0,T]$, that will transfer the state of the system to the desired target state.

The state and measurement system given by (8.1-1) and (8.1-2) is **observable** if and only if there exists a finite time T such that the initial state **x**(0) can be determined from the observation history **y**(t), $t \in [0,T]$, given the control **u**(t) $t \in [0,T]$.

Qualitatively, a system is controllable if every state variable can be affected independently, directly or indirectly, by the control vector **u**. Similarly, the system is observable if the measurement vector **y** is affected, directly or indirectly, by each of the state variables.

A particular system may not be controllable to a certain target state, even without bounds on the control, simply because the structure of the dynamical system does not allow the control to affect some portion of the state. For example, consider the system with state equations

$$\dot{x}_1 = x_2$$
$$\dot{x}_2 = -6x_1 + x_3 + u$$
$$\dot{x}_3 = -x_3.$$

The evolution of $x_3(t)$ cannot be affected by the control $u(t)$.

EXAMPLE 8.1-1   **Noncontrollable and Nonobservable System**

Consider the system given by

$$\begin{bmatrix} \dot{x}_1 \\ \dot{x}_2 \end{bmatrix} = \begin{bmatrix} 2 & 0 \\ -1 & 1 \end{bmatrix} \begin{bmatrix} x_1 \\ x_2 \end{bmatrix} + \begin{bmatrix} 1 \\ -1 \end{bmatrix} u$$

$$y = \begin{bmatrix} 1 & 1 \end{bmatrix} \begin{bmatrix} x_1 \\ x_2 \end{bmatrix}.$$

It would appear that the control affects both state variables and that the measurement senses both state variables. However, in fact, the control has no effect on the quantity

$$y = x_1 + x_2,$$

which is governed by the differential equation

$$\dot{y} = y.$$

The corresponding solution

$$x_1(t) + x_2(t) = [x_1(0) + x_2(0)]e^t$$

does not depend on $u(t)$. Thus, no matter what the control input is, the sum of the states will evolve according to this relationship. The system is clearly not controllable. Furthermore, suppose $x_1(0) + x_2(0) = 0$. Then the resulting measurement history $y(t) \equiv 0$ only tells us that $x_1(0) + x_2(0) = 0$. It does not allow us to determine $x_1(0)$ and $x_2(0)$ independently. Thus, the system is not observable.

Controllability and Observability Criteria

Example 8.1-1 illustrates the fact that controllability and observability may not be obvious from the system equations themselves. One way to check for controllability and observability, as indicated in Section 2.1, is by **diagonalization**. Let λ_i and $\boldsymbol{\xi}_i$, $i = 1, \ldots, N_x$, denote the eigenvalues and eigenvectors, respectively, for the matrix **A**, and let the eigenvector transformation matrix, whose columns are the eigenvectors, be designated by $\mathbf{M} = [\boldsymbol{\xi}_1, \ldots, \boldsymbol{\xi}_{N_x}]$. If the eigenvectors are linearly independent (for example, if the eigenvalues are distinct), then the transformation for the state vector **x** to the **eigenstate** vector

$$\mathbf{z} = \mathbf{M}^{-1}\mathbf{x} \tag{8.1-3}$$

yields the decoupled state equations

$$\dot{\mathbf{z}} = \hat{\mathbf{A}}\mathbf{z} + \mathbf{M}^{-1}\mathbf{B}\mathbf{u} \tag{8.1-4}$$

and the corresponding measurement equations

$$\mathbf{y} = \mathbf{CMz} + \mathbf{Du}, \tag{8.1-5}$$

where

$$\hat{A} = M^{-1}AM = \text{diag}[\lambda_1, \ldots, \lambda_{N_x}].$$

The **diagonalized controllability criterion** is that a system is controllable if and only if there are *no zero rows of the matrix* $M^{-1}B$. If this condition is satisfied, then the control **u** affects each component of the vector **z** and therefore each component of the state vector **x**, through the one-to-one correspondence given by (8.1-3).

Similarly, the **diagonalized observability criterion** is that a system is observable if and only if there are *no zero columns in the matrix* **CM**. If this condition holds, then the measurements **y** contain information from every component of **z** and therefore **x** through (8.1-3).

EXAMPLE 8.1-2   **Diagonalizing Example 8.1-1**

The **A** matrix in Example 8.1-1, given by

$$A = \begin{bmatrix} 2 & 0 \\ -1 & 1 \end{bmatrix},$$

has eigenvalues $\lambda_1 = 1$, $\lambda_2 = 2$. Corresponding eigenvectors yield the eigenvector transformation matrix

$$M = \begin{bmatrix} 0 & 1 \\ 1 & -1 \end{bmatrix} \Rightarrow M^{-1} = \begin{bmatrix} 1 & 1 \\ 1 & 0 \end{bmatrix}.$$

To check the diagonalized controllability criterion, we form

$$M^{-1}B = \begin{bmatrix} 1 & 1 \\ 1 & 0 \end{bmatrix}\begin{bmatrix} 1 \\ -1 \end{bmatrix} = \begin{bmatrix} 0 \\ 1 \end{bmatrix}.$$

Since $M^{-1}B$ contains a zero row, the controllability criterion is not satisifed. This agrees with our previous observation that this system is not controllable. To check the diagonalized observability criterion, we form

$$CM = \begin{bmatrix} 1 & 1 \end{bmatrix}\begin{bmatrix} 0 & 1 \\ 1 & -1 \end{bmatrix} = \begin{bmatrix} 1 & 0 \end{bmatrix}.$$

Since **CM** contains a zero column, the observability criterion is not satisfied, which agrees with our previous observation that this system is not observable.

Kalman Controllability and Observability Criteria

The diagonalized controllability and observability criteria required not only that we find the eigenvalues and eigenvectors of **A** but also that the eigenvectors be linearly independent. Although the previous conditions also hold for a system transformed to Jordon block diagonal form using generalized eigenvectors (Brogan, 1982, pp. 232–240), there are alternate tests (Kalman et al., 1963) that do not require computation of the eigenvalues and eigenvectors of **A**.

The **Kalman controllability criterion** is that the system (8.1-1), or equivalently the pair of matrices (**A, B**), is controllable if and only if

$$\text{rank}[\mathbf{P}] = N_x, \tag{8.1-6}$$

where **P** is the $N_x \times N_u N_x$ **controllability matrix**

$$\mathbf{P} = [\mathbf{B}, \mathbf{AB}, \mathbf{A}^2\mathbf{B}, \ldots, \mathbf{A}^{N_x-1}\mathbf{B}]. \tag{8.1-7}$$

To verify this result, recall from Section 3.3 that the $N_x \times N_x$ state transition matrix

$$\mathbf{\Phi}(t) = e^{\mathbf{A}t} = \mathbf{I} + \mathbf{A}t + \mathbf{A}^2\frac{t^2}{2!} + \mathbf{A}^3\frac{t^3}{3!} + \cdots \tag{8.1-8}$$

can be written, using the Cayley–Hamilton theorem, as a finite series given by (3.3-27), which we repeat here

$$\mathbf{\Phi}(t) = \alpha_0(t)\mathbf{I} + \alpha_1(t)\mathbf{A} + \alpha_2(t)\mathbf{A}^2 + \cdots + \alpha_{N_x-1}(t)\mathbf{A}^{N_x-1}. \tag{8.1-9}$$

For a particular initial state **x**(0) and control **u**(t), the solution to (8.1-1) can be written as

$$\mathbf{x}(t) = \mathbf{\Phi}(t)\mathbf{x}(0) + \int_0^t \mathbf{\Phi}(t - \tau)\mathbf{B}\mathbf{u}(\tau)\,d\tau. \tag{8.1-10}$$

For controllability we require that an arbitrary initial state be transferable to an arbitrary final state by some corresponding control **u**(t). For example, picking the final state as the origin, we require a final time $T < \infty$ such that

$$-\mathbf{x}(0) = \int_0^T \mathbf{\Phi}(-\tau)\mathbf{B}\mathbf{u}(\tau)\,d\tau = \sum_{k=0}^{N_x-1}\left[\mathbf{A}^k\mathbf{B}\int_0^T \alpha_k(-\tau)\mathbf{u}(\tau)\,d\tau\right], \tag{8.1-11}$$

where we have used the fact that $\Phi(t - \tau) = \Phi(t)\Phi(-\tau)$. Defining the $N_u \times 1$ vectors

$$\mu_k = \int_0^T \alpha_k(-\tau)\mathbf{u}(\tau)\,d\tau, \tag{8.1-12}$$

we have

$$-\mathbf{x}(0) = \sum_{k=0}^{N_x-1} \mathbf{A}^k\mathbf{B}\mu_k = [\mathbf{B}, \mathbf{AB}, \mathbf{A^2B}, \ldots, \mathbf{A}^{N_x-1}\mathbf{B}]\begin{bmatrix} \mu_0 \\ \mu_1 \\ \vdots \\ \mu_{N_x-1} \end{bmatrix}. \tag{8.1-13}$$

Therefore, for controllability every vector $\mathbf{x}(0)$ must be expressible as a linear combination of the columns of the controllability matrix $\mathbf{P}$. Thus, these columns must span an N_x-dimensional space, which requires rank[$\mathbf{P}$] = N_x.

In an analogous fashion, the **Kalman observability criterion** is that the system (8.1-1) and (8.1-2) is observable if and only if

$$\text{rank}[\mathbf{Q}] = N_x, \tag{8.1-14}$$

where $\mathbf{Q}$ is the **observability matrix**

$$\mathbf{Q} = \begin{bmatrix} \mathbf{C} \\ \mathbf{CA} \\ \mathbf{CA^2} \\ \vdots \\ \mathbf{CA}^{N_x-1} \end{bmatrix}. \tag{8.1-15}$$

EXAMPLE 8.1-3   **Using the Kalman Conditions**

The $\mathbf{P}$ matrix for Example 8.1-1 is given by

$$\mathbf{P} = \begin{bmatrix} 1 & \vdots \\ -1 & \vdots \end{bmatrix}\begin{bmatrix} 2 & 0 \\ -1 & 1 \end{bmatrix}\begin{bmatrix} 1 \\ -1 \end{bmatrix} = \begin{bmatrix} 1 & 2 \\ -1 & -2 \end{bmatrix}.$$

Since the second column is just two times the first column, the columns are not linearly independent and, hence, the rank of $\mathbf{P}$ must be less than 2. Since $\mathbf{P}$ is square, we can also verify that the controllability criterion is not satisfied by noting that $|\mathbf{P}| = 0$.

The **Q** matrix for Example 8.1-1 is given by

$$\mathbf{Q} = \left[\begin{array}{c} 1 \qquad\qquad 1 \\ \hline [1 \quad 1] \begin{bmatrix} 2 & 0 \\ -1 & 1 \end{bmatrix} \end{array} \right] = \begin{bmatrix} 1 & 1 \\ 1 & 1 \end{bmatrix}.$$

Again, the rows (or columns) are linearly dependent so that the rank of **Q** is less than 2. Because **Q** is square, we conclude that the observability criterion is not satisfied, since $|\mathbf{Q}| = 0$.

8.2   EIGENVALUE PLACEMENT

State Variable Feedback

Consider the state-space system given by (8.1-1) and (8.1-2) and assume that the output is the entire state vector $\mathbf{x}(t)$. This assumption will be relaxed in Section 8.3.

With every component of **x** measured, that is, $\mathbf{y} = \mathbf{x}$, we have $\mathbf{C} = \mathbf{I}$ and $\mathbf{D} = \mathbf{0}$. The design objective is to find a control law $\mathbf{u}(\mathbf{r},\mathbf{x})$ that will stabilize the system with respect to a command input $\mathbf{r}(t)$, within a prescribed accuracy, and with suitable transient response behavior. In particular, we would like to have complete control over the placement of the eigenvalues for the controlled system, and perhaps even some control over the eigenvectors which, when real, control the direction of approach to the equilibrium point in state space.

In a fashion analogous to the one-dimensional IO case where we used error feedback, let us consider a **state variable feedback** controller of the form

$$\mathbf{u} = \mathbf{F}\mathbf{r}(t) - \mathbf{K}\mathbf{x}(t), \tag{8.2-1}$$

where $\mathbf{r}(t)$ is an $N_r \times 1$ vector of **command inputs**, **F** an $N_u \times N_r$ **input transformation matrix**, and **K** an $N_u \times N_x$ **feedback gain matrix**. Note that the **F** matrix is required, since the dimension of the command inputs may not equal the dimension of the control vector. If we apply this controller to the system given by (8.1-1) and (8.1-2), we obtain the closed-loop system

$$\dot{\mathbf{x}} = \overline{\mathbf{A}}\mathbf{x} + \overline{\mathbf{B}}\mathbf{r} \tag{8.2-2}$$

$$\mathbf{y} = \mathbf{x}, \tag{8.2-3}$$

where

$$\overline{A} = A - BK \qquad (8.2\text{-}4)$$

$$\overline{B} = BF. \qquad (8.2\text{-}5)$$

Note that the resulting state-space system has the same form as the original system, namely, a linear constant-coefficient control system, except that the matrix $\overline{A}$ depends on a matrix of parameters K, and the "control" inputs are now the command inputs $r(t)$.

For the case of constant command inputs $r(t) \equiv \overline{r}$, the equilibrium solutions $x(t) \equiv \overline{x}$ are given by the solutions to

$$[A - BK]\overline{x} = -BF\overline{r}.$$

Thus, both the F and K matrices affect the location of the closed-loop equilibrium point(s). But the K matrix alone affects the eigenvalues and eigenvectors of the controlled system. In fact, if the system is completely controllable (satisfies one of the controllability criteria), then any desired set of eigenvalues for $\overline{A}$ can be chosen through an appropriate choice of K (Davison, 1968; Heymann, 1968; Wonham, 1967).

EXAMPLE 8.2-1   **Second-Order IO System**

Consider a second-order IO system of the form

$$\ddot{z} + p_1\dot{z} + p_0z = q_0u \qquad (8.2\text{-}6)$$

in which both z and $\dot{z}$ are measured and $q_0 \neq 0$. We use z instead of y, since we do not want to imply by (8.2-6) that only z is the output. By letting $x_1 = y_1 = z$ and $x_2 = y_2 = \dot{z}$, we obtain an equivalent state-space representation

$$\dot{x} = \begin{bmatrix} 0 & 1 \\ -p_0 & -p_1 \end{bmatrix} x + \begin{bmatrix} 0 \\ q_0 \end{bmatrix} u$$

$$y = x$$

in which the output is the entire state. Using the Kalman controllability criterion, we note that

$$|P| = -q_0^2.$$

Since $q_0 \neq 0$, the controllability criterion is satisfied. If we apply state feedback of the form (8.2-1),

$$u = K_s r - k_1 x_1 - k_2 x_2,$$

where in this case $F = K_s$ is simply a scaling factor and $\mathbf{K} = [k_1, \quad k_2]$ is a row vector. We obtain

$$\overline{\mathbf{A}} = \begin{bmatrix} 0 & 1 \\ -p_0 - k_1 q_0 & -p_1 - k_2 q_0 \end{bmatrix},$$

which has the characteristic equation

$$\lambda^2 + (p_1 + k_2 q_0)\lambda + (p_0 + k_1 q_0) = 0.$$

By appropriate choice for k_1 and k_2, we have complete control over the two coefficients in the characteristic equation. Thus, the eigenvalues of $\overline{\mathbf{A}}$ can be arbitrarily chosen. For example, to obtain a system with a characteristic equation of the form

$$\lambda^2 + 2\zeta\omega_n\lambda + \omega_n^2 = 0, \tag{8.2-7}$$

we can match coefficients to obtain

$$k_1 = \frac{\omega_n^2 - p_0}{q_0}, \qquad k_2 = \frac{2\zeta\omega_n - p_1}{q_0}. \tag{8.2-8}$$

Steady-State Response of Second-Order Systems

Since the second-order IO system studied in previous chapters is a basic building block for linear dynamical systems, it is of interest to examine, in a bit more detail, the system discussed in Example 8.2-1. Suppose that we have chosen feedback gains k_1 and k_2 to make this system stable. Then, under a constant step input $r(t) = \bar{r}$, we obtain the steady-state solution

$$\bar{x}_1 = \frac{K_s q_0 \bar{r}}{p_0 + k_1 q_0}$$

$$\bar{x}_2 = 0.$$

The output for our state-space system is both x_1 and x_2. Hence, the SISO concept of the output tracking the input is not directly applicable. In this case, we have one input variable and two output variables. In general, we cannot have any one of the state variables track any one of the input variables. In this instance, clearly, x_2 cannot track r, since its equilibrium value is always zero. However, x_1 can track r, since the equilibrium solution for x_1 contains $\bar{r}$. If this is our objective, then we note that in the steady state we can make $\bar{x}_1$ be any factor f times $\bar{r}$ by setting

$$K_S = \frac{(p_0 + k_1 q_0)f}{q_0}.$$

However, another way to get x_1 to track r and to eliminate any steady-state error due to uncertain bias inputs is to introduce **integral control** by means of an additional **error integral** state variable

$$x_3 = \int_0^t (fr - x_1)\, dt$$

in a third state equation. In this case, the total system including an uncertain input v is given by

$$\dot{x}_1 = x_2$$
$$\dot{x}_2 = -p_0 x_1 - p_1 x_2 + q_0(u + v) \qquad \text{(8.2-9)}$$
$$\dot{x}_3 = fr - x_1.$$

If we now choose the feedback control according to

$$u = fr - k_1 x_1 - k_2 x_2 - k_3 x_3,$$

then the resulting system is given by

$$\dot{x}_1 = x_2$$
$$\dot{x}_2 = -(p_0 + q_0 k_1)x_1 - (p_1 + q_0 k_2)x_2 - q_0 k_3 x_3 + f q_0 r + q_0 v$$
$$\dot{x}_3 = fr - x_1.$$

Under the step input $r(t) = \bar{r}$ and constant (bias) uncertain input $v(t) = \bar{v}$, we have the steady-state solution

$$\bar{x}_1 = f\bar{r}$$
$$\bar{x}_2 = 0$$
$$\bar{x}_3 = \frac{q_0 - p_0 - k_1 q_0}{k_3 q_0} f\bar{r} - \frac{\bar{v}}{k_3}.$$

This solution yields the desired steady-state relationship for $\bar{x}_1$ in spite of the bias input $\bar{v}$. This is at the expense of a nonzero equilibrium for $\bar{x}_3$. Since the state variable x_3 is just the integral of the error $e = fr - x_1$, it gives a measure of how well we track the input on the average.

The $\bar{A}$ matrix for the controlled system is given by

$$\bar{A} = \begin{bmatrix} 0 & 1 & 0 \\ -(p_0 + q_0 k_1) & -(p_1 + q_0 k_2) & -q_0 k_3 \\ -1 & 0 & 0 \end{bmatrix},$$

which has the characteristic equation

$$\lambda^3 + (p_1 + k_2 q_0)\lambda^2 + (p_0 + k_1 q_0)\lambda - k_3 q_0 = 0. \tag{8.2-10}$$

The three feedback gains k_1, k_2, and k_3 can now be chosen to give desired eigenvalues. For example, to obtain a system with a characteristic equation of the form

$$(\lambda + \beta)(\lambda^2 + 2\zeta\omega_n\lambda + \omega_n^2) = 0, \tag{8.2-11}$$

one can expand this equation and match coefficients with (8.2-10) to obtain

$$k_1 = \frac{\omega_n^2 + 2\zeta\omega_n\beta - p_0}{q_0}$$

$$k_2 = \frac{2\zeta\omega_n + \beta - p_1}{q_0} \tag{8.2-12}$$

$$k_3 = -\frac{\beta\omega_n^2}{q_0}.$$

In comparing the control design without integral error feedback [(8.2-7) and (8.2-8)] with the control design with integral error feedback [(8.2-11) and (8.2-12)], we observe that the complex roots can be kept the same in both instances. If we keep the root $\lambda = -\beta$ well to the left of the complex roots, then the dynamic response of the third-order system will closely match that of the second-order system. The effect of the root $\lambda = -\beta$ will be to provide tracking of the input with a zero steady-state error. With large β, a significant amount of control action may also be necessary and could result in control saturation in a practical situation.

To implement state variable feedback along with integral error feedback, the controller must be equivalent to the system

$$\dot{x}_3 = fr - x_1$$
$$u = fr - k_1 x_1 - k_2 x_2 - k_3 x_3.$$

According to the assumptions of state variable feedback, we have available x_1 and x_2 (either by direct measurement or from an observer to be discussed later). The input command is, of course, also available. If we use op amps to construct the controller, then the following procedure can be used. First, feed x_1 and the negative of r into a summing integrator, as in Figure 6.2-8, with $R_1 = 1$, $R_2 = 1/f$, and $C = 1$. The output will be x_3. Next, feed x_1, x_2, x_3, and the negative of r into a summing amplifier, as in Figure 6.2-5, with $R_1 = 1/k_1$, $R_2 = 1/k_2$, $R_3 = 1/k_3$, $R_4 = 1/f$, and $R_f = 1$. The output will be u. This control signal can now be used directly with the physical system through an appropriate interface such as a power amplifier.

Input-Output Tracking

The concept introduced in the previous section of having a state variable track the input can be generalized somewhat. Suppose that some (scalar) linear combination of the state variables and a scalar control, given by

$$\bar{y} = \overline{\mathbf{C}}\mathbf{x} + \bar{d}u = [\bar{c}_1, \ldots, \bar{c}_{N_x}]\mathbf{x} + \bar{d}u,$$

is required to track a scalar input. Note that $\bar{y}$ is not the output. Under the assumptions of this section, the vector output $\mathbf{y}$ is the state $\mathbf{x}$. Consider now introducing a new state variable x_{N_x+1}, satisfying

$$\dot{x}_{N_x+1} = fr - \bar{y} = fr - \overline{\mathbf{C}}\mathbf{x} - \bar{d}u.$$

The complete set of state equations can now be written in matrix form as

$$\begin{bmatrix} \dot{\mathbf{x}} \\ \dot{x}_{N_x+1} \end{bmatrix} = \begin{bmatrix} \mathbf{A} & 0 \\ \hline -\overline{\mathbf{C}} & 0 \end{bmatrix} \begin{bmatrix} \mathbf{x} \\ x_{N_x+1} \end{bmatrix} + \begin{bmatrix} \mathbf{B} \\ \hline -\bar{d} \end{bmatrix} u + \begin{bmatrix} 0 \\ \hline f \end{bmatrix} r.$$

Now introduce state variable feedback of the form

$$u = fr - \mathbf{K}\mathbf{x} - k_{N_x+1}x_{N_x+1},$$

where $\mathbf{K} = [k_1, \ldots, k_{N_x}]$. The controlled system equations become

$$\begin{bmatrix} \dot{\mathbf{x}} \\ \dot{x}_{N_x+1} \end{bmatrix} = \begin{bmatrix} \mathbf{A} - \mathbf{B}\mathbf{K} & -\mathbf{B}k_{N_x+1} \\ \hline -\overline{\mathbf{C}} + \bar{d}\mathbf{K} & \bar{d}k_{N_x+1} \end{bmatrix} \begin{bmatrix} \mathbf{x} \\ x_{N_x+1} \end{bmatrix} + \begin{bmatrix} \mathbf{B} \\ \hline 1 - \bar{d} \end{bmatrix} fr.$$

An equilibrium solution under a constant input $r(t) = \bar{r}$ is given by

$$\begin{bmatrix} \bar{\mathbf{x}} \\ \bar{x}_{N_x+1} \end{bmatrix} = \begin{bmatrix} \mathbf{A} - \mathbf{B}\mathbf{K} & -\mathbf{B}k_{N_x+1} \\ \hline -\overline{\mathbf{C}} + \bar{d}\mathbf{K} & \bar{d}k_{N_x+1} \end{bmatrix}^{-1} \begin{bmatrix} \mathbf{B} \\ \hline 1 - \bar{d} \end{bmatrix} f\bar{r}.$$

provided that the inverse exists. This latter condition effectively dictates which linear combinations of state and control variables can track a given input. Note, in particular, that when an equilibrium solution exists, then $\dot{x}_{N_x+1} = 0$ implies

$$\overline{\mathbf{C}}\mathbf{x} + \overline{d}u = f\overline{r},$$

which achieves the desired tracking, since $\overline{y} = \overline{\mathbf{C}}\mathbf{x} + \overline{d}u$.

Multiple-Input Systems

For controllable single-input systems, the eigenvalue placement procedure outlined in Example 8.2-1 will always provide a unique solution for the gain matrix **K**. However, with controllable multiple-input systems, the gain matrix **K** is not unique.

EXAMPLE 8.2-2   **A Single-Input System**

Consider a controllable single-input system of the form (8.1-1) with

$$\mathbf{A} = \begin{bmatrix} 0 & 1 \\ 1 & 0 \end{bmatrix}, \qquad \mathbf{B} = \begin{bmatrix} 0 \\ 1 \end{bmatrix}.$$

Suppose we wish to place the eigenvalues at $\lambda_1 = -2$ and $\lambda_2 = -3$ using state variable feedback of the form

$$u = r - \mathbf{K}\mathbf{x}.$$

The corresponding $\overline{\mathbf{A}}$ matrix is given by

$$\overline{\mathbf{A}} = \begin{bmatrix} 0 & 1 \\ 1 - k_1 & -k_2 \end{bmatrix}$$

with the characteristic equation

$$\lambda^2 + k_2\lambda + (k_1 - 1) = 0.$$

By substituting the desired eigenvalues into this equation, we obtain a system of two equations for the two unknowns, k_1 and k_2,

$$k_1 - 2k_2 = -3$$
$$k_1 - 3k_2 = -8,$$

which have the unique solutions

$$k_1 = 7$$
$$k_2 = 5.$$

EXAMPLE 8.2-3   **A Multiple-Input System**

Let a system have the same **A** matrix as in Example 8.2-2, but with a **B** matrix given by the identity matrix

$$\mathbf{B} = \begin{bmatrix} 1 & 0 \\ 0 & 1 \end{bmatrix}.$$

Since **P** is of rank 2, the system remains controllable. Using state variable feedback of the form

$$\mathbf{u} = \mathbf{r} - \mathbf{Kx},$$

that is,

$$\begin{bmatrix} u_1 \\ u_2 \end{bmatrix} = \begin{bmatrix} r_1 \\ r_2 \end{bmatrix} - \begin{bmatrix} k_{11} & k_{12} \\ k_{21} & k_{22} \end{bmatrix} \begin{bmatrix} x_1 \\ x_2 \end{bmatrix},$$

we obtain the [**A** − **BK**] matrix

$$[\mathbf{A} - \mathbf{BK}] = \begin{bmatrix} -k_{11} & 1 - k_{12} \\ 1 - k_{21} & -k_{22} \end{bmatrix}$$

with the characteristic equation

$$\lambda^2 + (k_{11} + k_{22})\lambda + k_{11}k_{22} - (k_{21} - 1)(k_{12} - 1) = 0.$$

If we now seek the same eigenvalues as in Example 8.2-2 by substituting $\lambda_1 = -2$ and $\lambda_2 = -3$ into this equation, we can solve for two of the feedback gains in terms of the other two. Hence, the determination of the feedback gains is not unique. In this example we have

$$k_{22} = 5 - k_{11}$$

$$k_{21} = 1 + \frac{(5 - k_{11})k_{11} - 6}{k_{12} - 1}.$$

One advantage of the nonuniqueness is that the control system designer may be able to provide a more **robust controller**, which works even in some off-design conditions, by choosing one set of solutions for the parameters versus another. In particular, it may be possible to choose the k_{ij} not only to yield the desired eigenvalues but also such that, if certain of the feedback elements fail in operation, the resulting system does not become unstable. In our example, suppose we choose $k_{11} = 1$ and $k_{12} = 2$. Then, for the nominal closed-loop system to have the desired eigenvalues, we need $k_{21} = -1$ and $k_{22} = 4$. Now, suppose the k_{11} feedback element fails, so that $k_{11} = 0$. With the other k_{ij} fixed at their nominal design values, the characteristic equation becomes

$$\lambda^2 + 4\lambda + 2 = 0$$

and the control system remains stable.

A General Procedure for Eigenvalue Placement

An eigenvalue placement method does exist (Wonham, 1967, 1985; Brogan, 1982) that not only has the advantage of allowing for a systematic specification of the $\mathbf{K}$ matrix for the multiple-input case but also can be readily adapted for numerical solution. The basic problem is to find elements of the feedback gain matrix $\mathbf{K}$ such that specified eigenvalues λ_1, $\ldots$, λ_{N_x} are solutions to the characteristic equation for the $\overline{\mathbf{A}}$ matrix defined by (8.2-4). In other words, the eigenvalues must satisfy

$$|\lambda \mathbf{I}_{N_x} - (\mathbf{A} - \mathbf{BK})| = 0, \tag{8.2-12}$$

where $\mathbf{I}_{N_x}$ is the N_x-dimensional identity matrix.

For systems with only a few state and control variables, the matrix in (8.2-12) can simply be constructed as a function of the elements of $\mathbf{K}$. Then the elements of $\mathbf{K}$ can be chosen to yield a zero determinant in (8.2-12), as in Example 8.2-3. However, there is a modification of this approach that yields an explicit closed-form solution for $\mathbf{K}$. The approach is suitable for any number of state and control variables and is advantageous when the number of control variables is less than the number of state variables.

Let $\overline{P}(\lambda)$ be the characteristic equation for the $[\mathbf{A} - \mathbf{BK}]$ matrix

$$\overline{P}(\lambda) = |\lambda \mathbf{I}_{N_x} - (\mathbf{A} - \mathbf{BK})| \tag{8.2-13}$$

and $P(\lambda)$ be the characteristic equation for the $\mathbf{A}$ matrix

$$P(\lambda) = |\lambda \mathbf{I}_{N_x} - \mathbf{A}|. \tag{8.2-14}$$

Denote the inverse of the matrix $[\lambda \mathbf{I}_{N_x} - \mathbf{A}]$ by

$$\Phi(\lambda) \triangleq [\lambda \mathbf{I}_{N_x} - \mathbf{A}]^{-1}. \tag{8.2-15}$$

We can now write $\overline{P}(\lambda)$ as

$$\overline{P}(\lambda) = |\lambda \mathbf{I}_{N_x} - \mathbf{A} + [\lambda \mathbf{I}_{N_x} - \mathbf{A}][\lambda \mathbf{I}_{N_x} - \mathbf{A}]^{-1}\mathbf{BK}|$$

or

$$\overline{P}(\lambda) = |[\lambda \mathbf{I}_{N_x} - \mathbf{A}][\mathbf{I}_{N_x} + \mathbf{\Phi}(\lambda)\mathbf{BK}]|.$$

Since the two matrices in brackets are square, we can write

$$\overline{P}(\lambda) = |\lambda \mathbf{I}_{N_x} - \mathbf{A}| \, |I_{N_x} + \mathbf{\Phi}(\lambda)\mathbf{BK}|$$

or

$$\overline{P}(\lambda) = P(\lambda)|I_{N_x} + \mathbf{\Phi}(\lambda)\mathbf{BK}|. \qquad \textbf{(8.2-16)}$$

We can now use the identity

$$|\mathbf{I}_n + \mathbf{DE}| = |\mathbf{I}_m + \mathbf{ED}|,$$

where $\mathbf{D}$ is an $n \times m$ matrix and $\mathbf{E}$ an $m \times n$ matrix, to write (8.2-16) as

$$\overline{P}(\lambda) = P(\lambda)|\mathbf{I}_{N_u} + \mathbf{K}\mathbf{\Phi}(\lambda)\mathbf{B}|. \qquad \textbf{(8.2-17)}$$

Of course, (8.2-17) is just another way of writing the characteristic equation for the $[\mathbf{A} - \mathbf{BK}]$ matrix. However, in this form, there are obvious ways of choosing $\mathbf{K}$ to satisfy $\overline{P}(\lambda) = 0$. In particular, one can choose $\mathbf{K}$ so that a column of the matrix

$$[\mathbf{I}_{N_u} + \mathbf{K}\mathbf{\Phi}(\lambda)\mathbf{B}] \qquad \textbf{(8.2-18)}$$

is zero. This will guarantee that the corresponding determinant is zero and, hence,

$$\overline{P}(\lambda) = 0. \qquad \textbf{(8.2-19)}$$

EXAMPLE 8.2-4   **Redesigning the Single-Input System from Example 8.2-2**

We will use (8.2-17) to rework Example 8.2-2. We have

$$[\lambda \mathbf{I}_{N_x} - \mathbf{A}] = \begin{bmatrix} \lambda & -1 \\ -1 & \lambda \end{bmatrix}$$

$$\mathbf{\Phi}(\lambda) = \frac{1}{\lambda^2 - 1}\begin{bmatrix} \lambda & 1 \\ 1 & \lambda \end{bmatrix}.$$

Thus,

$$\mathbf{K}\Phi(\lambda)\mathbf{B} = \frac{1}{\lambda^2 - 1} [k_1 \quad k_2] \begin{bmatrix} \lambda & 1 \\ 1 & \lambda \end{bmatrix} \begin{bmatrix} 0 \\ 1 \end{bmatrix}$$

$$= \frac{k_1 + k_2\lambda}{\lambda^2 - 1}.$$

Since $\mathbf{I}_{N_u} = 1$, the matrix (8.2-18) is made zero by choosing k_1 and k_2 such that

$$1 + \frac{k_1 + k_2\lambda}{\lambda^2 - 1} = 0.$$

In particular, with $\lambda_1 = -2$ and $\lambda_2 = -3$, we obtain

$$k_1 - 2k_2 = -3$$
$$k_1 - 3k_2 = -8,$$

which are the same set of equations we obtained in Example 8.2-2.

EXAMPLE 8.2-5   **Redesigning the Multiple-Input System from Example 8.2-3**

Reconsider the multiple-input system in Example 8.2-3. In this case,

$$\mathbf{I}_{N_u} + \mathbf{K}\Phi(\lambda)\mathbf{B} = \begin{bmatrix} 1 & 0 \\ 0 & 1 \end{bmatrix} + \frac{1}{\lambda^2 - 1} \begin{bmatrix} k_{11} & k_{12} \\ k_{21} & k_{22} \end{bmatrix} \begin{bmatrix} \lambda & 1 \\ 1 & \lambda \end{bmatrix} \begin{bmatrix} 1 & 0 \\ 0 & 1 \end{bmatrix}$$

$$= \begin{bmatrix} 1 + \dfrac{k_{11}\lambda + k_{12}}{\lambda^2 - 1} & \dfrac{k_{11} + k_{12}\lambda}{\lambda^2 - 1} \\ \dfrac{k_{21}\lambda + k_{22}}{\lambda^2 - 1} & 1 + \dfrac{k_{21} + k_{22}\lambda}{\lambda^2 - 1} \end{bmatrix}.$$

We can now choose feedback gains to make the determinant of this matrix zero for both eigenvalues $\lambda_1 = -2$ and $\lambda_2 = -3$. This can be done, for example, by making the first column zero when $\lambda = -2$ and the second column zero when $\lambda = -3$, which yields

$$\mathbf{K} = \frac{1}{5} \begin{bmatrix} 9 & 3 \\ 8 & 16 \end{bmatrix}.$$

Note that the specific set of feedback gains obtained in this example resulted from a particular choice for the determinant column to be zeroed for each eigenvalue. An alternate set of feedback gains are obtained if the gains are chosen to make the second column zero with the first eigenvalue and the first column zero with the second eigenvalue. One other combination is possible in this case, in that the gains may be chosen to make the second column zero with both eigenvalues.

The general procedure can be made computationally more efficient by actually solving for $\mathbf{K}$ directly. Let $\mathbf{e}_j$ be the jth column of $\mathbf{I}_{N_u}$, that is,

$$
\mathbf{e}_1 = \begin{bmatrix} 1 \\ 0 \\ 0 \\ \vdots \\ 0 \end{bmatrix}, \qquad \mathbf{e}_2 = \begin{bmatrix} 0 \\ 1 \\ 0 \\ \vdots \\ 0 \end{bmatrix}, \dots
$$

and let

$$
\mathbf{\Psi}(\lambda_i) \triangleq \mathbf{\Phi}(\lambda_i)\mathbf{B}. \tag{8.2-20}
$$

For the jth column of (8.2-18) to be zero, we require

$$
\mathbf{K}\mathbf{\Psi}_j(\lambda_i) = -\mathbf{e}_j, \tag{8.2-21}
$$

where $\mathbf{\Psi}_j$ is the jth column of $\mathbf{\Psi}$. If the desired eigenvalues are distinct, then we can find N_x linearly independent columns of $\mathbf{\Psi}(\lambda_i)$, $i = 1, \dots, N_x$, to form

$$
\mathbf{G} \triangleq [\mathbf{\Psi}_{j_1}(\lambda_1), \dots, \mathbf{\Psi}_{j_{N_x}}(\lambda_{N_x})] \tag{8.2-22}
$$

and then solve for

$$
\mathbf{K} = -[\mathbf{e}_{j_1}, \dots, \mathbf{e}_{j_{N_x}}]\mathbf{G}^{-1}, \tag{8.2-23}
$$

where $j_1, \dots, j_{N_x}$ stand for the corresponding linearly independent columns in $\mathbf{I}_{N_u}$.

Original Roots

In some design situations, we may wish to keep some of the original eigenvalues of $\mathbf{A}$. If λ is an eigenvalue of $\mathbf{A}$ and is also to be an eigenvalue of $\overline{\mathbf{A}} = \mathbf{A} - \mathbf{BK}$, then we use $\lambda + \epsilon$ in computing $\mathbf{G}$ and let $\epsilon \to 0$ after computing $\mathbf{G}^{-1}$.

EXAMPLE 8.2-6   **Using an Original Root**

Consider a controllable dynamical system in which

$$\mathbf{A} = \begin{bmatrix} -1 & 0 & 1 \\ 0 & 2 & 0 \\ 0 & 0 & -1 \end{bmatrix}, \qquad \mathbf{B} = \begin{bmatrix} 0 & 1 \\ 1 & 0 \\ 0 & 1 \end{bmatrix}.$$

The eigenvalues for the **A** matrix are given by

$$\lambda_1 = -1$$
$$\lambda_2 = 2$$
$$\lambda_3 = -1.$$

Let us design a feedback control law of the form (8.2-1), using (8.2-23), such that the overall system will have the eigenvalues

$$\lambda_1 = -1$$
$$\lambda_2 = -2$$
$$\lambda_3 = -4.$$

We obtain from (8.2-15)

$$\Phi(\lambda) = [\lambda\mathbf{I} - \mathbf{A}]^{-1} = \begin{bmatrix} \dfrac{1}{\lambda + 1} & 0 & \dfrac{1}{(\lambda + 1)^2} \\ 0 & \dfrac{1}{\lambda - 2} & 0 \\ 0 & 0 & \dfrac{1}{\lambda + 1} \end{bmatrix}$$

and from (8.2-21)

$$\Psi(\lambda) = \Phi(\lambda)\mathbf{B} = \begin{bmatrix} 0 & \dfrac{\lambda + 2}{(\lambda + 1)^2} \\ \dfrac{1}{\lambda - 2} & 0 \\ 0 & \dfrac{1}{\lambda + 1} \end{bmatrix}. \qquad \text{(8.2-24)}$$

Thus,

$$\Psi(-1 + \epsilon) = \begin{bmatrix} 0 & \dfrac{1 + \epsilon}{\epsilon^2} \\ \dfrac{1}{\epsilon - 3} & 0 \\ 0 & \dfrac{1}{\epsilon} \end{bmatrix} = [\Psi_1(\lambda_1 + \epsilon), \quad \Psi_2(\lambda_1 + \epsilon)]$$

$$\Psi(-2) = \begin{bmatrix} 0 & 0 \\ -\dfrac{1}{4} & 0 \\ 0 & -1 \end{bmatrix} = [\Psi_1(\lambda_2), \quad \Psi_2(\lambda_2)]$$

$$\Psi(-4) = \begin{bmatrix} 0 & -\dfrac{2}{9} \\ -\dfrac{1}{6} & 0 \\ 0 & -\dfrac{1}{3} \end{bmatrix} = [\Psi_1(\lambda_3), \quad \Psi_2(\lambda_3)]$$

To form **G** as defined by (8.2-22), we need to choose three linearly independent columns, one each from $\Psi(\lambda_1)$, $\Psi(\lambda_2)$, and $\Psi(\lambda_3)$. For example, we can use $\Psi_1(\lambda_1 + \epsilon)$, $\Psi_2(\lambda_2)$, and $\Psi_2(\lambda_3)$, yielding

$$\mathbf{G} = [\Psi_1(\lambda_1 + \epsilon), \quad \Psi_2(\lambda_2), \quad \Psi_2(\lambda_3)] = \begin{bmatrix} 0 & 0 & -\dfrac{2}{9} \\ \dfrac{1}{\epsilon - 3} & 0 & 0 \\ 0 & -1 & -\dfrac{1}{3} \end{bmatrix},$$

which has the inverse

$$\mathbf{G}^{-1} = \frac{9(\epsilon - 3)}{2} \begin{bmatrix} 0 & \dfrac{2}{9} & 0 \\ \dfrac{1}{3(\epsilon - 3)} & 0 & \dfrac{-2}{9(\epsilon - 3)} \\ \dfrac{-1}{\epsilon - 3} & 0 & 0 \end{bmatrix}.$$

In the limit as $\epsilon \rightarrow 0$, we obtain

$$
G^{-1} = \begin{bmatrix} 0 & -3 & 0 \\ \frac{3}{2} & 0 & -1 \\ -\frac{9}{2} & 0 & 0 \end{bmatrix}.
$$

Since we used columns 1, 2, and 2 from the three Ψ matrices, respectively, we employ the corresponding columns from the 2×2 identity matrix $I_{N_u} = I_2$ to calculate K, from (8.2-23), as

$$
K = -[e_1 \quad e_2 \quad e_2]G^{-1} = -\begin{bmatrix} 1 & 0 & 0 \\ 0 & 1 & 1 \end{bmatrix} \begin{bmatrix} 0 & -3 & 0 \\ \frac{3}{2} & 0 & -1 \\ -\frac{9}{2} & 0 & 0 \end{bmatrix},
$$

which yields

$$
K = \begin{bmatrix} 0 & 3 & 0 \\ 3 & 0 & 1 \end{bmatrix},
$$

and the reader can verify that $A - BK$ has the desired eigenvalues $\lambda = -1, -2, -4$.

Complex Eigenvalues

If the controlled system is to have complex eigenvalues, then these eigenvalues must be specified in terms of complex conjugate pairs. In forming G, the same column of $\Psi(\lambda)$ is used for a given complex pair.

EXAMPLE 8.2-7   **Using Complex Eigenvalues**

Let the A and B matrices be the same as in Example 8.2-6. We are to design a feedback control law of the form (8.2-1), using (8.2-20)–(8.2-23), such that the overall system will have the eigenvalues

$$\lambda_1 = -3$$
$$\lambda_2 = -2 + i$$
$$\lambda_3 = -2 - i.$$

Since $\mathbf{A}$ and $\mathbf{B}$ are the same as in the previous example, $\mathbf{\Psi}(\lambda)$ will again be given by (8.2-24). Let us now select

$$\mathbf{G} = [\mathbf{\Psi}_1(-3), \quad \mathbf{\Psi}_2(-2 + i), \quad \mathbf{\Psi}_2(-2 - i)],$$

where the last two columns of $\mathbf{G}$ are chosen from the same column of $\mathbf{\Psi}(\lambda)$. Thus,

$$\mathbf{G} = \begin{bmatrix} 0 & -\frac{1}{2} & -\frac{1}{2} \\ -\frac{1}{5} & 0 & 0 \\ 0 & \dfrac{-1-i}{2} & \dfrac{-1+i}{2} \end{bmatrix}$$

with the inverse

$$\mathbf{G}^{-1} = \frac{7}{2i} \begin{bmatrix} 0 & -5 & 0 \\ -1-i & 0 & i \\ -1+i & 0 & -i \end{bmatrix}.$$

We can now calculate $\mathbf{K}$ from (8.2-23), which yields

$$\mathbf{K} = \begin{bmatrix} 0 & 5 & 0 \\ 2 & 0 & 0 \end{bmatrix}.$$

We note in this example that even though $\mathbf{G}$ and $\mathbf{G}^{-1}$ are complex, $\mathbf{K}$ remains real. This is the reason for selecting the same column of $\mathbf{\Psi}(\lambda)$ twice (once for a desired complex root and once for its conjugate).

Repeated Roots

If repeated roots are desired, then the previous procedure may or may not yield N_x linearly independent $\mathbf{\Psi}_j(\lambda_i)$ columns, and a modified procedure may be necessary (Brogan, 1982). An additional independent equation for forcing $\overline{P}(\lambda_i) = 0$ is given by differentiating (8.2-21), yielding

$$\mathbf{K} \frac{d\mathbf{\Psi}_j}{d\lambda}\bigg|_{\lambda = \lambda_i} = \mathbf{0}. \tag{8.2-25}$$

EXAMPLE 8.2-8   **Using Repeated Roots**

Consider a system with the same **A** and **B** matrices used in Example 8.2-6. This time, we wish to design a feedback control law of the form (8.2-1), using (8.2-23) and (8.2-25), such that the overall system will have eigenvalues $\lambda_1 = -3$, $\lambda_2 = \lambda_3 = -4$.

The $\mathbf{\Psi}(\lambda)$ matrix is again given by (8.2-24). Thus,

$$\mathbf{\Psi}(-3) = \begin{bmatrix} 0 & -\frac{1}{4} \\ -\frac{1}{5} & 0 \\ 0 & -\frac{1}{2} \end{bmatrix} = [\mathbf{\Psi}_1(\lambda_1), \quad \mathbf{\Psi}_2(\lambda_2)]$$

$$\mathbf{\Psi}(-4) = \begin{bmatrix} 0 & -\frac{2}{9} \\ -\frac{1}{6} & 0 \\ 0 & \frac{1}{3} \end{bmatrix} = [\mathbf{\Psi}_1(\lambda_2), \quad \mathbf{\Psi}_2(\lambda_2)]$$

$$\frac{d\mathbf{\Psi}}{d\lambda}\bigg|_{\lambda = -4} = \begin{bmatrix} 0 & -\frac{1}{27} \\ -\frac{1}{36} & 0 \\ 0 & -\frac{1}{9} \end{bmatrix} = \left[\frac{d\mathbf{\Psi}_1(\lambda_2)}{d\lambda}, \quad \frac{d\mathbf{\Psi}_2(\lambda_2)}{d\lambda}\right].$$

We need to choose three linearly independent columns. We use column $\mathbf{\Psi}_1(\lambda_1)$, $\mathbf{\Psi}_2(\lambda_2)$, and $d\mathbf{\Psi}_2(\lambda_2)/d\lambda$ and compute **K** according to

$$\mathbf{K} = -[\mathbf{e}_1 \quad \mathbf{e}_2 \quad \mathbf{0}]\left[\mathbf{\Psi}_1(\lambda_1), \quad \mathbf{\Psi}_2(\lambda_2), \quad \frac{d\mathbf{\Psi}_2(\lambda_2)}{d\lambda}\right]^{-1},$$

so that

$$\mathbf{K} = -\begin{bmatrix} 1 & 0 & 0 \\ 0 & 1 & 0 \end{bmatrix} \begin{bmatrix} 0 & -\frac{2}{9} & -\frac{1}{27} \\ -\frac{1}{5} & 0 & 0 \\ 0 & \frac{1}{3} & -\frac{1}{9} \end{bmatrix}^{-1},$$

which yields

$$\mathbf{K} = \begin{bmatrix} 0 & 5 & 0 \\ 3 & 0 & -3 \end{bmatrix}.$$

8.3 OBSERVER DESIGN FOR LINEAR SYSTEMS

In Section 8.2 we demonstrated how to compute a feedback gain matrix $\mathbf{K}$ so that a state variable feedback controller of the form

$$\mathbf{u} = \mathbf{F}r(t) - \mathbf{K}\mathbf{x} \qquad (8.3\text{-}1)$$

will provide specified eigenvalues for the controlled system

$$\dot{\mathbf{x}} = (\mathbf{A} - \mathbf{B}\mathbf{K})\mathbf{x} + \mathbf{B}\mathbf{F}r. \qquad (8.3\text{-}2)$$

It was assumed that the output was the entire state, that is,

$$\mathbf{y} = \mathbf{x}. \qquad (8.3\text{-}3)$$

For general control system design, condition (8.3-3) may be impractical or not even possible. It may be impractical because of the large number of sensors required, or impossible simply because some of the states cannot be measured. Fortunately, the methods associated with state variable feedback design can still be applied, even if the output is not the full state vector given by (8.3-3). In fact, we can continue to use state variable feedback methods as long as the system is observable. The reason for this is that for observable systems, it is possible to build a device called a **Luenberger observer** (Luenberger, 1971) that will approximately reconstruct any states that are not measured directly.

Even if the entire state vector is not measured, we first proceed with state variable feedback design as discussed in Section 8.2. After the gain matrix $\mathbf{K}$ has been determined, we then direct attention to the design of an observer. The observer (state estimator) becomes a part of the overall feedback scheme, as depicted in Figure 1.4-3.

Identity Observer

Even if the states are not all directly measurable, it is possible to build an observer for the system as long as the system is observable. An observer that reconstructs the entire state vector is referred to as an **identity observer**.

The idea behind an observer is to use the system model to generate an estimate for the state $\mathbf{x}$. Since we know the input $\mathbf{u}$ to the system, one might consider as an observer the state $\hat{\mathbf{x}}$ obtained by integrating (on-line) the equations

$$\dot{\hat{\mathbf{x}}} = \mathbf{A}\hat{\mathbf{x}} + \mathbf{B}\mathbf{u}. \qquad (8.3\text{-}4)$$

Such a device could be built using integrating operational amplifiers, as discussed in Section 6.2. However, an observer based on (8.3-4) would not, in general, be a good one. To see this, let us rewrite the actual system as

$$\dot{\mathbf{x}} = \mathbf{A}\mathbf{x} + \mathbf{B}\mathbf{u} + \mathbf{v}(t), \qquad (8.3\text{-}5)$$

where $\mathbf{v}$ is an $N_x \times 1$ vector that represents all of the system uncertainties. These can include inaccuracies in $\mathbf{A}$ and $\mathbf{B}$, nonlinear effects, and higher-order effects, as well as uncertain inputs. Consider now the effect of $\mathbf{v}(t)$ on the **error** defined by

$$\mathbf{e} = \hat{\mathbf{x}} - \mathbf{x}. \qquad (8.3\text{-}6)$$

If we differentiate (8.3-6) and substitute in (8.3-4) and (8.3-5), we obtain

$$\dot{\mathbf{e}} = \mathbf{A}\mathbf{e} - \mathbf{v}(t). \qquad (8.3\text{-}7)$$

There are two fundamental problems with the observer given by (8.3-4). First, we observe from (8.3-7) that unless all of the eigenvalues of $\mathbf{A}$ have negative real parts, the error equation will be unstable. Even if $\mathbf{A}$ has all eigenvalues with negative real parts, the return time for (8.3-7) may be large. A system with a large return time will be much more sensitive to $\mathbf{v}(t)$ than a system with a small return time (see Section 4.4). A second problem is the fact that no output information is used in (8.3-4). A properly designed observer should be able to use this information in estimating the state.

Toward this end, consider adding a negative feedback term to (8.3-4), proportional to the error in the output, to yield an observer of the form

$$\dot{\hat{\mathbf{x}}} = \mathbf{A}\hat{\mathbf{x}} + \mathbf{B}\mathbf{u} - \mathbf{G}[\hat{\mathbf{y}} - \mathbf{y}], \qquad (8.3\text{-}8)$$

where $\hat{\mathbf{x}}(0) \overset{\Delta}{=} \mathbf{0}$, $\mathbf{G}$ is an $N_x \times N_y$ matrix to be determined, and

$$\hat{\mathbf{y}} = \mathbf{C}\hat{\mathbf{x}} + \mathbf{D}\mathbf{u} \qquad (8.3\text{-}9)$$

is the **predicted measurement** in accordance with (8.1-2). The structure of the state estimator (8.3-8)–(8.3-9) is the same as a Kalman filter (Kalman, 1960), which is used for state estimation in the presence of Gaussian random noise inputs.

Substituting (8.3-9) into (8.3-8) and rearranging terms yields the **identity observer**

$$\dot{\hat{\mathbf{x}}} = [\mathbf{A} - \mathbf{G}\mathbf{C}]\hat{\mathbf{x}} + [\mathbf{B} - \mathbf{G}\mathbf{D}]\mathbf{u} + \mathbf{G}\mathbf{y}, \qquad (8.3\text{-}10)$$

with $\hat{\mathbf{x}}(0) \overset{\Delta}{=} \mathbf{0}$. The components of the $\mathbf{G}$ matrix can now be chosen so that the error equation

$$\dot{\mathbf{e}} = \dot{\hat{\mathbf{x}}} - \dot{\mathbf{x}} \qquad (8.3\text{-}11)$$

will have desirable stability properties. In particular, substituting (8.3-5) and (8.3-8) into (8.3-11) yields

$$\dot{\mathbf{e}} = \mathbf{A}\hat{\mathbf{x}} - \mathbf{G}\hat{\mathbf{y}} + \mathbf{G}\mathbf{y} - \mathbf{A}\mathbf{x} - \mathbf{v}(t). \qquad (8.3\text{-}12)$$

Substituting for $\mathbf{y}$ and $\hat{\mathbf{y}}$ using (8.1-2) and (8.3-9), we can rewrite (8.3-12) as

$$\dot{\mathbf{e}} = [\mathbf{A} - \mathbf{G}\mathbf{C}]\mathbf{e} - \mathbf{v}(t), \qquad (8.3\text{-}13)$$

where (8.3-6) has been used to eliminate $\hat{\mathbf{x}} - \mathbf{x}$. Note that $[\mathbf{A} - \mathbf{G}\mathbf{C}]$ is an $N_x \times N_x$ matrix with $\mathbf{G}$ to be chosen so that the error as given by (8.3-13) will approach a small neighborhood of the origin despite the uncertain input $\mathbf{v}(t)$. For example, the return time of the error equation should be chosen so that it is not only faster than the return time of the controlled system but also fast enough to satisfy performance requirements on the steady-state error resulting from bounds on the expected inputs for $\mathbf{v}$.

An observer based on (8.3-10) overcomes the two fundamental problems mentioned previously. Both the input $\mathbf{u}$ and the output $\mathbf{y}$ are now "inputs" to the observer equation. The stability properties of the error equation are completely under the designer's control, provided that one can choose a $\mathbf{G}$ matrix in (8.3-13) to give the desired eigenvalues for $[\mathbf{A} - \mathbf{G}\mathbf{C}]$. This will always be possible, provided that the system is observable. More precisely, if and only if $\mathbf{Q}$, as defined by (8.1-15), is of rank N_x can the eigenvalues of $[\mathbf{A} - \mathbf{G}\mathbf{C}]$ be arbitrarily determined by an appropriate choice of $\mathbf{G}$ (Luenberger, 1971). In general, $\mathbf{G}$ is chosen so that $[\mathbf{A} - \mathbf{G}\mathbf{C}]$ has eigenvalues different from those of $[\mathbf{A} - \mathbf{B}\mathbf{K}]$. In particular, they should be chosen to the "left" of the controlled system eigenvalues so that the return time of the observer will be faster than that for the controlled system.

State-Space Design Using an Identity Observer

Suppose that a system given by (8.1-1) and (8.1-2) is both controllable and observable and that only some (or perhaps some linear combination) of the states are measured directly. A state variable feedback controller can still be designed for this system in the form of (8.2-1) by choosing $\mathbf{K}$ so that $\overline{\mathbf{A}}$ in (8.2-2) has the desired eigenvalues. However, we must now use an observer to supply part of the missing state information. In particular, if we estimate the entire state, then the controller will be of the form

$$\mathbf{u} = \mathbf{Fr} - \mathbf{K}\hat{\mathbf{x}} \qquad (8.3\text{-}14)$$

with $\hat{\mathbf{x}}$ determined from (8.3-10), where $\mathbf{G}$ has been chosen so that $[\mathbf{A} - \mathbf{GC}]$ has eigenvalues to the "left" of $[\mathbf{A} - \mathbf{BK}]$. The total controlled system will then be governed by

$$\dot{\mathbf{x}} = \mathbf{Ax} - \mathbf{BK}\hat{\mathbf{x}} + \mathbf{BFr} \qquad (8.3\text{-}15)$$

and Equation (8.3-10). A block diagram for this system is depicted in Figure 8.3-1. In practice, the control signal $\mathbf{u}$ will be available for the observer, since this same signal must be input into the system to be controlled.

If we substitute for $\mathbf{u}$ in (8.3-10), we obtain

$$\dot{\hat{\mathbf{x}}} = [\mathbf{A} - \mathbf{GC} - (\mathbf{B} - \mathbf{GD})\mathbf{K}]\hat{\mathbf{x}} + \mathbf{Gy} + [\mathbf{B} - \mathbf{GD}]\mathbf{Fr}. \qquad (8.3\text{-}16)$$

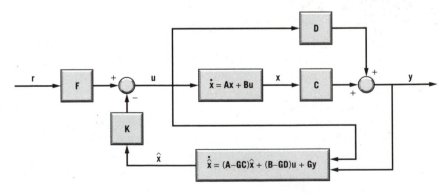

Figure 8.3-1 State-space feedback using an identity observer.

It would appear from the coupled equations (8.3-15) and (8.3-16) that the observer might change the closed-loop stability properties of the system intended with the choice of **K**. However, for controllable and observable systems, observers do not change the closed-loop eigenvalues of the controlled system when they are introduced, but simply adjoin their own eigenvalues. That is, the eigenvalues for the system (8.3-15) and (8.3-16) are those associated with $[\mathbf{A} - \mathbf{BK}]$ and $[\mathbf{A} - \mathbf{GC}]$, respectively.

EXAMPLE 8.3-1 **Second-Order IO System**

Consider the second-order IO system of the form

$$\ddot{y} + p_1\dot{y} + p_0 y = q_0 u.$$

We wish to use state variable feedback methods to design a controller. We first convert the system to a state-space notation. By letting $x_1 = y$ and $x_2 = \dot{y}$, we obtain the equivalent system

$$\dot{x}_1 = x_2$$
$$\dot{x}_2 = -p_0 x_1 - p_1 x_2 + q_0 u$$
$$y = x_1,$$

which is of the form (8.1-1) and (8.1-2) in terms of the matrices

$$\mathbf{A} = \begin{bmatrix} 0 & 1 \\ -p_0 & -p_1 \end{bmatrix}, \quad \mathbf{B} = \begin{bmatrix} 0 \\ q_0 \end{bmatrix}, \quad \mathbf{C} = [1 \quad 0], \quad \mathbf{D} = \mathbf{0}.$$

The system is observable ($|\mathbf{Q}| = 1$) and, provided that $q_0 \neq 0$, it is controllable ($|\mathbf{P}| = -q_0^2$). Using a state variable feedback controller of the form (8.2-1), in particular,

$$u = r - k_1 x_1 - k_2 x_2,$$

we obtain

$$\overline{\mathbf{A}} = \mathbf{A} - \mathbf{BK} = \begin{bmatrix} 0 & 1 \\ -(p_0 + q_0 k_1) & -(p_1 + q_0 k_2) \end{bmatrix}$$

with the characteristic equation

$$\lambda^2 + (p_1 + q_0 k_2)\lambda + (p_0 + q_0 k_1) = 0. \tag{8.3-17}$$

Since this is a single-input system, we can uniquely determine k_1 and k_2 by matching coefficients. Let us design the controller so that the controlled system has a characteristic equation of the form

$$\lambda^2 + 2\zeta\omega_n\lambda + \omega_n^2 = 0, \tag{8.3-18}$$

with $0 < \zeta < 1$ and $\omega_n > 0$. Matching coefficients between (8.3-17) and (8.3-18) yields the same results we obtained in Example 8.2-1 as given by (8.2-8).

In this case, since only x_1 is measured, we really only need to obtain an estimate for x_2 from the observer. However, the identity observer may still be used by simply discarding the estimate for x_1. If we use an identity observer, then we must find $\mathbf{G} = [g_1 \quad g_2]^T$ so that $[\mathbf{A} - \mathbf{GC}]$ has proper eigenvalues. We have

$$\mathbf{A} - \mathbf{GC} = \begin{bmatrix} 0 & 1 \\ -p_0 & -p_1 \end{bmatrix} - \begin{bmatrix} g_1 \\ g_2 \end{bmatrix} \begin{bmatrix} 1 & 0 \end{bmatrix},$$

which reduces to

$$\mathbf{A} - \mathbf{GC} = \begin{bmatrix} -g_1 & 1 \\ -(p_0 + g_2) & -p_1 \end{bmatrix}.$$

The characteristic equation for $[\mathbf{A} - \mathbf{GC}]$ is given by

$$\lambda^2 + (p_1 + g_1)\lambda + (p_1 g_1 + p_0 + g_2) = 0. \tag{8.3-19}$$

Let us choose g_1 and g_2 so that the observer has a characteristic equation of the form

$$\lambda^2 + 2\bar{\zeta}\bar{\omega}_n\lambda + \bar{\omega}_n^2 = 0, \tag{8.3-20}$$

where, when compared with (8.3-18), $\bar{\zeta}\bar{\omega}_n > \zeta\omega_n$ so that the observer eigenvalues are to the left of the controlled system eigenvalues. Matching coefficients between (8.3-19) and (8.3-20), we have

$$g_1 = -p_1 + 2\bar{\zeta}\bar{\omega}_n, \qquad g_2 = -p_0 - p_1 g_1 + \bar{\omega}_n^2. \tag{8.3-21}$$

The observer equations (8.3-10) for this case can be written as

$$\begin{aligned} \dot{\hat{x}}_1 &= \hat{x}_2 - (\hat{x}_1 - y)g_1 \\ \dot{\hat{x}}_2 &= -p_0\hat{x}_1 - p_1\hat{x}_2 - (\hat{x}_1 - y)g_2 + q_0 u. \end{aligned} \tag{8.3-22}$$

These equations may be solved on-line to yield $\hat{x}_1$ and $\hat{x}_2$. The resulting controller is

$$u = r - k_1\hat{x} - k_2\hat{x}_2. \tag{8.3-23}$$

Note that we could use the actual measurement $y = x_1$ in the controller (8.3-23). However, if the measurement y contains any noise we can filter out the noise, to a large extent, by using the full state estimate $\hat{x}$.

In general, both the observer (8.3-10) and controller (8.3-1) may be implemented using operational amplifiers (see Chapter 6). The advantage of using operational amplifiers for the observer is that the output estimates from the observer are "on-line" and can be fed directly into the controller whether the controller is digital or analog. The initial conditions for the integrators may not be known. However, in most situations this should not present difficulties with a properly designed observer. Even if all initial conditions are set equal to zero, any error in doing so will rapidly tend to zero according to the error equation. In other words, initially, there may be an error in the estimate of the state, but this error will quickly become small, since the return time for the observer is quicker than that for the controlled system. Note that any step input for r will have an effect similar to that for the initial condition error.

Reduced-Order Observer

For the system given by (8.1-1)–(8.1-2), which we repeat here for reference,

$$\dot{\mathbf{x}} = \mathbf{Ax} + \mathbf{Bu} \qquad (8.3\text{-}24)$$

$$\mathbf{y} = \mathbf{Cx} + \mathbf{Du}, \qquad (8.3\text{-}25)$$

suppose that $\mathbf{y} \neq \mathbf{x}$, but that rank $[\mathbf{C}] = N_y < N_x$. That is, we do not measure the entire state, but the measurements that we do take are linearly independent. If we use an identity observer to estimate $\hat{x}$, then the observer is a dynamical system of order N_x. Since we have N_y linearly independent measurements $\mathbf{y}$, there is some redundancy in the observer, which leads us to consider a design based on an alternate **reduced-order observer**, of order $N_x - N_y$. This observer will incorporate the $\mathbf{y}$ measurements for part of the state and estimate only the missing state information instead of the full state vector.

If the system is observable and rank$[\mathbf{C}] = N_y$, then not only can such a reduced-order observer be designed but also the eigenvalues for the observer error can be set to any desired values. The results that we present are due to Luenberger (1979, p. 304).

In general, the first step in the process is to make a state-space coordinate transformation so that the measurements $\mathbf{y}(t)$ become part of the new state vector. Then we design an observer for the remainder of the new state vector.

Choose any $(N_x - N_y) \times N_x$ matrix $\mathbf{T}$ such that

$$\mathbf{M}^{-1} \triangleq \begin{bmatrix} \mathbf{T} \\ \mathbf{C} \end{bmatrix} \qquad (8.3\text{-}26)$$

is nonsingular. Such a matrix $\mathbf{T}$ exists, since rank $[\mathbf{C}] = N_y$. Then the new state vector

$$\mathbf{z} = \mathbf{M}^{-1}\mathbf{x} = \begin{bmatrix} \mathbf{T}\mathbf{x} \\ \mathbf{C}\mathbf{x} \end{bmatrix} \triangleq \begin{bmatrix} \boldsymbol{\omega} \\ \boldsymbol{\eta} \end{bmatrix} \qquad (8.3\text{-}27)$$

is partitioned to define an **unmeasured state** $\boldsymbol{\omega}$ and a **measured state** $\boldsymbol{\eta} \triangleq \mathbf{y} - \mathbf{D}\mathbf{u}$. This transformation, applied to (8.3-24), yields new state equations $\dot{\mathbf{z}} = \tilde{\mathbf{A}}\mathbf{z} + \tilde{\mathbf{B}}\mathbf{u}$, with $\tilde{\mathbf{A}} = \mathbf{M}^{-1}\mathbf{A}\mathbf{M}$ and $\tilde{\mathbf{B}} = \mathbf{M}^{-1}\mathbf{B}$, in the partitioned form

$$\begin{bmatrix} \dot{\boldsymbol{\omega}} \\ \dot{\boldsymbol{\eta}} \end{bmatrix} = \begin{bmatrix} \mathbf{A}_{11} & \mathbf{A}_{12} \\ \mathbf{A}_{21} & \mathbf{A}_{22} \end{bmatrix} \begin{bmatrix} \boldsymbol{\omega} \\ \boldsymbol{\eta} \end{bmatrix} + \begin{bmatrix} \mathbf{B}_1 \\ \mathbf{B}_2 \end{bmatrix} \mathbf{u}. \qquad (8.3\text{-}28)$$

Now, given $\mathbf{y}(t)$ and $\mathbf{u}(t)$ and therefore $\boldsymbol{\eta}(t)$, define the **reduced state**

$$\boldsymbol{\gamma}(t) \triangleq \boldsymbol{\omega}(t) - \mathbf{G}\boldsymbol{\eta}(t), \qquad (8.3\text{-}29)$$

where $\mathbf{G}$ is an $(N_x - N_y) \times N_y$ **reduced-order gain matrix** to be determined. Knowledge of an estimate $\hat{\boldsymbol{\gamma}}(t)$ for $\boldsymbol{\gamma}(t)$, along with the measured $\boldsymbol{\eta}(t)$, would determine an estimate $\hat{\boldsymbol{\omega}}(t)$ for $\boldsymbol{\omega}(t)$, via (8.3-29) and, hence, an estimate $\hat{\mathbf{z}}(t)$ for the transformed state $\mathbf{z}(t)$. Then from (8.3-27) we could construct an estimate $\hat{\mathbf{x}}(t)$ for the original state vector as

$$\hat{\mathbf{x}}(t) = \mathbf{M} \begin{bmatrix} \hat{\boldsymbol{\omega}}(t) \\ \boldsymbol{\eta}(t) \end{bmatrix}. \qquad (8.3\text{-}30)$$

Differentiating (8.3-29) and using the transformed state equations (8.3-28) yields

$$\dot{\boldsymbol{\gamma}} = \dot{\boldsymbol{\omega}} - \mathbf{G}\dot{\boldsymbol{\eta}} = [\mathbf{A}_{11} - \mathbf{G}\mathbf{A}_{21}]\boldsymbol{\omega} + [\mathbf{A}_{12} - \mathbf{G}\mathbf{A}_{22}]\boldsymbol{\eta} + [\mathbf{B}_1 - \mathbf{G}\mathbf{B}_2]\mathbf{u}.$$

Substituting for $\boldsymbol{\omega}$ from (8.3-29), we obtain

$$\dot{\boldsymbol{\gamma}} = [\mathbf{A}_{11} - \mathbf{G}\mathbf{A}_{21}]\boldsymbol{\gamma}$$
$$+ [(\mathbf{A}_{12} - \mathbf{G}\mathbf{A}_{22}) + (\mathbf{A}_{11} - \mathbf{G}\mathbf{A}_{21})\mathbf{G}]\boldsymbol{\eta} + [\mathbf{B}_1 - \mathbf{G}\mathbf{B}_2]\mathbf{u}.$$

Given $\mathbf{y}(t)$ and $\mathbf{u}(t)$, and hence $\boldsymbol{\eta}(t)$, this differential equation would determine $\boldsymbol{\gamma}(t)$, except that we do not know $\boldsymbol{\gamma}(0)$. As an estimate $\hat{\boldsymbol{\gamma}}(t)$ for

$\gamma(t)$, we choose an arbitrary value for $\hat{\gamma}(0)$ and a copy of the differential equation, yielding the **reduced-order observer** given by

$$\dot{\hat{\gamma}} = [\mathbf{A}_{11} - \mathbf{GA}_{21}]\hat{\gamma}$$
$$+ [(\mathbf{A}_{12} - \mathbf{GA}_{22}) + (\mathbf{A}_{11} - \mathbf{GA}_{21})\mathbf{G}]\eta + [\mathbf{B}_1 - \mathbf{GB}_2]\mathbf{u}, \quad (8.3\text{-}31)$$

where the resulting state estimate $\hat{\mathbf{x}}(t)$ is computed from (8.3-30), with

$$\hat{\omega}(t) = \hat{\gamma}(t) + \mathbf{G}\eta(t) \qquad (8.3\text{-}32)$$

$$\eta(t) = \mathbf{y}(t) - \mathbf{Du}(t). \qquad (8.3\text{-}33)$$

In terms of the error $\mathbf{e}(t) \overset{\Delta}{=} \hat{\omega}(t) - \omega(t) = \hat{\gamma}(t) - \gamma(t)$, we have

$$\dot{\mathbf{e}} = [\mathbf{A}_{11} - \mathbf{GA}_{21}]\mathbf{e}. \qquad (8.3\text{-}34)$$

If the original state-space system is observable, we can pick the elements of the $(N_x - N_y) \times N_x$ gain matrix $\mathbf{G}$ so that the error equation matrix $[\mathbf{A}_{11} - \mathbf{GA}_{21}]$ has any desired eigenvalues (Luenberger, 1971). Furthermore, if a full state feedback controller of the form $\mathbf{u} = \mathbf{Fr} - \mathbf{Kx}$ has been designed for a particular set of eigenvalues for $[\mathbf{A} - \mathbf{BK}]$, the controller implemented as $\mathbf{u} = \mathbf{Fr} - \mathbf{K}\hat{\mathbf{x}}$ will have the desired eigenvalues, independent of those chosen for the reduced-order observer. That is, as in the case of an identity observer, the design of the feedback controller can be separated from the design of the reduced-order observer, provided that the system is controllable and observable.

EXAMPLE 8.3-2   Second-Order IO System

We will use the same system as in Example 8.3-1

$$\dot{x}_1 = x_2$$
$$\dot{x}_2 = -p_0 x_1 - p_1 x_2 + q_0 u$$
$$y = x_1.$$

The output is $y = x_1$ and we need only an estimate for x_2. To conform to our notation, where the measured states were the last N_y components of the new transformed state vector $\mathbf{z}$, we swap the order of the states by using the transformation $(z_1, z_2) = (x_2, y) = (\omega, \eta)$. However, in this simple case, we will use the notation (x_2, y) instead of (ω, η). Therefore, we have

$$\dot{x}_2 = -p_1 x_2 - p_0 y + q_0 u = a_{11} x_2 + a_{12} y + b_1 u$$
$$\dot{y} = x_2 \qquad\qquad\qquad = a_{21} x_2 + a_{22} y + b_2 u.$$

Thus,

$$a_{11} = -p_1, \quad a_{12} = -p_0, \quad a_{21} = 1, \quad a_{22} = 0, \quad b_1 = q_0, \quad \text{and} \quad b_2 = 0.$$

With $\gamma = \omega - G\eta = x_2 - gy$, (8.3-31) yields the reduced-order observer

$$\dot{\gamma} = -(p_1 + g)\hat{\gamma} - (p_0 + p_1 g + g^2)y(t) + q_0 u(t), \qquad \textbf{(8.3-35)}$$

with

$$\hat{x}_1 = y \qquad\qquad\qquad\qquad\qquad\qquad \textbf{(8.3-36)}$$

$$\hat{x}_2 = \hat{\gamma} + gy. \qquad\qquad\qquad\qquad\qquad \textbf{(8.3-37)}$$

The error $e = \hat{x}_2 - x_2 = \hat{\gamma} - \gamma$ is given by

$$\dot{e} = -(p_1 + g)e, \qquad\qquad\qquad\qquad\qquad \textbf{(8.3-38)}$$

where g is chosen so that $p_1 + g > 0$ and the error eigenvalue is to the left of the eigenvalues for the state-space system. That is, the coefficient g is selected so that the return time for the error equation (8.3-38) is faster than the return time for the controlled system. Once this has been specified, Equation (8.3-35) can be integrated on-line (e.g., with operational amplifiers) using the "inputs" y and u to yield $\hat{\gamma}$. The estimate for $\hat{x}_2$ is then obtained directly from (8.3-37).

8.4   DIGITAL IMPLEMENTATION OF STATE-SPACE CONTROLLERS

Sampled-Data Systems

In the control system design situations discussed thus far, we have assumed that the sensors, controllers, and actuators all behave in a continuous-time fashion. This will be true for many applications where continuous-time devices are used at each stage of the control process; for example, when strain gages, motor tachometers, and so on, are used for measurement, operational amplifiers are used for control system logic, and servomotors are employed as the actuators. However, in many modern applications of control theory, the measurement or logic devices are digital. For example, optical encoders can be used to measure

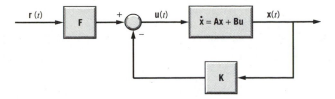

Figure 8.4-1 Continuous-time control of a state-space system.

displacement and mircroprocessors can be used to carry out the logic of a feed-back control design. We will briefly examine how continuous control design can be modified for use in systems where digital devices make up some or all of the control components.

Zero-Order Hold

Suppose that we have used state variable feedback methods to design a controller for the system (8.1-1) and (8.1-2), where $C = I$ and $D = 0$, as illustrated in Figure 8.4-1. If digital devices are used, then the control signal actually applied to the system will be a series of piecewise constant controls. The use of **discrete-time** devices has the effect of introducing a **zero-order hold** (also called a **sample and hold**) in the system, as illustrated in Figure 8.4-2.

A zero-order hold is a device whose output is a stepwise approximation of the input, as illustrated in Figure 8.4-3. Of course, we do not actually place a zero-order hold in the circuit. It simply represents the effect of using digital devices. The digital system illustrated in Figure 8.4-2 is said to be equivalent to the continuous system illustrated in Figure 8.4-1 if the responses of the two systems closely match for the same input and initial conditions.

Analysis and Design of Sampling Time Interval

Let us first calculate the response of the continuous-time system at the discrete-time intervals, $t = 0, T, 2T, \ldots$, using a state feedback

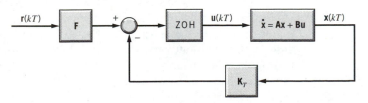

Figure 8.4-2 Discrete-time control of a state-space system.

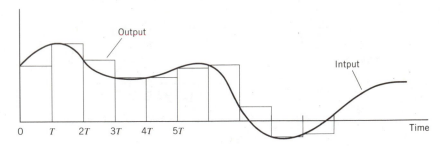

Figure 8.4-3 A comparison of the input and output for a zero-order hold.

controller of the form

$$\mathbf{u} = \mathbf{Fr} - \mathbf{Kx}. \tag{8.4-1}$$

The state $\mathbf{x}(t)$ in the continuous-time system will evolve according to the differential equation

$$\dot{\mathbf{x}}(t) = [\mathbf{A} - \mathbf{BK}]\mathbf{x}(t) + \mathbf{BFr}(t). \tag{8.4-2}$$

In terms of the state transition matrix $\overline{\mathbf{\Phi}}$ (obtained using $[\mathbf{A} - \mathbf{BK}]$) for this system, we have according to the methods of Section 4.2

$$\mathbf{x}(t) = \overline{\mathbf{\Phi}}(t - t_0)\mathbf{x}(t_0) + \int_{t_0}^{t} \overline{\mathbf{\Phi}}(t - \tau)\mathbf{BFr}(\tau)\, d\tau,$$

so that in terms of any "initial" time kT, the state at time $kT + T$ is given by

$$\mathbf{x}(kT + T) = \overline{\mathbf{\Phi}}(T)\mathbf{x}(kT) + \int_{kT}^{kT+T} \overline{\mathbf{\Phi}}(kT + T - \tau)\mathbf{BFr}(\tau)\, d\tau. \tag{8.4-3}$$

The state $\hat{\mathbf{x}}(t)$ of the digital discrete-time system starting at kT will evolve over the sampling period T according to

$$\dot{\hat{\mathbf{x}}}(t) = \mathbf{A}\hat{\mathbf{x}}(t) + \mathbf{B}[\mathbf{Fr}(kT) - \hat{\mathbf{K}}\hat{\mathbf{x}}(kT)], \tag{8.4-4}$$

where the state feedback controller is in the same form as (8.4-1), but with a different feedback gain $\hat{\mathbf{K}}$. Note that we are assuming that both the output (the state $\hat{\mathbf{x}}$) and the input $\mathbf{r}$ pass through sample-and-hold devices and, thus, in particular, $\hat{\mathbf{x}}(kT)$ cannot be combined with $\mathbf{x}(t)$, as was done in (8.4-2). If the command signal is continuous-time, rather than being given by a digital device, the following analysis is still valid, provided that

$r(t) \approx r(kT)$ over one sampling period. The coefficient of $\mathbf{B}$ in (8.4-4) represents a step input into a continuous-time system. We can find the evolution of the state of this system from the state transition matrix $\boldsymbol{\Phi}$ (obtained using $\mathbf{A}$) as

$$\hat{\mathbf{x}}(t) = \boldsymbol{\Phi}(t - t_0)\hat{\mathbf{x}}(t_0) + \int_{t_0}^{t} \boldsymbol{\Phi}(t - \tau)\mathbf{B}[\mathbf{Fr}(kT) - \hat{\mathbf{K}}\hat{\mathbf{x}}(kT)]\,d\tau.$$

In terms of the "initial" time kT, the state at time $kT + T$ is given by

$$\hat{\mathbf{x}}(kT + T) = \boldsymbol{\Phi}(T)\hat{\mathbf{x}}(kT) + \int_{kT}^{kT+T} \boldsymbol{\Phi}(kT + T - \tau)\mathbf{BFr}(kT)\,d\tau$$

$$- \int_{kT}^{kT+T} \boldsymbol{\Phi}(kT + T - \tau)\mathbf{B}\hat{\mathbf{K}}\hat{\mathbf{x}}(kT)\,d\tau. \tag{8.4-5}$$

The fundamental problem in the digital implementation of a continuous-time controller is to find $\hat{\mathbf{K}}$ so that the states $\mathbf{x}(kT + T)$ as determined by (8.4-3) are as close as possible to the states $\hat{\mathbf{x}}(kT + T)$ determined by (8.4-5), given $\mathbf{x}(kT) = \hat{\mathbf{x}}(kT)$. Thus, the following matching conditions must hold:

$$\overline{\boldsymbol{\Phi}}(T)\mathbf{x}(kT) = \boldsymbol{\Phi}(T)\hat{\mathbf{x}}(kT) - \int_{kT}^{kT+T} \boldsymbol{\Phi}(kT + T - \tau)\mathbf{B}\hat{\mathbf{K}}\hat{\mathbf{x}}(kT)\,d\tau$$

and

$$\int_{kT}^{kT+T} \overline{\boldsymbol{\Phi}}(kT + T - \tau)\mathbf{BFr}(\tau)\,d\tau = \int_{kT}^{kT+T} \boldsymbol{\Phi}(kT + T - \tau)\mathbf{BFr}(kT)\,d\tau.$$

By using these conditions, with the additional assumption that $r(t) \approx r(kT)$ [or equivalently, $r(t)$ is applied in a stepwise fashion], it is possible to show (Kuo, 1980) that, for small T,

$$\hat{\mathbf{K}} \approx \mathbf{K} + \frac{T}{2}\mathbf{K}[\mathbf{A} - \mathbf{BK}]. \tag{8.4-6}$$

Thus, given the original design for $\mathbf{K}$ based on the continuous-time system, the appropriate design for $\hat{\mathbf{K}}$ to be used in the digital system can be readily calculated from (8.4-6), provided that the delay time T caused by the digital elements can be determined and is not too large. According to the **Nyquist sampling theorem**, to obtain the proper frequency representation, one must sample at a frequency that is, at least, twice the highest system frequency. It should be noted, however, that the time plot of a signal sampled at just twice the highest frequency of the signal will not be smooth. To obtain a smooth time plot, one must sample at an even higher frequency.

8.5 EXERCISES

8.5-1 Carry out all the details to work Examples 8.1-2 and 8.1-3.

8.5-2 Using state-space methods, design a controller for the inverted pendulum. The approximate model (see Section 1.3) is given by

$$\dot{x}_1 = x_2$$

$$\dot{x}_2 = -a_1 x_1 - a_2 x_2 + bu$$

$$y = x_1.$$

The particular system of interest has

$$a_1 = -17.627/\text{sec}^2$$

$$a_2 = 0.187/\text{sec}$$

$$b = 0.6455/\text{sec}^2\text{-V}.$$

The controller is to be of the form

$$u = r - k_1 x_1 - k_2 x_2.$$

Determine k_1 and k_2 so that the controlled system will have a damping ratio $\zeta = 0.5$ and an undamped natural frequency $\omega_n = 5$ rad/sec.

8.5-3 A new controller is to be designed for the inverted pendulum defined in Exercise 8.5-2 using state feedback plus integral error control. Let the error difference between the input r and angle $\theta = x_1$ be given by $e = r - \theta$. It is required that the eigenvalues be placed at $\lambda_{1,2} = -1 \pm i$ and $\lambda_3 = -2$. A unit step is applied at $t = 0$ with the system at the equilibrium point $\mathbf{x} = \mathbf{0}$. What is the error $e(t)$ when $t = 1/2$ sec. Under these specifications, what is the bandwidth of the system?

8.5-4 Consider the problem of designing a controller to stabilize the origin for the system

$$\dot{x}_1 = x_2 + u$$

$$\dot{x}_2 = -u$$

with measurement output $y = x_1$.
(a) Before attempting the design, determine if such a controller can be developed, by checking controllability.
(b) For a feedback controller of the form $u = -\mathbf{Kx}$, determine the range of values for the elements of the matrix $\mathbf{K}$ such that the closed-loop system is asymptotically stable.
(c) Design a feedback controller $u = -\mathbf{Kx}$ such that the closed-

loop system has a dominant eigenvalue $\lambda = -2$ and a corresponding eigenvector $\xi = [9 \quad -6]^T$.

(d) For the system resulting from (c), determine the remaining eigenvalues and eigenvectors and sketch the trajectories in (x_1, x_2) space.

8.5-5 A system has the transfer function

$$G_P(s) = \frac{\theta(s)}{U(s)} = G_1(s)G_2(s),$$

where

$$G_1(s) = \frac{\Gamma(s)}{U(s)} = \frac{1}{1 + 0.1s}, \qquad G_2(s) = \frac{\theta(s)}{\Gamma(s)} = \frac{1}{s^2 - 4}.$$

(a) Develop a state-space model with $x = [\theta, \dot{\theta}, \Gamma]^T$ and $y = \theta$. Determine the matrices A, B, C, and D in (8.1-1)–(8.1-2).

(b) Design a feedback controller $u = fr - Kx$ so that the closed-loop system has eigenvalues $\lambda = -1, -2, -4$ and the steady-state output $y_\infty = \bar{r}$ in response to a step input $r(t) = \bar{r}$.

8.5-6 Using the general matrix procedure for eigenvalue placement, design a state variable feedback controller of the form

$$u = r - Kx$$

for a state space system $\dot{x} = Ax + Bu$ defined by

$$A = \begin{bmatrix} -3 & 1 \\ 2 & -2 \end{bmatrix}, \qquad B = \begin{bmatrix} 1 & 0 \\ 0 & 2 \end{bmatrix}.$$

The controlled system is to have damping ratio $\zeta = 1/2$ and undamped natural frequency $\omega_n = 4$ rad/sec.

8.5-7 Repeat Exercise 8.5-6, except that the controlled system is to have the eigenvalues $\lambda = -2, -3$ and

(a) $u(r, x)$ contains only u_1, with the remaining control variables identically zero.

(b) $u(r, x)$ contains only u_2, with the remaining control variables identically zero.

8.5-8 For the feedback control system in Exercise 8.5-4:

(a) Determine whether or not a full-state estimator can be developed, by checking controllability.

(b) Develop a full-state estimator such that the state estimation error differential equations have eigenvalues $\lambda = -6, -8$.

8.5-9 For the state-space system in Exercise 8.5-5(a), design an identify observer so that the observer error equations have eigenvalues

$\lambda = -5 \pm i, -5$. Write the observer dynamical equations in matrix form with u and y as inputs.

8.5-10 For the system in Exercise 8.5-2, the state vector is to be estimated using an identity observer. Determine **G** so that the observer will have a damping ratio $\bar{\zeta} = 0.5$ and an undamped natural frequency $\bar{\omega}_n = 6$ rad/sec.

8.5-11 For the system described by

$$\ddot{y} + 2y = u,$$

(a) Obtain an equivalent state-space representation in companion form.

(b) Is the system controllable?

(c) Using state variable feedback of the form

$$u = r - k_1 x_1 - k_2 x_2,$$

determine k_1 and k_2 so that the controlled system will have eigenvalues at $\lambda = -2 \pm i$.

(d) Is the system observable?

(e) Determine values for the **G** vector used with the identity observer so that the observer eigenvalues are placed at $\lambda = -3 \pm i$.

(f) Draw a wiring diagram using operational amplifiers to illustrate the design of the complete control system including the identity observer.

8.5-12 Recall that the linearized differential equation describing the motion of a magnetically suspended rotor is given by

$$\ddot{y} - \beta_1 y = \beta_2 u,$$

where $\beta_1 = k/m$ and $\beta_2 = 1/m$, as previously defined in Section 1.3. Suppose that $k = m = 1$.

(a) Using state-space methods, design a controller for this system of the form

$$u = r - k_1 x_1 - k_2 x_2.$$

The controlled system is to have eigenvalues $\lambda = -5 \pm 5i$.

(b) Design an identity observer for the system with observer eigenvalues placed at $\lambda = -6 \pm 5i$.

(c) The observer is to be implemented using operational amplifiers. Draw a wiring diagram that illustrates this implementation.

(d) Design a reduced-order observer that will provide the same return time as the identity observer.

8.5-13 Using a reduced-order observer to obtain $\hat{x}_2$, rework Exercises
  (a) 8.5-10
  (b) 8.5-11
  Place the eigenvalues of the observer to the left of the controlled
  system eigenvalues.

8.5-14 The controller designed in Exercise 8.5-2 is to be implemented by
  using a PC equipped with A/D and D/A ports so that data can be
  read in and control signals sent out. Three different languages were
  used to implement the control law: C, FORTRAN, and BASIC.
  The time delay associated with each language was $T = 6$ msec,
  $T = 15$ msec, and $T = 40$ msec, respectively. Using digital
  redesign methods, determine the state variable feedback gains to
  be used in each case. The observer was still implemented
  employing operational amplifiers.

References

Ash, R.H., and Ash, G.R., "Numerical Computation of Root Loci Using the Newton–Raphson Technique," *IEEE Transactions on Automatic Control*, **AC-13** (5), Oct. 1968:576–582.

Barmish, B.R., and Leitmann, G., "On Ultimate Boundedness Control of Uncertain Systems in the Absence of Matching Conditions," *IEEE Transactions on Automatic Control*, **AC-27** (2), Feb. 1982:153–158.

Bose, N.K., "A System-Theoretic Approach to Stability of Sets of Polynomials," *Contemporary Mathematics*, **47**, 1985:25–34.

Brogan, W.L., *Modern Control Theory*, Prentice Hall, Englewood Cliffs, N.J., 1982.

Chung, C.S., Kong, C.D., and Yum, Y.H., "A Study on the Dynamic Characteristics of the Korean Yi-Dynasty Bell Type Structure," *Korean Society of Mechanical Engineers Journal*, **1** (2), Dec., 1987:133–139.

Coddington, E.A., and Levinson, N., *Theory of Ordinary Differential Equations*, McGraw-Hill, New York, 1955.

Corless, M., and Leitmann, G., "Continuous State Feedback Guaranteeing Uniform Ultimate Boundedness for Uncertain Dynamical Systems," *IEEE Transactions on Automatic Control*, **AC-26** (5), Oct. 1981:1139–1144.

Davison, E.J., "On Pole Assignment in Multivariable Linear Systems," *IEEE Transactions on Automatic Control*, **AC-13** (6), Dec. 1968:747–748.

Doeblin, E.O., *Measurement Systems*, McGraw-Hill, New York, 1966.

Evans, W.R., "Graphical Analysis of Control Systems," *Transactions of the AIEE*, **67**, Part I, 1948:547–551.

Gayek, J.E. and Vincent, T.L., "On the Asymptotic Stability of Boundary Trajectories," *International Journal of Control*, **41** (4), April 1985:1077–1086.

Gayek, J.E. and Vincent, T.L., "An Existence Theorem for Boundary Trajectories," *Journal of Mathematical Analysis and Applications*, **132** (1), May 1988:290–299.

Grantham, W.J., and Vincent, T.L., "A Controllability Minimum Principle," *Journal of Optimization Theory and Applications*, **17** (1–2), Oct. 1975:93–114.

Gutman, S., "Uncertain Dynamical Systems—Lyapunov Min-Max Approach," *IEEE Transactions on Automatic Control*, **AC-24** (3), June 1979:437–443.

Heymann, M., "Comments on Pole Assignment in Multi-Input Controllable Linear Systems," *IEEE Transactions on Automatic Control*, **AC-13** (6), Dec. 1968:748–749.

Hildebrand, F.B., *Advanced Calculus with Applications*, Prentice Hall, Englewood Cliffs, N.J., 1962.

Kalman, R.E., "A New Approach to Linear Filtering and Prediction Problems," *Transactions of the ASME, Journal of Basic Engineering*, **82** (1), March 1960: 34–45.

Kalman, R.E., Ho, Y.C., and Narendra, K.S., "Controllability of Linear Dynamical Systems," *Contributions to Differential Equations,* **I** (2), 1963:189–213.

Kharitonov, V.L., "Asymptotic Stability of an Equilibrium Position of a Family of Systems of Linear Differential Equations," *Differential Equations,* **14** (11), Nov. 1978:1483–1485.

Kreyszig, E., *Advanced Engineering Mathematics,* 5th ed., Wiley, New York, 1983.

Kuo, B.C., *Digital Control Systems,* Holt, Rinehart, and Winston, New York, 1980.

Kuo, B.C., *Automatic Control Systems,* 4th ed., Prentice Hall, Englewood Cliffs, N.J., 1982.

Leitmann, G., "On the Efficacy of Nonlinear Control in Uncertain Linear Systems," *Journal of Dynamic Systems, Measurement, and Control,* **103** (2), June 1981: 95–102.

Luenberger, D.G., "An Introduction to Observers," *IEEE Transactions on Automatic Control,* **AC-16** (6), Dec. 1971:596–602.

Luenberger, D.G., *Introduction to Dynamic Systems,* Wiley, New York, 1979.

Meditch, J.S., *Stochastic Optimal Linear Estimation and Control,* McGraw-Hill, New York, 1969.

Nayfeh, A.H., and Mook, D.T., *Nonlinear Oscillations,* Wiley, New York, 1979.

Press, W.H., Flannery, B.P., Teukolsky, S.A., and Vetterling, W.T., *Numerical Recipes in C,* Cambridge University Press, New York, 1988.

Ryan, E.P., Leitmann, G., and Corless, M., "Practical Stabilization of Uncertain Dynamical Systems, Application to Robotic Tracking," *Journal of Optimization Theory and Applications,* **47** (2), Oct. 1985:235.

Takahashi, Y., Rabins, M.J., and Auslander, D.M., *Control and Dynamic Systems,* Addison-Wesley, Reading, Mass., 1970.

Wonham, W.M., "On Pole Assignment in Multi-Input Controllable Linear Systems," *IEEE Transactions on Automatic Control,* **AC-12** (6), Dec. 1967:660–665.

Wonham, W.M., *Linear Multivariable Control,* 3rd ed., Springer-Verlag, New York, 1985.

Appendix

Answers to Even-Numbered Exercises

CHAPTER 1

1.5-2 (a) $x = Q \qquad a = \dfrac{-1}{RC}$

$\qquad u = E \qquad b = \dfrac{1}{R}$

(b) $u = E \qquad a = \dfrac{-1}{RC} \qquad b = \dfrac{1}{RC}$

1.5-4 $a_{11} = 0 \qquad a_{12} = 1 \qquad b_1 = 0$

$\qquad a_{21} = \dfrac{-1}{LC} \qquad a_{22} = \dfrac{-R}{L} \qquad b_2 = \dfrac{1}{L}$

1.5-6 Whenever the function and its first- and second-order partial derivatives exist and are continuous.

1.5-8 (a) $\overline{X}_1 = \dfrac{\mu g m_2}{K} \qquad \overline{U} = \mu g(m_1 + m_2)$

(b) $\dot{x}_1 = x_2 - x_3$

$\qquad \dot{x}_2 = -\dfrac{k}{m_1} x_1 + \dfrac{u}{m_1}$

$\qquad \dot{x}_3 = \dfrac{k}{m_2} x_1.$

1.5-10 **(a)** Given

(b) Given

(c) For $\mathbf{x} = [r, \dot{r}, \theta, \dot{\theta}]^T$ (with all components small), $u = \Gamma$, and $\mathbf{y} = [r, \theta]^T$, the linearized equations are $\dot{\mathbf{x}} = \mathbf{A}\mathbf{x} + \mathbf{B}u$, $\mathbf{y} = \mathbf{C}\mathbf{x}$,

where

$$\mathbf{A} = \begin{bmatrix} 0 & 1 & 0 & 0 \\ -\dfrac{mg}{\rho}\left[\dfrac{J}{R} + m(R+h)\right] & 0 & \dfrac{mg}{\rho}\left(\dfrac{hJ}{R} - I\right) & 0 \\ 0 & 0 & 0 & 1 \\ -\dfrac{mg}{\rho}\left(\dfrac{J}{R^2} + m\right) & 0 & \dfrac{mg}{\rho}\dfrac{hJ}{R^2} & 0 \end{bmatrix}$$

$$\mathbf{B} = \begin{bmatrix} 0 \\ \dfrac{1}{\rho}\left[\dfrac{J}{R} + m(R+h)\right] \\ 0 \\ \dfrac{1}{\rho}\left(\dfrac{J}{R^2} + m\right) \end{bmatrix}, \qquad \mathbf{C} = \begin{bmatrix} 1 & 0 & 0 & 0 \\ 0 & 0 & 1 & 0 \end{bmatrix},$$

with

$$\rho = m\left(I + \dfrac{h^2}{R^2}J\right) + \dfrac{IJ}{R^2}.$$

1.5-12 $\dot{x}_1 = x_2$

$$\dot{x}_2 = -\left(\beta_2 \dfrac{T}{m} + \beta_1 x_5^2 + \beta_2\beta_4 x_5^2\right)x_1 - \beta_2 x_5 x_2 + \beta_3 x_5^2 u$$

$$\dot{x}_3 = \beta_2 x_5 x_1$$

$$\dot{x}_4 = x_5 x_3$$

$$\dot{x}_5 = \dfrac{T}{m} - \beta_4 x_5^2.$$

1.5-14 $\dfrac{L_F}{K_A}\dddot{J\theta} + \dfrac{R_F}{K_A}J\ddot{\theta} - \dfrac{L_F}{K_A}mg\ell\dot{\theta}\cos\theta - \dfrac{R_F}{K_A}mg\ell\sin\theta = E_F,$

where K_A is a constant, with $\Gamma = K_A I_A$ being the motor torque.

1.5-16 One approach is given by

1. With a sufficiently heavy pendulum attached, apply a constant voltage E. Then at equilibrium, use (1.3-50) to determine k_Γ.

2. Remove the pendulum, leaving just the rotor and shaft inertia. In (1.3-48), with $\Gamma_{ex} = 0$, apply a constant voltage $E \neq 0$. Use (1.3-48) at equilibrium to determine k_B.

CHAPTER 2

2.5-2 (a) $\dot{z} = \begin{bmatrix} 2 & 0 \\ 0 & -2 \end{bmatrix} z + \begin{bmatrix} 15.75 \\ -15.75 \end{bmatrix} u$

$y = \begin{bmatrix} 1 & 1 \end{bmatrix} z$

(b) $\dot{z} = \begin{bmatrix} 4i & 0 \\ 0 & -4i \end{bmatrix} z + \frac{1}{4}\begin{bmatrix} -i & 2(1-i) \\ i & 2(1+i) \end{bmatrix} u$

$y = \begin{bmatrix} 1 & 1 \\ 2i & -2i \end{bmatrix} z$

2.5-4 (a) $\ddot{y} + 4\dot{y} + 3y = 13u + 2\dot{u}$
(b) $\ddot{y} + \dot{y} = 2u$
(c) Observability condition not satisfied.
(d) $\ddot{y} - 4\dot{y} + 3y = -u + \dot{u}$.

2.5-6 Follows from $CA^k = k$th row of A, $k = 1, \ldots, N_x - 1$.

2.5-8 System is controllable.

2.5-10 Given.

2.5-12 (a) $G(s) = \dfrac{2 + s}{s^2 + 2s + 3}$

(b) $A = \begin{bmatrix} 0 & 1 \\ -3 & -2 \end{bmatrix}, \quad B = \begin{bmatrix} 1 \\ 0 \end{bmatrix}, \quad C = \begin{bmatrix} 1 & 0 \end{bmatrix}$

(c) $G(s) = \dfrac{2 + s}{s^2 + 3s + 5}$.

2.5-14 Given.

2.5-16 (a) $G_1(s) = \dfrac{0.5}{0.5 + (s + 2)(1 + Ts)}$

(b) $G_2(s) = \dfrac{20(1 + \tau s)}{0.5 + (s + 2)(1 + Ts)}$

(c) $G(s) = \begin{bmatrix} \dfrac{0.5}{P(s)} & \dfrac{20(1 + Ts)}{P(s)} \end{bmatrix}, \quad P(s) = 0.5 + (s + 2)(1 + Ts)$

(d) With $x = \begin{bmatrix} x_1 & x_2 \end{bmatrix}^T$ and $u = \begin{bmatrix} u_1 & u_2 \end{bmatrix}^T$, one diagonal representation is given by

$$\dot{x} = \begin{bmatrix} -2.6834 & 0 \\ 0 & -9.3166 \end{bmatrix} x + \begin{bmatrix} 0.7538 & 22.0605 \\ -0.7538 & -2.0605 \end{bmatrix} u$$

$$y = \begin{bmatrix} 1 & 1 \end{bmatrix} x.$$

CHAPTER 3

3.6-2 (a) $y = e^{-3t}$
(b) $y = 3e^{-2t} - 2e^{-3t}$
(c) $y = 3e^{-2t} - 2e^{-3t}$
(d) $y = e^{-t} + \dfrac{2}{\sqrt{7}} e^{-t/2} \sin \dfrac{\sqrt{7}}{2} t.$

3.6-4 (a) $\Phi(t) = \begin{bmatrix} 2 & 2 \\ -1 & -1 \end{bmatrix} e^{-t} + \begin{bmatrix} -1 & -2 \\ 1 & 2 \end{bmatrix} e^{-2t}$

(b) $\Phi(t) = \begin{bmatrix} \cos 2t + 8 \sin 2t & 26 \sin 2t \\ -2.5 \sin 2t & \cos 2t - 8 \sin 2t \end{bmatrix}$

(c) $\Phi(t) = e^{-t} \begin{bmatrix} \cos 2t + \frac{1}{2} \sin 2t & \frac{1}{2} \sin 2t \\ -\frac{5}{2} \sin 2t & \cos 2t - \frac{1}{2} \sin 2t \end{bmatrix}.$

3.6-6 $x(t) = \begin{bmatrix} 2 \\ 1 \end{bmatrix} e^{3t} + \begin{bmatrix} -1 \\ -1 \end{bmatrix} e^{2t}.$

3.6-8 For $a \neq 0$, the origin is a saddle point, as in Figure 3.4-1e. For $a = 0$, trajectories are as in Figure 3.4-1f.

3.6-10 Stability requires $\beta_2 k_2 > 0$ and $\beta_2 k_1 - \beta_1 > 0$.

3.6-12 Stable for $a > 0$, $b > 0$, and $ab - c > 0$. Unstable if $a = 0$ or $b = 0$. Stable if $c = 0$ with $a > 0$ and $b > 0$.

CHAPTER 4

4.5-2 (a) $y = x_2 = e^{-t} - e^{-2t}$
(b) $y = x_2 = -4 + 4 \cos 2t + 0.5 \sin 2t$
(c) $y = x_2 = 0.5 e^{-t} \sin 2t$

4.5-4 $\ddot{y} + 2\zeta\omega_n\dot{y} + \omega_n^2 y = \omega_n^2 u,$ $\zeta = 0.3579,$ $\omega_n = 672.88$ rad/sec.

4.5-6 $\zeta = 0.8892.$

4.5-8 (a) For Example 4.4-1,

$$\lambda_{1,2} = -\frac{1}{2} \pm i\frac{\sqrt{3}}{2}, \qquad \bar{x} = \begin{bmatrix} 1 \\ 0 \end{bmatrix}.$$

(b) For example 4.4-2,

$$\lambda_{1,2} = -1, -2, \qquad \bar{x} = \begin{bmatrix} \frac{3}{2} \\ -2 \end{bmatrix}.$$

CHAPTER 5

5.5-2 (a) $y(t) = -3 + (3 + 9t)e^{-t}$,   as $t \to \infty$, $y(t) \to y_r = -3$

(b) $y(t) = \frac{1}{25}[(24 - 90t)e^{-t} + 57 \sin 2t - 24 \cos 2t]$,   as $t \to \infty$,

$y(t) \to y_r(t) = \frac{57}{25} \sin 2t - \frac{24}{25} \cos 2t$.

5.5-4 $y_r(t) = Y \sin(\omega t + \varphi)$

(a) $Y = \dfrac{2}{\sqrt{(2 - \omega^2)^2 + 9\omega^2}}$, $\varphi = \tan^{-1}\left[\dfrac{3\omega}{\omega^2 - 2}\right]$

(b) $Y = \dfrac{52}{|4 - \omega^2|}$, $\varphi = \begin{cases} 0 & \text{if } 0 < \omega < 2 \\ \pi & \text{if } \omega > 2 \end{cases}$

(c) $Y = \dfrac{1}{\sqrt{(5 - \omega^2)^2 + 4\omega^2}}$, $\varphi = \tan^{-1}\left[\dfrac{2\omega}{\omega^2 - 5}\right]$.

5.5-6 Low frequency: $|G(i\omega)|_{db} \to 0$ as $\omega \to 0$

High frequency: $|G(i\omega)|_{db} \to -40 \log \dfrac{\omega}{\omega_n}$ as $\omega \to \infty$

Intersect: $|G(i\omega)|_{db} = 0$, $\dfrac{\omega}{\omega_n} = 1$,   where $\omega_n = 4$.

5.5-8 Given.

5.5-10 Given.

CHAPTER 6

6.6-2 (a) $G_p(s) = \dfrac{H(s)}{Q_i(s)} = \dfrac{1}{As + k_1}$

(b) $G(s) = \dfrac{K}{As + k_1 + K}$

(c) $h(t) = \frac{1}{3}[2 + e^{-3t/2}]$

(d) $h_\infty = \frac{2}{3}$.

6.6-4 (a) $\zeta = \frac{1}{5}$, $\omega_n = 5$

(b) $y(t) = 1 - 1.0103\, e^{-t} \sin(4.899t + 1.3694)$.

6.6-6 (a) $K = 3.5$

(b) $\zeta = 0.025$

(c) For $\omega = 2$, $|y(t)| \to |G(i\omega)| = 17.5$ ft.

6.6-8 (a) $K_S = 0.1,\ K = 5,\ \tau_i = 0.2$

(b) -20 db; $+20$ db/decade at $\omega = 1$; -20 db/decade at $\omega = \sqrt{5}$

(c) $\omega_b = 7$ rad/sec, $T = \frac{1}{2}$.

6.6-10 $K = 20,\ a = 0.4.$

6.6-12 (a) $G_P(s) = \dfrac{mg/J}{s^2} = \dfrac{0.1963}{s^2}$

(b) Unstable

(c) $G(s) = \dfrac{K_S K \dfrac{mg}{J}\left(s + \dfrac{1}{\tau_a}\right)}{s^3 + \dfrac{1}{\tau_a}s^2 + K\dfrac{mg}{J}\left(\dfrac{\tau_e}{\tau_a}\right)s + \left(\dfrac{K}{\tau_a}\right)\dfrac{mg}{J}}$

(d) $K = 5.093,\ \tau_e = 1.8,\ \tau_a = 0.2$

(e) $K_S = 0.1.$

6.6-14 (a) $K > 27.308$

(b) $K = 88.467.$

CHAPTER 7

7.5-2 (a) $\theta_m = \pm 60°,\ 180°$

(b) $\sigma = -\frac{4}{3}$

(c) $\theta_d = -63.43°$

(d) $s = \pm i\sqrt{5}$

(e) $s_b = -1$

(f) $s_b = -\frac{5}{3}.$

7.5-4 (a) Real axis: $-\infty < \sigma \le 0$

Asymptotes: centroid at $\sigma = -2$, angles $\theta_m = \pm 60°,\ 180°$

No breakpoints

Departure from $s = -3 + i2$: $\theta_d = -56.3°$

Imaginary-axis intercepts at $s = \pm i3.6$

Asymptotically stable for $0 < K < 78$

(b) Real axis: $-4 \le \sigma \le -1$

Asymptotes: centroid at $\sigma = -\frac{3}{2}$, angles $\theta_m = \pm 90°$

No breakpoints

Departure from $s = -3 - i2$: $\theta_d = 18.43°$

No imaginary-axis crossings

Asymptotically stable for $K > 0.$

7.5-6 (a) Real axis: $-4.2936 \leq \sigma \leq 4.1063$
   Asymptotes: centroid at $\sigma = -0.0936$,   angles $\theta_m = \pm 90°$
   Breakpoint: $s_b = -0.0936$
   Imaginary-axis intercept: $s = 0$
   Asymptotically stable for $K > 27.312$

(b) Real axis: $-4.2936 \leq \sigma \leq -0.3$, $0 \leq \sigma \leq 4.1063$
   Asymptotes: centroid at $\sigma = 0.0564$,   angles $\theta_m = \pm 90°$
   Breakpoint: $s_b = 1.2116$
   No imaginary-axis crossings
   Unstable for $K > 0$

(c) Real axis: $-\infty < \sigma \leq 0$
   Asymptotes: centroid at $\sigma = -0.0624$,   angles $\theta_m = \pm 60°$,
   $180°$
   No breakpoints
   Imaginary-axis intercepts: $s = \pm i4.7435$
   Asymptotically stable for $\tau_i > 9.5274$

(d) Real axis: $-10 \leq \sigma \leq -4.2936$, $0 \leq \sigma \leq 4.1063$
   Asymptotes: centroid at $\sigma = 4.9064$,   angles $\theta_m = \pm 90°$
   Breakpoint: $s_b = 2.204$
   No imaginary-axis crossings
   Unstable for $K > 0$.

7.5-8 Center at $s = -2 + i0$,   radius $r = 2$.

7.5-10 Guaranteed asymptotically stable for $0 < K < 0.7421$ or $K > 61.7579$.

7.5-12 Given.

7.5-14 $\zeta = 0.5$ and $\zeta = 0.3$.

7.5-16 GM $= 10.3$ db, with phase cross-over at $\omega_\varphi = 7.2$ rad/sec
   PM $= 36°$, corresponding to gain cross-over at $\omega_k = 3.2$ rad/sec.

7.5-18 (a) GM $= 25$ db, PM $= 57°$, stable

(b) $F(s) = \dfrac{1 + 0.1s}{s(1 + 0.5s)(1 + 0.2s)}$.

CHAPTER 8

8.5-2 $k_1 = 66.042$, $k_2 = 7.456$.

8.5-4 (a) Controllable $\Rightarrow$ yes

(b) $k_2 < k_1 < 0$

(c) $k_2 = \frac{3}{2}k_1 - 2$. For example, choose $\lambda_2 = -4$. Then $\mathbf{K} = [k_1 \quad k_2] = [-8 \quad -14]$.

(d) for $\lambda_2 = -4$, $\boldsymbol{\xi}_2 = [1 \quad -0.8]^T$. Stable node.

8.5-6 $\mathbf{K} = \begin{bmatrix} -2 & 14 \\ 0 & 0 \end{bmatrix}$.

8.5-8 (a) Observable $\Rightarrow$ yes

(b) $\mathbf{G} = [14 \quad 48]^T$.

8.5-10 $\mathbf{G} = [5.8128 \quad 52.542]^T$.

8.5-12 (a) $k_1 = 51$, $k_2 = 10$

(b) $\mathbf{G} = [12 \quad 62]^T$

(c) Identity observer is given by

$$\dot{\hat{\mathbf{x}}} = \begin{bmatrix} -12 & 1 \\ -61 & 0 \end{bmatrix} \hat{\mathbf{x}} + \begin{bmatrix} 0 \\ 1 \end{bmatrix} u + \begin{bmatrix} 12 \\ 62 \end{bmatrix} y.$$

(d) Reduced-order observer is given by

$$\dot{\omega} = -6\omega - 35y + u$$
$$\hat{x}_2 = \omega + 6y.$$

8.5-14 For $T = 0.006$ sec, $\hat{\mathbf{K}} = [65.45 \quad 7.546]^T$

For $T = 0.015$ sec, $\hat{\mathbf{K}} = [64.64 \quad 7.676]^T$

For $T = 0.040$ sec, $\hat{\mathbf{K}} = [62.31 \quad 8.035]^T$.

Index